Lecture Notes in Bioinformatics 3615

Subseries of Lecture Notes in Computer Science

Bertram Ludäscher   Louiqa Raschid (Eds.)

Data Integration in the Life Sciences

Second International Workshop, DILS 2005
San Diego, CA, USA, July 20-22, 2005
Proceedings

 Springer

Series Editors

Sorin Istrail, Celera Genomics, Applied Biosystems, Rockville, MD, USA
Pavel Pevzner, University of California, San Diego, CA, USA
Michael Waterman, University of Southern California, Los Angeles, CA, USA

Volume Editors

Bertram Ludäscher
University of California at Davis, Department of Computer Science
One Shields Ave, Davis, CA 95616, USA
E-mail: ludaesch@ucdavis.edu

Louiqa Raschid
University of Maryland, Department of Computer Science
A.V. Williams Building, College Park, MD 20742, USA
E-mail: louiqa@umiacs.umd.edu

Library of Congress Control Number: 2005928957

CR Subject Classification (1998): H.2, H.3, H.4, J.3

ISSN   0302-9743
ISBN-10   3-540-27967-9 Springer Berlin Heidelberg New York
ISBN-13   978-3-540-27967-9 Springer Berlin Heidelberg New York

Springer is a part of Springer Science+Business Media

springeronline.com

© Springer-Verlag Berlin Heidelberg 2005
Printed in Germany

Typesetting: Camera-ready by author, data conversion by Scientific Publishing Services, Chennai, India
Printed on acid-free paper   SPIN: 11530084   06/3142   5 4 3 2 1 0

Preface

The explosion in the number and size of life science data resources, and the rapid growth in the variety and volume of laboratory data has been fueled by world-wide research activity and the emergence of new technologies. The modeling, management and analysis of this data often requires a comprehensive integration of heterogeneous and typically semistructured data, distributed across many possibly data sources. Recent interoperability standards such as XML and WSDL solve some (easy) problems, but data and process integration often remain time-consuming and error-pone manual tasks. The difficulty of these tasks is compounded by the high degree of semantic heterogeneity across data sources, varying data quality, and other domain-specific application requirements.

DILS 2005 was the 2nd International Workshop on Data Integration in the Life Sciences, following a successful first DILS workshop, March 2004 in Leipzig, Germany. For a specialized workshop, the DILS 2005 call for papers created a large interest (over 50 abstracts and eventually 42 paper submissions; an increase of over 20% over DILS 2004), out of which the international Program Committee selected 15 full papers, as well as 5 short papers, and 8 posters/demonstrations, which are all included in this volume. They cover a wide spectrum of theoretical and practical issues including scientific/clinical workflows, ontologies, tools and systems, and integration techniques. DILS 2005 also featured keynotes by Dr. Peter Buneman, Professor at the School of Informatics, University of Edinburgh, and Dr. Shankar Subramaniam, Professor at the Department of Bioengineering and Chemistry, UC San Diego. The program also included 6 invited presentations and reports on ongoing research activities in academia and industry and a panel organized by the AMIA Geomics Working Group.

The workshop was organized by the San Diego Supercomputer Center (SDSC) and took place July 20–22, 2005 at the University of California, San Diego. Additional sponsors included Microsoft Research, the American Medical Informatics Association (AMIA), the UC Davis Genome Center, and the University of Maryland Center for Bioinformatics and Computational Biology.

As the workshop co-chairs and editors of this volume, we thank all authors who submitted papers and the Program Committee members and external reviewers for their excellent work. Special thanks also go to Amarnath Gupta who served as workshop general chair, and his team, especially Donna Turner, Jon Meyer, and LInda Ferri, all at SDSC. We thank Chani Johnson and the Microsoft CMT Team for the excellent support of their paper management system. Finally, we thank Alfred Hofmann, Erika Siebert-Cole, and the team from Springer for their cooperation and help in putting this volume together.

June 2005   Bertram Ludäscher and Louiqa Raschid

2nd International Workshop on Data Integration in the Life Sciences (DILS)

University of California, San Diego
July 20–22, 2005

DILS 2005 Co-chairs

Amarnath Gupta (General Chair) University of California, San Diego, USA
Bertram Ludäscher (PC Co-chair) University of California, Davis, USA
Louiqa Raschid (PC Co-chair) University of Maryland, USA

Program Committee

Vineet Bafna	University of California, San Diego, USA
Chitta Baral	Arizona State University, USA
Judith Blake	Jackson Laboratory, USA
Shawn Bowers	University of California, Davis, USA
Terence Critchlow	Lawrence Livermore National Laboratory, USA
Alin Deutsch	University of California, San Diego, USA
Barbara Eckman	IBM Life Sciences, USA
Christoph Freytag	Humboldt University, Berlin, Germany
Floris Geerts	University of Edinburgh, UK
Carole Goble	University of Manchester, UK
Amarnath Gupta	University of California, San Diego, USA
Michael Gribskov	Purdue University, USA
Ralf Hofestaedt	University of Bielefeld, Germany
Hasan Jamil	Wayne State University, USA
Matthew Jones	University of California, Santa Barbara, USA
Jessie Kennedy	Napier University, Edinburgh, UK
Zoé Lacroix	Arizona State University, USA
Ulf Leser	Humboldt University Berlin, Germany
Felix Naumann	Humboldt University Berlin, Germany
Frank Olken	Lawrence Berkeley National Laboratory, USA
Jignesh Patel	University of Michigan, USA
Erhard Rahm	University of Leipzig, Germany
Julia Rice	IBM Life Sciences, USA
Peter Tarczy-Hornoch	University of Washington, USA
Limsoon Wong	Institute for Infocomm Research, Singapore
Aidong Zhang	State University of New York at Buffalo, USA

Additional Reviewers

Alexander Bilke	Humboldt University Berlin, Germany
Jens Bleiholder	Humboldt University Berlin, Germany
Hong-Hai Do	University of Leipzig, Germany
Antoon Goderis	University of Manchester, UK
Woo-Chang Hwang	State University of New York at Buffalo, USA
Daxin Jiang	State University of New York at Buffalo, USA
Toralf Kirsten	University of Leipzig, Germany
Peter Li	University of Newcastle, UK
Phillip Lord	University of Manchester, UK
Hervé Ménager	Arizona State University, USA
Peter Mork	University of Washington, USA
HweeHwa Pang	Institute for Infocomm Research, Singapore
Pengjun Pei	State University of New York at Buffalo, USA
Benjamin Prins	University of Bielefeld, Germany
Robert Stevens	University of Manchester, UK
Thoralf Töpel	University of Bielefeld, Germany
Silke Trißl	Humboldt University Berlin, Germany
Chris Wroe	University of Manchester, UK
Xian Xu	State University of New York at Buffalo, USA

Sponsors

Microsoft Research	research.microsoft.com
San Diego Supercomputer Center	www.sdsc.edu
American Medical Informatics Association (AMIA)	www.amia.org
UC Davis Genome Center	genomics.ucdavis.edu
U Maryland Institute for Advanced Computer Studies	www.umiacs.umd.edu
University of California, San Diego	www.ucsd.edu

Organization Committee

Amarnath Gupta	San Diego Supercomputer Center
Jon C. Meyer	San Diego Supercomputer Center
Donna Turner	San Diego Supercomputer Center
Linda Ferri	San Diego Supercomputer Center

Website

For more information on the workshop please visit the workshop website at www.sdsc.edu/dils05.

Table of Contents

Data Integration I–IV

Potpourri

Posters and Demonstrations

Invited Briefings

Challenges in Biological Data Integration in the Post-genome Sequence Era

(Keynote Talk)

Shankar Subramaniam

University of California, San Diego
shankar@sdsc.edu

Abstract. We are witnessing the emergence of the "data rich" era in biology. The myriad data in biology ranging from sequence strings to complex phenotypic and disease-relevant data pose a huge challenge to modern biology. The standard paradigm in biology that deals with "hypothesis to experimentation (low throughput data) to models" is being gradually replaced by "data to hypothesis to models and experimentation to more data and models". And unlike data in physical sciences, that in biological sciences is almost guaranteed to be highly heterogeneous and incomplete. In order to make significant advances in this data rich era, it is essential that there be robust data repositories that allow interoperable navigation, query and analysis across diverse data, a plug-and-play tools environment that will facilitate seamless interplay of tools and data and versatile user interfaces that will allow biologists to visualize and present the results of analysis in the most intuitive and user-friendly manner. This talk will address several of the challenges posed by enormous need for scientific data integration in biology with specific exemplars and strategies. The issues addressed will include:

- Architecture of Data and Knowledge Repositories
- Databases: Flat, Relational and Object-Oriented; what is most appropriate?
- The imminent need for Ontologies in biology
- The Middle Layer: How to design it?
- Applications and integration of applications into the middle layer
- Reduction and Analysis of Data: the largest challenge!
- How to integrate legacy knowledge with data?
- User Interfaces: web browser and beyond

The complex and diverse nature of biology mandates that there is no "one solution fits all" model for the above issues. While there is a need to have similar solutions across multiple disciplines within biology, the dichotomy of having to deal with the context, which is everything in some cases, poses severe design challenges. For example, can a system that describes cellular signaling also describe developmental genetics? Can the ontologies that span different areas (e.g. anatomy, gene and protein data, cellular biology) be compatible and connective? Can the detailed biological knowledge accrued painstakingly over decades be easily integrated with high throughput data? These are only few of the questions that arise in designing and building modern data and knowledge systems in biology.

B. Ludäscher and L. Raschid (Eds.): DILS 2005, LNBI 3615, p. 1, 2005.
© Springer-Verlag Berlin Heidelberg 2005

Curated Databases

(Keynote Talk)

Peter Buneman

School of Informatics and Digital Curation Centre,
University of Edinburgh
opb@inf.ed.ac.uk

Abstract. Measured in dollars per byte, the cost of data in some bio-logical data sets exceeds that of "big science" data by several orders of magnitude. This somewhat pointless observation does at least underline the fact that biological databases are constructed and maintained with a very great deal human effort—they are *curated*. So what are the issues with curated data, and how well does current database technology serve them?

In this talk I shall describe some of the new challenges to database research that arise from curated databases and what my colleagues and I are doing to tackle them. They include annotation, data provenance, database archiving, data publishing and security. I shall also attempt to summarise the work of the recently formed Digital Curation Centre, which is concerned not only with these database-related issues but also with the larger problems of ensuring that our scientific and scholarly data is understandable not only by current users but is "curated" in the sense that it will be usable in the future.

B. Ludäscher and L. Raschid (Eds.): DILS 2005, LNBI 3615, p. 2, 2005.
© Springer-Verlag Berlin Heidelberg 2005

A User-Centric Framework for Accessing Biological Sources and Tools*

Sarah Cohen-Boulakia[1], Susan Davidson[2], and Christine Froidevaux[1]

[1] LRI, CNRS UMR 8023, Université Paris-Sud, Orsay, France
. {cohen, chris}@lri.fr
[2] Department of Computer and Information Science,
University of Pennsylvania, USA
susan@cis.upenn.edu

Abstract. Biologists face two problems in interpreting their experiments: the integration of their data with information from multiple heterogeneous sources and data analysis with bioinformatics tools. It is difficult for scientists to choose between the numerous sources and tools without assistance. Following a thorough analysis of scientists' needs during the querying process, we found that biologists express *preferences* concerning the sources to be queried and the tools to be used. Interviews also showed that the querying process itself – the *strategy* followed – differs between scientists. In response to these findings, we have introduced a user-centric framework allowing to specify various querying processes. Then we have developed the BioGuide system which helps the scientists to choose suitable sources and tools, find complementary information in sources, and deal with divergent data. It is generic in that it can be adapted by each user to provide answers respecting his/her *preferences*, and obtained following his/her *strategies*.

Availability: http://www.lri.fr/~cohen/bioguide/bioguide.html

1   Introduction

Life sciences are continuously evolving so that the number and size of new sources providing specialized information in biological sciences have increased exponentially in the last few years,[1] as well as the number of tools required to carry out bioinformatics tasks. Scientists are therefore frequently faced with the problem of selecting sources and tools when interpreting their data. The diversity of sources and tools available makes it increasingly difficult to make this selection without assistance.

We firstly introduce a framework allowing to specify various querying processes. Our work was developed following a thorough study of scientists' needs during querying and data management. After interviewing scientists working in

---

* This work was supported in part by the European Project HKIS IST-2001-38153, the Fulbright Program as well as a Hitachi Chair at INRIA.
[1] See the annual Nucleic Acids Research database issue (January).

B. Ludäscher and L. Raschid (Eds.): DILS 2005, LNBI 3615, pp. 3–18, 2005.

various domains, we found that they expressed *preferences* concerning the sources queried and the tools used. Moreover, this study emphasized the fact that the process of querying itself – the *strategy* – varies from one scientist to another. We have then designed the BioGuide system, which provides scientists with support during the querying process. BioGuide assists the scientist with data searches within sources, providing information concerning the sequences of sources to be consulted and the tools to be used: the *paths* between sources to be followed.

We first describe the method used to assess scientists' requirements, and present the needs identified (section 2). We then describe the notion of strategy (section 3) and the way in which we propose to manage preferences (section 4). Section 5 introduces the formal framework and presents the general architecture of BioGuide, explaining how it provides support for the querying process. The biological significance of the results obtained will be presented in section 6. Section 7 compares our work to previous work and concludes the paper.

2   User Requirements

2.1   Process: Interviews and Questionnaire

We started with a thorough study of user requirements (cf. BioGuide site). We investigated the way in which scientists query sources and perform bioinformatics tasks (in the spirit of [18] and [6]), paying particular attention to determining why biologists query one source rather than another (*preferences*) and identifying the steps of their querying process (*strategies*).

A questionnaire was developed based on lists of user requirements in three kinds of documents: (i) survey articles [11] and reports of workshops on biological source querying (ii) studies on data quality [14], [4], [15] and (iii) studies on user guidance during the querying process, involving BioMediator [12], BioNavigation [9] and DSS [2]. The questionnaire comprised 28 questions and was constructed according to standard guidelines. As an illustration, four questions are provided:

- Choose a particular context from your own area of study and list some biological queries that you frequently make.
- If several sources yield answers for your query, do you access all of them or only few? If you query only a few, how do you proceed?
- In your mind, what is a "high-quality" source/tool?
- When you look for data related to two linked entities (e.g. a gene and the protein it encodes), how do you proceed (sources accessed, way of correlating information, etc.)?

After collecting responses to the questionnaire, we conducted *interviews* according to classical techniques. We sent questionnaires to 20 individuals, including both biologists and bioinformatics specialists. Their research interests fell into three main domains: *studies of diseases*, *functional* and *structural genomics*.

From the questionnaire, we identified 156 common queries. Some had almost identical structures (e.g. the search for genes involved in *breast* or in *bladder* cancer) and we grouped them together, giving a total of 119 distinct queries.

2.2 Transparent Queries and Traceability

In most cases, neither the sources to access nor the tools to be used were specified by the biologists in their queries. Instead, their queries involved only biological **entities** and **relationships** between entities. An example of such queries is *"Return all contigs that map 'close' to marker M on chromosome 19"* which includes the biological entities CONTIG, MARKER and CHROMOSOME and includes the relationships "maps close to" and "(located) on". We conclude that scientists find it very useful not to have to specify the sources and tools that is, to make **transparent** queries [10].

Follow-up interviews showed that scientists want to ask transparent queries while being aware of the **origin of the answers** obtained. They want to know the *why-provenance* [1] that is, which sources and/or which tools have been used to calculate the data they obtain. Traceability is particularly important for verifying results, drawing conclusions and testing biological hypotheses [19].

2.3 Source and Tool Requirements

A more complex step in the querying process is the **assembly** of information between entities. From the sample queries, we observed that relationships between entities are either explicitly **stored** in the sources or **calculated** by a bioinformatics tool. For example, in the query *"Return all contigs that map 'close' to marker M on chromosome 19"*, the fact that Marker M is on chromosome 19 must be *stored* in the data sources queried by the biologist. Conversely, the relationship of "close mapping" can be *calculated* (e.g. using *Blastn*). For each calculated relationship between entities, we also determined which tools were used to achieve it (e.g. *Blastn*) based on the interview information.

Different kinds of **links** between sources may therefore be distinguished: *internal links* (within the same source), *cross-references* (between different sources) and *tool-links*. *Internal links* may be seen as a way of obtaining information on one entity from another entity within the same source. *Cross-references* are hypertext links from an entity in one source to complementary information in another source, and are not necessarily symmetric (e.g. there are an increasing number of specialized sources which crossreference GenBank but are not referenced in return). Finally, *tool-links* are services provided by a source, yielding links with entities in other sources. Each source may provide several different services achieving a given relationship. For example, GenBank provides different tools (e.g. *Blastx, tBlastn*) to enable users to carrying out "similarity searches" between the genes of GenBank and proteins of various sources.

It is also clear from interviews that scientists have **preferences** concerning entities in sources and tools. One of the key issues facing bioinformaticians is therefore to help the scientists to evaluate their confidence in sources and tools, and to make use of this confidence in a semi-automatic querying process. We return to this in section 4.

3   Strategies

Interviews revealed that each scientist followed paths between sources and
queried the sources by first considering each entity for which information was
sought and then by linking information about entities by means of cross-
references or tools. Since information is collected entity by entity, each entity
is treated exactly once. However, the scientists differed considerably in other aspects of querying, in particular whether or not (i) they followed an order on the
entities, (ii) they were willing to explore other entities, and (iii) they were willing
to visit a source more than once. We term these query criteria *Ord* (*Ordered*),
OnlyGE (*OnlyGivenEntities*) and *SourceOFA* (*SourceOnceForAll*), respectively,
and call the combination of criteria the query **strategy**.

3.1   Querying Entities by Following an Order

The first criterion, *Ord*, determines whether the entities of interest are searched
in the given order or whether all orderings of the entities are considered. It is typically chosen when the scientists know that the desired information is provided
by the given ordering, as opposed to when they want to get as much information
as possible[2]. For example, if the scientists search for the chromosomal location
of the sequence of a given BAC (Bacterial Artificial Chromosome), they may access a few sources containing BAC information and may follow cross-references
to sources providing information about chromosomal location. In this situation,
the scientists order the entities so as to start from the known entity and end
with the entity sought; only links from BAC to CHROMOSOME are followed.
However, if the information sought is not available in the data sources, the biologists may browse the sources to obtain as much information as possible. The
two entities are therefore also considered in reverse order (from CHROMOSOME
to BAC). Thus, they consider all the permutations between entities (from BAC
to CHROMOSOME and from CHROMOSOME to BAC).

3.2   Querying Only Given Entities

The second criterion, *OnlyGE*, determines whether the scientists are interested
in finding information using only the given entities, or whether they are willing
to explore *additional* entities that are *biologically linked* to the entities explicitly
sought. As an illustration, consider the previous example of scientists interested
in finding data on the chromosomal location of a given BAC b. If the scientists
do not find any information about the BAC b by querying sources for entities
BAC and CHROMOSOME, they may consult sources providing information on
other entities, such as GENE, and try to determine the location of genes known
to be present on b. This makes it possible to determine the location of the
BAC b.

---

[2] Note that if the entities are not ordered the non-symmetric aspect of links between
sources can be resolved.

3.3   Querying a Source Once for All

The third strategy criterion, *SourceOFA*, determines whether or not a given source can be visited more than once. The second approach is primarily adopted by scientists who wish to validate information already obtained. Visiting a given source multiple times makes it possible for the biologist to check whether the information obtained - and to which further information has been added via the browsing of several sources - has remained coherent. This process is particularly important when the data reflects expertise, as experts may disagree, resulting in divergent data. Continuing with our example, the scientists may query the source MapView to obtain data about a given BAC and follow a cross-reference to GenBank to find the chromosomal location of that BAC. GenBank is queried here because it contains all the available genomic data. However, GenBank is a large public data repository, containing information originating from many different laboratories; therefore, some of the data it contains may be erroneous. The biologists then follow links from localization information in GenBank to the same kind of information in MapView to compare the results.

3.4   Combining the Criteria

Interestingly, criteria may be combined, generating a wide variety of querying processes. Scientists typically adopt the simple strategy where the criteria *Ord*, *OnlyGE*, *SourceOFA* are chosen. If the results obtained are not satisfactory, the scientists may then drop one of these criteria, e.g. allow the entities to be queried in any order. Section 6 shows how following strategies allows the scientists to find complementary data and to deal with divergent data. We will also see how allowing them to choose his/her **strategy** represents a real challenge in the development of systems providing support for the querying process.

4   Management of Preferences

Our goal is to get as much information as possible from the sources using alternative paths that follow the chosen strategy. Unfortunately, the number of alternative paths may be very large. BioGuide therefore allows users to state **preferences** to filter and rank the paths considered.

4.1   Initializing Preferences

Responses to our questionnaire showed that the reason why a source or tool is preferred varies between scientists. Interviews revealed that about 30 criteria determine preferences (e.g. reliability, completeness and ease of use), mainly in association with entities in sources and links between them. Some users even base their preferences for tool-links on the sources which provide them. We thus asked and helped the users to quantify the confidence that they have in the components of each path, i.e. entities in sources and links between them. To guide the user, initial confidence values for components of a path can be automatically generated using information such as the *average speed* of a tool,

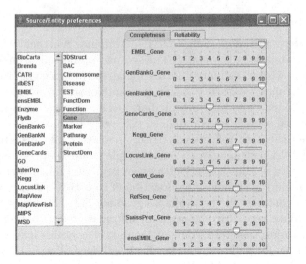

Fig. 1. Initializing Preferences

or the *source-entity cardinality* (i.e. an estimate of the number of instances of an entity in a source) [9]. These initial values may then be improved, adjusted or rectified by comparing the values obtained for all the source-entities related to a given set of entities and/or to a given set of sources. BioGuide provides a user-friendly interface (Fig. 1) through which the user can adjust the improved initial values.

4.2   Using Values of Preferences

Firstly, we introduce the notion of **level of filter preference** and distinguish three different levels: (i) *global*, (ii) *intermediate* and (iii) *local*. The *global level* corresponds to a filter on a path, i.e. on the sequence of sources and links taken as a whole. Filters at the *intermediate level* focus on a given entity or relationship. At the *local level*, filters relate to a given source or a given link, allowing the biologist to name the source/tool to use. Section 6 will illustrate this notion.

If the number of alternative paths is still too large, we can **sort** them according to the biologist's preferences [2], [9]. To do this, we must associate a value with each path. The way in which the global value of a path is computed from the confidence assigned to its components (source-entities and links), i.e. the *sort-operation* used (e.g. the *weighted sum*), can vary (cf. BioGuide site).

5   BioGuide: Querying According to Strategies

In this section we introduce the architecture of BioGuide (see Fig. 2) and then describe more precisely its two main modules: *EntityPathsGenerator*(EPG) and *SourceEntityPathTranslator* (SEPT).

5.1   Architecture

From a query expressed in natural language (Q_{nat}), the scientist first has to extract the underlying biological entities and the relationships between them (Q_{entRel}). In BioGuide, this pre-process is performed by the user, but could easily be automated, as described by [16]. BioGuide supports biologists in this task by providing a graph of entities (described in the next subsection).

The steps (I) to (IV) of the BioGuide process are shown in Fig. 2. (I) The *initial user's query* Q consists of (i) Q_{entRel}, the entities and relationships underlying the user's query; and (ii) the choice of the user concerning entity related strategy criteria (*Ord* and *OnlyGE*). (II) From Q, the *EPG* module yields P_e, the set of paths in the graph of entities generated according to the entity related strategy criteria. (III) The *extended user's query* Q_{se} consists of (a) P_e, the output of the *EPG* module, (b) the choice of the user concerning the strategy criterion *SourceOFA*, and (c) the user's preferences. (IV) Using Q_{se} and the source-entities graph, the *SEPT* module generates the list L_{pse} of paths between source-entities that can be used to retrieve the data.

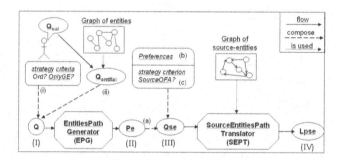

Fig. 2. BioGuide architecture

5.2   *EntityPathsGenerator*: Transparency and Strategies

We now present how the *EPG* module processes and we describe its components.

Graph of Entities: We extracted entities and relationships from the collected queries and used the answers given during interviews to build the *graph of entities*. The nodes are the biological entities and the edges are the biological relationships between them (see Fig. 3). This graph expresses biological knowledge (e.g. *proteins are encoded by genes*), bioinformatics knowledge about tools (e.g. *proteins and genes may be similar*) and knowledge about sources (e.g. *information on disease often cross-reference information on 3D-structure*). Labels on the edges specify the kind of semantic relationship between these entities. The users can make use of this graph to build questions by selecting entities and, possibly, relationships between these entities. Moreover, if they do not want to only consider the given entities of their query, they may characterize the *additional entities* and relationships that they would like to consider or to avoid. This can

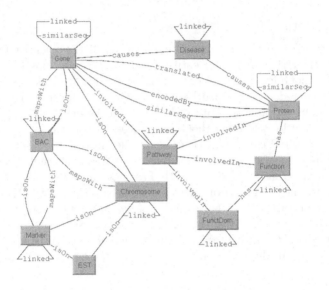

Fig. 3. Graph of Entities (Subpart)

be done by explicitly referring to entities and relationships or by specifying the kind of relationships (e.g. those achieved by tools) used to reach these *additional* entities. We now present more formally the notion of *initial query*.

Input of the EPG Module: Q. The *initial user's query* is Q={L_{Ent}, S_{Rel}, StrategyE, S_{noEnt}, S_{noRel}, PropertiesRel} where L_{Ent} and S_{Rel} denote the list of entities and the set of names of relationships (possibly empty), respectively; StrategyE contains the choice of the user concerning the strategy criteria *Ord* and *OnlyGE*; if *OnlyGE* is not chosen by the user then (a) s/he may specify which entities (or relationships) s/he wishes to avoid, by adding them to the set S_{noEnt} (or S_{noRel}) and (b) PropertiesRel is a conjunction of properties expressing which *kinds* of relationships can be used to reach *additional* entities.

As an illustration, consider the previous example in which the user wishes to find information connecting a given BAC and its Chromosomal location (L_{Ent}=[BAC, CHROMOSOME]) without choosing an order between entities and considering not only the given entities of his/her query (StrategyE ={}). The user may wish to avoid distant entities such as EST (S_{noEnt}={EST}) and may choose to follow only non-tool relationships(S_{noEnt}={}, S_{noRel}={}, Properties-Rel=OnlyNonTool).

The *EPG* module is based on an **algorithm** which aims at calculating from Q the corresponding set of paths in the graph of entities. As an illustration, the following paths are returned by *EPG* from the previous query: (BAC isOn CHROMO)[3], (CHROMO isOn BAC), (BAC isOn GENE), (GENE isOn CHROMO)[4].

---

[3] CHROMO will stand for CHROMOSOME.
[4] Relationships between entities are symmetric.

Output of the EPG Module: P_e. More formally, the *EPG* module calculates P_e, the set of paths in the graph of entities which respect the following four properties. (1) Each path in P_e contains all the entities and relationships specified by the user and visits each entity once only. Moreover, (2) if the user has chosen the strategy criterion *Ord* then the entities in each path must be considered in the order indicated in the list L_{Ent}, and (3) if the user has chosen the criterion *OnlyGE* then each entity of each path must belong to L_{Ent}. Conversely, (4) if *OnlyGE* has not been chosen, paths may consider *additional* entities and relationships (i.e. not specified in L_{Ent} and S_{Rel}). In this case, these entities and relationships must be different from those in S_{noEnt}, S_{noRel} and the edges followed must satisfy conditions expressed in PropertiesRel.

The EPG algorithm is **sound and complete** with respect to these properties.

5.3   *SourceEntityPathTranslator*: Preferences and Strategies

The next step involves finding the sources containing entities and the links giving relationships, which is the aim of the SEPT module that we present with its main components here-after.

The Graph of Source-Entities: After carrying out a thorough study of the sources and tools mentioned in interviews, we designed a graph of source-entities (see Fig. 4). Each node represents an entity in a source. Arrows indicate the links between a given entity in a source and another entity (in the same source or another source). Labels on arrows specify the kind of link. *CrossRef* and *Internal* labels indicate cross-reference and internal links, respectively. Other labels (such as *Blast*) refer to tools.

More formally, let E be the finite set of biological entities (e.g. BAC, GENE), and R be the set of pairs of entities linked by relationships. Let Lab_r be the finite set of labels of relationships between entities (e.g. *SimilarTo*), S be a finite set of data sources (e.g. *GenBank*), N⊆SxE be the set of pairs (source,entity) (e.g.

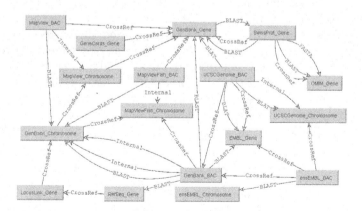

Fig. 4. Graph of Source-Entities (Subpart: only source-entities relating to BAC, CHROMOSOME and GENE)

*(GenBank,*GENE*))*, A be the set of directed links (arrows) between (source,entity) pairs, and Lab$_l$ be the finite set of labels of links (e.g. *CrossRef, Blast*) between (source,entity) pairs. Lab$_l$ contains the names of the links achieving relationships, the names of which are in Lab$_r$. In the rest of the paper we will use the following abbreviations to mention sources: GB, LL, RF, MV, MVF and UG stand for GenBank, RefSeq, LocusLink, MapView, MapViewFish and UCSCGenome.

Definition 1. *The **GraphOfSourceEntities** is a directed labelled graph given by the 3-tuple (N,A,f$_{labl}$), where (1) N is the set of nodes given as (source,entity) (2) A ⊆ N x N is the set of arrows (directed links between nodes) (3) f$_{labl}$: A → Lab$_l$ provides the label of each arrow.*

Definition 2. *A **path in GraphOfSourceEntities** is a sequence of pairs of arrows and labels, $(a_1, l_1), (a_2, l_2), ..., (a_k, l_k)$ such that, for i $(1 \leq i \leq k)$, a_i is an arrow from the node n_{i-1} to the node n_i (adjacent arrows) and such that $n_i \neq n_j$ (no cyclic path), for $i \neq j$, $(0 \leq i, j \leq k)$. The **length** of the path is k, the number of arrows.*

The GraphOfSourceEntities is constructed so that: (i) (s,e) is a node if and only if the source s contains the entity e and (ii) a=(s,e) l (s',e') is an arrow if and only if (1) the source s provides a link labelled by l from entity e to entity e' of source s' and (2) there is a relationship r in the graph of entities between e and e' such that l achieves the relation r.

Using the GraphOfSourceEntities the users can specify their filter preferences. In this step, the users may also define their sort preferences and select whether or not they wish to consider each source once for all. We present more formally the notion of *extended query* based on the graph of source-entities.

Input of the SEPT Module: Q$_{se}$. The **extended query** of the user (cf. Figure 2 step (III)) is Q$_{se}$ ={P$_e$,PrefCond, L$_{rank}$,Op$_{rank}$, StrategyS} where P$_e$ is the set of paths in the graph of entities obtained from Q (cf. section 5.1); PrefCond is a boolean formula expressing filter preferences on paths of source-entities (cf. section 4.1); L$_{rank}$ is a list of pairs (entity, preference criterion) used to rank the paths; Op$_{rank}$ is the *sort-operation* chosen to calculate the value of the preference on each path from the value of preference criteria for its components (pairs of source-entities and links); and StrategyS describes the choice of the user concerning the criterion *SourceOFA* (cf. section 3.3).

The *SEPT* module is based on an algorithm which aims at calculating from Q$_{se}$ the corresponding list of paths in the graph of source-entities, L$_{pse}$. An example of path in L$_{pse}$ is p$_{se}$=(GB,BAC) $\overset{BlastN\_NCBI}{\rightarrow}$ (RS,GENE) $\overset{CrossRef}{\rightarrow}$ (LL, GENE) $\overset{CrossRef}{\rightarrow}$ (GB, CHROMO). Let us mention that this path have been generated using the path p$_e$=BAC *mapsWith* GENE *isOn* CHROMO of P$_e$.

Definition 3. *Let us consider $p_e = e_1 r_1 ... r_{t-1} e_t$ a path of Pe, $p_{se} = (s_1, e_1)$ $l_1 (s_2, e_2) ... l_{n-1} (s_n, e_n)$ a path of P$_{se}$ and m the number of entities in the query. p_{se} **corresponds to** p_e if and only if (1) the set of entities of p_{se} is equal to the set of entities in p_e and entities in p_{se} appear in the same order as in p_e ;*

(2) several source-entities concerning the same entity are possible in p_{se} ($m \leq n$) but they must be consecutive and linked by cross-references ; (3) let (s_i, e_i) l_i (s_{i+1}, e_{i+1}) be an arrow of p_{se} ($1 \leq i \prec n$), if e_i and e_{i+1} are occurrences of two distinct entities, x and y, there must be an arrow x r y in p_e such that l_i achieve r ($\exists j, 1 \leq j \prec m$, $x = x_i$, $y = x_{j+1}$ and $r = r_j$).

Let us return to our example. The path p_{se} corresponds to p_e since: (1) entities are the same and are in the same order; (2) the source-entities related to the GENE entity are consecutive and linked with cross-references; and (3) the BlastN_NCBI tool and a cross-reference achieve the relationships *mapsWith* and *isOn*.

Output of the SEPT Module: L_{pse}. From Q_{se} the SEPT module yields L_{pse} a list of paths in the graph of source-entities. These paths satisfy the three following properties: (1) Paths of L_{pse} *correspond to* paths of P_e according to the previous definition; (2) each path in L_{pse} satisfies the preference filters; (3) the list of paths in L_{pse} is ranked following sort-preferences specified in Op_{rank} and L_{rank}. The SEPT algorithm is **correct and complete** with respect to these properties.

5.4   Towards a Meaning for Source-Entities Paths

We provide below the meaning of paths between source-entities from a relational database perspective: (i) each node (s,e) in the graph of source-entities is a **view** over the source s of the entity e (represented by a table s_e); and (ii) each link is a kind of *join*. More precisely, tool-links are mapped to a particular conditional join, the **similarity join**, in which data are joined if and only if they are very similar [17]. We considered several similarity functions based on those used by tools (Blast etc.). Furthermore, *internal* and *cross-reference* links are mapped to a **link-join**. A *link-join* between two tables si_ek and sj_ek' (respectively related to source-entities (si, ek) and (sj, ek')), with id as identifier (primary key), is defined by using the table *Link(IdBeg, SourceBeg, IdEnd, SourceEnd)* as follows si_ek ⋈ $_{(si\_ek.id= Link.idBegin)}$ *Link* ⋈ $_{(sj\_ek'.id= Link.idEnd)}$ sj_ek'. *Link* contains internal and cross-reference links. A tuple (i_1, s_1, i_2, s_2) is in Link if there is a cross-reference (or internal link) from a biological data identified by i_1 in s_1 to another data identified by i_2 in s_2.

Consequently, depending on whether the **Ord** criterion is chosen or not, different paths are generated. Consider two ordered entities e_1 and e_2: if only one tuple of the form (i_2, s_2, i_1, s_1) concerns s_1 and s_2 in the *Link* table, then no path between s_1 and s_2 is generated. Conversely, if *Ord* is dropped then the path $(s_2, e_2) \rightarrow (s_1, e_1)$ is generated. Furthermore, if the criterion **OnlyGE** is dropped, new data may be found due to the ability to introduce new entities. Conversely, if **SourceOFA** is chosen then some links may be missed. With three entities, paths of the form $(s_1, e_1) \rightarrow (s_2, e_2) \rightarrow (s_1, e_3)$ cannot be calculated.

5.5   Complexity

The **complexity** of BioGuide is related to the number of source-entities paths generated. The worst case occurs when the graphs of entities and source-entities are complete. Table 1 gives the number of entities paths generated by *EPG* according to the strategy followed. q is the number of entities of the query, n+q is the number of entities in the graph of entities.

In any strategy where *Ord* is dropped (cases b and d), all permutations between the q entities of the user's query are considered. In the case where OGE^5 is dropped and *Ord* is taken (c), all the paths with at most i additional entities between q entities are considered (n is the upper bound of i), the first entity and the last one staying fixed. Then, for each entity e, the maximal number of paths of source-entities only focused on e (i.e. each source-entity concerns e) generated is given by the following formula: $\sum_{k=1}^{nbse} \frac{nbse!}{(nbse-k)!}$ where *nbse* is the number of sources that contain the entity e (k is the number of sources involved in the paths of source-entities).

In the worst case, the time complexity is very high. However, the queries identified in this study consider only a small number of entities at the same time (only 8 % of the queries had more than three entities) and the source-entities paths desired by the user rarely exceed 6 source-entities. Moreover, BioGuide generates paths that are shorter than 15 source-entities long in less than 1 second.

Table 1. Number of paths depending on the criteria combination

a. {OGE, Ord}	b. {OGE}	c. {Ord}	d. *no criteria*
1	q!	$\sum_{i=0}^{n} \frac{(i+q-2)!}{(q-2)!}$	$q(q-1)\sum_{i=0}^{n}(i+q-2)!$

6   Results

6.1   Using Strategies

The ability to use different strategies and alternative ways of retrieving data across sources, combined with the ability to use tools and take user preferences into account, was considered very useful by the biologists interviewed. A knowledge of which **tools** may be used for a particular bioinformatics task was considered important in a variety of domains, such as the annotation of newly acquired genome with sequence similarity search and 3D-structure analysis with structure comparison. Moreover, all of the biologists questioned used strategies where they do not **limit them to query the entities** of their query. For example, in cancer studies knowledge about PROTEINs and PATHWAYs is obtained using FUNCTION as an additional entity. In protein-protein docking studies, biologists may use STRUCTURALDOMAINs to link PROTEIN and 3D-STRUCTURE. In annotation projects, the CHROMOSOMAL location of INTRONs is found using data about ESTs. Furthermore, more than half the interviewees frequently

---

[5] OGE stands for OnlyGE.

adopted strategies where no **order** is fixed between entities. Only when the goal of the search was to find very high-quality data did biologists adopt strategies with a fix order between entities. This is the case when searching for samples for expensive experiments (e.g. crystallization of PROTEINs). Finally, strategies where a **source is queried once for all** are adopted by biologists for only a very small number of sources in which they have a high level of confidence. In most cases, strategies where sources are queried several times are adopted to ensure that the results obtained are reliable.

6.2   Example of CGH Analysis

A principal example of the use of BioGuide concerns the task of positioning genomic BAC clones on the draft of the human genome sequence [2]. In CGH (Comparative Genomic Hybridization) array experiments, BACs are used to identify new cancer-related genes and it is of the utmost importance to know the precise position of BACs on the genome sequence. We will study the following query: "Where are **all the BACs** of my CGH array located on the genome sequence?" where the underlying entities are BAC and CHROMOSOME.[6] We initially assumed that the scientist adopted a *simple* strategy choosing all of the criteria (*Ord*, *OnlyGE*, and *SourceOFA*). As for preferences, we assumed that the user indicated the following filters: no source with low completeness whatever the entity is (*global* level), no source providing CHROMOSOME with a medium reliability (*intermediate* level), and the ensEMBL source should not be queried (*local* level). The user also indicated that the results should be sorted by considering *completeness* for BAC and *reliability* for CHROMOSOME. The sort-operation is the weighted sum. Based on these filters and strategy criteria, BioGuide yielded seven source-entities paths. Instantiated data have been got using BioGuide within the HKIS platform[7] [2].

The results given by these paths are complementary, providing information on different instances of BACs. They also give complementary results concerning single instances of BACs. For example, the path (MVF,BAC) $\overset{Internal}{\rightarrow}$ (MVF, CHROMO) localizes BAC RP11-89F21 on chromosome band 17p11.2 whereas the path (UG,BAC) $\overset{Internal}{\rightarrow}$ (UG,CHROMO) is more precise, giving the exact position of this BAC on the chromosome sequence (15,021,683-15,022,225). More globally, these source-entities paths yield the location of about 80% of the BACs.

Let us assume that the user then tries to obtain information about the 20% missing BACs by adopting a more complex strategy without *OnlyGE*. The user also chooses to follow relationships achieved by tools, and not to consider MARKER as an additional entity. A new path of entities is generated with GENE as an additional entity. In the corresponding source-entities paths, all the missing BACs can now be located. For example, due to the path (GB,BAC) $\overset{BlastN\_NCBI}{\rightarrow}$ (RS,GENE) $\overset{CrossRef}{\rightarrow}$ (LL,GENE) $\overset{CrossRef}{\rightarrow}$ (GB,CHROMO) the chromosomal location

---

[6] Sources were queried on January 3, 2005; more details on this example are available from the BioGuide web site.

[7] http://www.hkis-project.com

of BAC RP11-782H1 was found. More precisely, this BAC (entry AC025749 in GB) mapped with (using the BlastN tool from NCBI) the gene P85B (entry NM_005027 in RS, which cross-refers entry 5296 in LL), which is is located on chromosome 19 (in GB PIK3R2 entry).

Finally, let us assume that the scientist then analyzed the results obtained. Several **divergent** locations were produced by these paths for the BACs CTD-2012D15 and CTD-2008I6. Indeed, BAC CTD-2012D15 may be considered to be located on chromosome X or 11. As sources locating the BAC on chromosome X (GB and MV) are considered less reliable than those locating the BAC on chromosome 11 (UG and MVF), the user is likely to consider it more probable that BAC CTD-2012D15 is located on chromosome 11 [2]. Conversely, the sources involved in the paths which locate the BAC CTD-2008I6 on chromosome 3 or 17 (UG and MVF) are considered to be equally reliable. The biologist must therefore explore new paths to correlate these pieces of information, and does it by adopting a strategy without *SourceOFA* and by considering tools-relationships between BAC and CHROMOSOME. Consequently a new path is generated: (UG,BAC) $\overset{Blat\_UCSC}{\rightarrow}$ (UG,CHROMO). The results provided allow the user to conclude that BAC CTD-2008I6 is duplicated in the genome, and is present on both chromosomes 3 and 17.

Due to its multiple-strategies approach, BioGuide enables the users to make the most of the available data and guides them to deal with divergent data.

7   Discussion and Conclusion

Based on a thorough study of scientists' needs, we have designed a **user-centric framework** to specify the notions of *queries*, *preferences* and *strategies*. From this framework we have proposed and implemented the BioGuide system which calculates the *paths* between source-entities. Then, we have presented the architecture of BioGuide and have provided a very easy-to-use **implementation**.

Over the last few years, three systems considering paths between sources have been developed: Biomediator [12], Bionavigation [8] [9] and DSS [2]. We sum-up the differences between our approach and these works. Firstly, the underlying query languages of [9] and [12] [13] are formal query languages: a *regular expressions based query language* and an XML-based path language called PQL, respectively. Following our user-centric approach we have proposed a user-friendly **graphical query language**. This language enables to express the strategy criteria which came out of the user requirements. Any query with a strategy combining the presence/absence of the *OnlyGE* and *Ord* criteria can be expressed using the query languages of [13] and [9]. Note that writing such queries may be a complex task (e.g. if *Ord* is dropped then the user has to enumerate all the possible orders between entities of his/her query). Finally, [12] and [9] require the *SourceOFA* criterion to be present ([12] and [9] do not provide a way of visiting a given source several times in a given path). In DSS, there is only one available strategy where the *OnlyGE* criterion is present and the other criteria are dropped.

Furthermore, each of these systems considers user preferences at different levels: [2] considers only global preferences whereas [9] considers both global and intermediate preferences (called meta-data in [9]). Only BioGuide considers all levels of preferences as far as it allows to deal with local preferences (sources can be named) too. Last but not least, BioGuide differs from the previous works in that it is based on labelled-graphs (graphs of entities and source-entities) in which two given entities (resp. source-entities) may be related by several biological relationships (resp. links like cross-references or tools). Therefore BioGuide yields many more alternative paths.

BioGuide thus provides a framework which is general enough to take into account all the abilities (strategies and preferences) of current systems and enables to specify new preferences and strategies. Its implementation allows these abilities to be managed in a simple yet unified and graphical way. We have shown the benefit of BioGuide by highlighting the biological relevance of the alternative paths obtained, through examples in various biological domains. BioGuide has been implemented and is very flexible allowing users to adapt the graphs and the preferences according to his/her needs. It is available for use at http://www.lri.fr/~cohen/bioguide/bioguide.html.

We are currently adding methods to filter and rank the paths in the spirit of [9]. Moreover, as BioGuide is architecture-independent we are studying its use in different integration systems: browsers (SRS [7]) but also mediators (K2 [3]).

Acknowledgments. We thank Olivier Biton for his help in the implementation of BioGuide. For the interviews, we are very grateful to biologists of IGM, Curie Institute, CIRAD, IBP, MIG, and IBBMC.[8]

References

1. Buneman, P., Khanna, S., Tan, W.: Why and Where: A Characterization of Data Provenance, *Proc. of Int. Conf. on Database Theory (ICDT),316-330*, 2001.
2. Cohen-Boulakia, S., Lair, S., Stransky, N., Graziani, S., Radvanyi, F., Barillot, E., Froidevaux, C.: Selecting biomedical data sources according to user preferences, *Bioinformatics, Proc. ISMB/ECCB04,* **20**, i86-i93, 2004.
3. Davidson, S., Crabtree, J., Brunk, B., Schug, J., Tannen, V., Overton, C., Stoeckert, C.: K2/Kleisli and GUS: Experiments in integrated access to genomic data sources *IBM Systems Journal,* 2001.
4. De Santis, L., Scannapieco, M., Catarci, T.: Trusting Data Quality in Cooperative Information Systems, *Proc. of CoopIS/DOA/ODBASE 2003,* 354-369, 2003.
5. Donelson, L., Tarczy-Hornoch, P., Mork, P., Dolan, C., Mitchell, J., Barrier, M., Mei, H.: The BioMediator System as a Data Integration Tool to Answer Diverse Biologic Queries, *Proc. of MedInfo, IMIA (in CDROM)*, 2004.
6. Ely, J.W., Osheroff, J.A., Gorman, P.N., Ebell, M.H., Chambliss, M.L., Pifer, E.A., Stavri, P.Z.: A taxonomy of generic clinical questions: classification study, *British Medical Journal BMJ* **321 (7258)**, 429-432, 2000.

---

[8] The exhaustive list of interviewed biologists is available on the Web site.

7. Etzold, T., Ulyanov, A. and Argos, P.: SRS: information retrieval system for molecular biology data banks. *Methods Enzymol*, **266**, 114-128, 1996.
8. Lacroix,Z., Parekh K., Raschid,L., Vidal,M.: Navigating through the Biological Maze, *Proc. Int. IEEE Computational Systems Bioinformatics (CSB)*, 594-595, 2004.
9. Lacroix,Z., Raschid,L., Vidal,M.: Efficient Techniques to Explore and Rank Paths in Life Science Data Sources, *Proc. Data Integration in the Life Sciences*, 187-202, 2004.
10. Levy, A.Y.: Combining Artificial Intelligent and Databases for Data Integration *Artificial Intelligence Today*, 249-268, 1999.
11. Lord, P., Bechhofer, S., Wilkinson, M.D., Schiltz, G., Gessler, D., Hull, D., Goble, C., Stein, L.: Applying Semantic Web Services to Bioinformatics: Experiences Gained, Lessons Learnt *Proc. of Semantic Web Conference (ISWC2004)*, 350-364, 2004.
12. Mork, P., Halevy, A., Tarczy-Hornoch, P.: A model for data integration systems of biomedical data applied to online genetic databases, *AMIA Symp*, 473-477, 2001.
13. Mork, P., Shaker, A., Halevy, A., Tarczy-Hornoch, P.: PQL: A declarative query language over dynamic biological schemata, *Proc. AMIA Symp*, 533-537, 2002.
14. Muller, H., Naumann, F.: Data Quality in Genome Databases, *Proc. Int. Conf. on Information Quality*, 269-284, 2003.
15. Naumann, F., Leser, U., Freytag, J.C.: Quality-driven Integration of Heterogenous Information Systems, *Proc. Int. Conf. Very Large DataBases (VLDB)*, 447-458, 1999.
16. Samsonova, M., Pisarev, A., Blagov, M.: Processing of natural language queries to a relational database, *Bioinformatics*,19, i241-i249, 2003.
17. Schallehn, E., Sattler, K-U., Saake, G.: Efficient similarity-based operations for data integration, *Data and Knowledge Engineering*,48, 361-387, 2003.
18. Stevens, R.D., Goble, A.C., Baker, P.G., Brass,A.: A classification of tasks in bioinformatics, *Bioinformatics*, **17(1)**, 180-188, 2001.
19. Zhao, J., Wroe, C., Goble, C., Stevens, R., Quan, D. and Greenwood, M.: Using Semantic Web Technologies for Representing e-Science Provenance *Proc of Semantic Web Conference (ISWC2004)*, 92-106, 2004.

BioLog: A Browser Based Collaboration and Resource Navigation Assistant for BioMedical Researchers

P. Singh[1], R. Bhimavarapu[1], H. Davulcu[1], C. Baral[1], S. Kim[2],
H. Liu[1], M. Bittner[2], and IV Ramakrishnan[3]

[1] Dept of CSE, Arizona State University, Tempe AZ 85287
[2] Translational Genomics Research Institute, Phoenix AZ 85004
[3] CS Dept, Stony Brook University, Stony Brook NY 11794
{prabhdeep.singh, ravi, hdavulcu,
Chitta, skim, hliu}@asu.edu
mbittner@tgen.org, ram@cs.sunysb.edu

Abstract. We often realize that communicating with other colleagues who are studying similar topics helps to identify information relevant to our area of study, which otherwise may not have been found. We wish to accelerate acquisition of collective knowledge in a defined area by identifying specific spheres of inquiry. Such spheres correspond to groups of people who are experts in a field. In this paper we provide a systematic way to gain knowledge from their online search activity, and enable them to organize and share their search findings for further analysis. We have built a prototype system, BioLog, to help biomedical researchers share this implicit knowledge among their peers and store their access patterns into a central system for reuse. BioLog has been deployed it in two labs within TGen as a pilot study. The data has been gathered and analyzed by preliminary text-mining and collaborative filtering methods.

1 Introduction

We often realize that communicating with other colleagues who are studying similar topics helps to identify information relevant to our area of study, which otherwise may not have been found. Hence, there have been many organizational efforts and a variety of tools produced to support sharing of knowledge, as well as data, within communities of shared research areas. The collective knowledge of sets of experts is different from the massive, general, text archives of information that we typically rely on since it is limited to a particular realm of findings. It is further different in that it reflects the experts' current models of what that field suggests and it is dynamic, and constantly changing as a result of researchers search activity. While data sharing among experts is improving constantly, model sharing has not improved. We wish to accelerate acquisition of collective knowledge in well defined areas by identifying specific spheres of inquiry and corresponding groups of people. We also provide a systematic way to gain knowledge from their online search activity, and enable them to organize and share their findings for further analysis.

B. Ludäscher and L. Raschid (Eds.): DILS 2005, LNBI 3615, pp. 19–30, 2005.

One place where experts' models are evident for further analysis and inferencing is their interaction logs with archived information sources. For example, PubMed [1] is a well-known repository of biological literature and serves as an invaluable biological repository. It is frequently used as a first stage tool in creating and refining new hypotheses. An expert's prior understanding of the biological relationships and their emerging models will be implicit in their search patterns of PubMed and other such biomedical resources.

Biologists go to PubMed when they have a model with some supporting evidence but want to seek further support. It is also sought when they have a incomplete model with some missing elements or a fragmented model with missing relationships. They type in keywords and PubMed retrieves a list of keyword matching abstracts. Researchers scan through the list and identify a subset of abstracts that might be relevant to their model – most likely based on the titles and the authors of articles. Once they identify the subset of articles, they follow-up on those articles and read the corresponding abstracts. Sometimes, they home in on their by iteratively narrowing down their keyword searches. However, they find it more informative to talk to their expert colleagues, who are studying similar subjects, to obtain recommendations and leads about other relevant articles that might contain missing links, as illustrated in Figure 1.

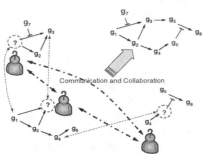

Fig. 1. Communication and collaboration among biologists to combine knowledge of missing links

One problem is that researchers often do not know whom to talk to. It could be someone in their lab or someone at another institute. A precursor to collaboration is to first find whom to work with or ask for help.

In most cases a biologist has some 'handles' (such as a set of nucleotide sequences or gene names) and he or she searches the repositories using those handles. For example, a biologist trying to figure out (parts of) a pathway that explains a particular phenomena may start with a list of gene and protein names as handles. Starting with one of those names, when one searches a repository like PubMed, it is likely that a large number of matches will be found. For example, the search term `g-protein' leads to 51,286 matches in PubMed. The researcher is then faced with the problem of narrowing down the articles that are relevant to his topic of investigation by adding additional keywords or trying alternative keywords. The time it takes to find the right matches plays a huge role in the overall timely success of the quest. A biomedical researcher would benefit tremendously if the various resources would rank the links in a way that matches her own priority. The situation here is closer to recommender systems such as the ones used in Netflix.com or Amazon.com where the system recommends movies and books respectively based on the users' past interaction with the system, the users' feedback (in terms of ratings in case of Netflix.com) and the global knowledge extracted from the web log of all the users as well as the corresponding web content.

We have built a helper application, named BioLog, to archive scientists' access pattern of PubMed of NIH/NCBI as well as the client software that allows users to browse through group specific archives. The system logs the user identity, search keywords used, list of matching articles, set of followed articles, and the amount of time spent on each abstract. We also extract list of gene names using a state-of-the art gene/protein extractor, the Abner [17] system, from each abstract. We developed preliminary recommender algorithms based on gene-to-gene, abstract-to-abstract and user-to-user relevance networks by using a combination of collaborative filtering and content-based filtering techniques. BioLog system automatically recommends alternative lists of genes, articles and other researchers upon each keyword search.

In this paper we propose a recommendation algorithm based upon a clustering technique. Clustering is a technique to group items or data points that are similar in a given context. It has been widely used for many quantitative studies, including gene expression data analysis [9,10]. This is a natural choice of approach to find relevant or similar set of articles or genes given co-observations of genes and articles. A similar set of articles may represent a specific research subject, and a similar set of genes may indicate members of a regulatory network. However, in the context of high dimensional datasets such as those relating PubMed articles, genes, and users, where the datasets are wide and sparse, with many irrelevant dimensions, it is difficult to find relationships that exist in subspace of the dataset. *Subspace clustering* [11] is a form of unsupervised machine learning that seeks to uncover groups of objects that are related in terms of only a subset of the attributes (dimensions) in the dataset. In our effort to identify similar articles or genes, when the number of genes runs over tens of thousands, the number of users in tens of thousands and the number of articles in millions, but the number of users in a group who access articles being relatively rather small, we demonstrate that subspace clustering is useful and effective.

The rest of the paper is structured as follows. Section 2 presents the related work. Section 3 is the system flow. Section 4 is the system design. Section 5 presents relevance networks. Section 6 presents the BioLog's recommendation algorithm. Finally, Section 7 presents our preliminary pilot studies.

2   Related Work

Collaborative filtering (or recommender systems) predicts products or topics a new user might like by using a database about other users past preferences. These systems are popular for their use on e-commerce web sites, where the systems use input about a customer's interests to generate a list of recommended items.

In Memory Based Algorithms [2] the task of collaborative filtering is to predict the votes/interests of the active user from a database of user votes from a sample or population of other users. The strategies mentioned in the memory based algorithms can be used in our current problem of recommending abstracts and users. The user database, which is the log of browsing history in our case, contains information of the various abstracts accessed by the users in the system. We can construct a user-abstract preference/access table, which is analogous to the user-item information mentioned earlier. Based on this information, we could compute the similarity between pairs of users. Based on the similarity, other un-accessed abstracts could be recommended. Using either of the similarity based metrics, similar users can be recommended too. The user-abstract table/matrix constructed from the log would be very sparse since each user would have accessed an insignificant percentage of the total number of abstracts (from PubMed). The Pearson's correlation based or the vector based similarity [3] would not yield good measures if there are very few abstracts in common between two users. Another major pitfall of this approach is in regard with its scalability. Recommendations at runtime for the active user would require the system to scan over the complete database to compute the similarity metrics between the active user and the other set of users and then uses the weights over the common set of abstracts for the selected users.

Probabilistic Cluster Models [4] is a model based method, in which the learning phase can be done offline. Quick recommendations can be given in real time, thereby making the recommendation system scalable. A crucial pitfall in this approach is the Bayesian assumption that the conditional probabilities of the variables given the class are independent. This may well not be the case in our domain. The probabilities of the occurrence of genes given the class, in a given cluster might not be independent with respect to each other. In fact, genes identified in a cluster might be strongly correlated. On the other hand, evaluation results given by the authors for this approach do not seem to be impressive. Other approaches based on correlation outperform this model on most of the datasets.

Clustering is a technique to group items or data points that are similar in a given context. It has been widely used for many quantitative studies, including gene expression data analysis [9,10]. This is a natural choice of approach to find relevant or similar set of abstracts or genes given co-observations of genes and abstracts. A similar set of abstracts may represent a specific research subject, and a similar set of genes may indicate members of a regulatory network.

As datasets become larger and more complex, clustering performance often degrades due to the curse of dimensionality [12, 13]. In high dimensional data, clusters often exist in subspaces [14], and many of the dimensions are often irrelevant. These irrelevant dimensions confuse clustering algorithms by hiding clusters in noisy data. In very high dimensions it is common for all of the instances in a dataset to be nearly equidistant from each other, completely masking the clusters. Feature transformation and feature selection techniques have been used to address the difficulties in clustering high dimensional datasets [11]. However, neither of these techniques is suitable for finding subspace clusters. Feature transformation such as Principle Components Analysis (PCA) attempt to summarize the data by creating new attributes which are combinations of the original attributes in the dataset. Since relative distances are preserved, the effects of the irrelevant dimension remain. Also, the new attributes can be very difficult to interpret. Feature selection techniques attempt to select the most relevant attributes over the whole dataset. While successful at removing noisy attributes [15], feature selection does not allow us to discover clusters that exist in different subspaces. *Subspace clustering* is a form of unsupervised machine learning approach that we utilize in this paper to uncover groups of objects that are related in terms of only a subset of the attributes (dimensions) in the dataset. In our effort to identify similar abstracts or genes, when the number of genes runs over tens of thousands, the number of users in tens of thousands and the number of articles in millions, but the number of users in a group who access articles being relatively rather small, subspace clustering is useful and effective.

Instead of matching the active user to similar customers, item-to-item based approach matches each of the user's purchased and rated items to similar items, and then combines those similar items into a recommendation list. To determine the most-similar match for a given item, the algorithm builds a similar-items table by finding items that customers tend to purchase together. Unlike the traditional collaborative filtering techniques, this algorithm's online computation scales independently of the number of customers and number of items in the product catalog. The above mentioned algorithm can be modified, replacing items with abstracts. This way, we can build up a *similar-abstracts* table by finding abstracts that users tend to look together. As more users tend to access a set of related articles, their pair wise similarity scores go up. Using the similar-abstracts table, related articles can be recommended. As mentioned earlier, this method's online computation scales independently to the number of abstracts and the set the genes, since we would be computing the similarity tables offline. Unlike traditional collaborative filtering techniques, the algorithm also reportedly performs well with limited user data, producing high-quality user data, producing high-quality recommendations. The offline computation of the similarity tables is extremely time intensive, with $O(N^2M)$ as worst case, where N is the number of abstracts/genes and M is the number of users/abstracts respectively for the two above mentioned adaptations to the domain.

3   System Flow

As shown in Figure 5, a biologist initially goes to PubMed types in a keyword search query and PubMed will fetch a list of articles matching the keyword. The biologist scans through the list and identifies a subset of articles that might be relevant to their inquiry, most likely based on the titles and the authors of articles. Once they find the articles of high relevance, they will click on one of the articles and read the abstract to make sure if it is really useful to what they are looking for. Biolog tracks these Web pages in a database log and archives them in a central cache repository with all relevant meta information. Currently we are using a MySQL backend but the module has been built to be database independent. The cached documents are also indexed using a high performance text search engine in order to support keyword searching in the cached documents. Next, gene-to-gene and abstract-to-abstract relevance networks are computed and the recommendation system uses these models.

4 Biolog System Design

We have built a helper application for Internet Explorer® (IE) to archive scientist's accessing pattern of vast archive of biomedical literatures at PubMed of NIH/NCBI. The archival process consists of a logger, which is responsible for capturing web pages during browsing based on domains which are to be tracked. The capturing of data is in terms of logging Meta information in the database as well as caching of web pages in a central repository.

In Figure 2 below the logger uses browser helper objects (BHO) [5] to store html pages in the file system cache as well as all relevant meta information such as machine name, URL, time-stamp etc to the database.

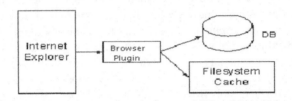

Fig. 2. Logger Architecture

Browser Helper Objects are components — specifically, in-process Component Object Model (COM) components — that Internet Explorer will load each time it starts up. Such objects run in the same memory context as the browser and can perform any action on the available windows and modules. Further, a new instance of the BHO is created each time a new browser window is created. In its simplest form, a BHO is a COM in-process server registered under a certain registry's key. Upon start up, Internet Explorer looks up that key and loads all the objects whose CLSID is stored there.

Logging of dynamic data on the Web has been a problem. By dynamic data we mean the data input by the user at run time during filling of form elements. We planted our logging module into the IE browser and this architecture can be imported to any other browser with plug-in support. The problem of trapping the dynamic data can be tackled during the pre-navigation step, which is, as soon as the dynamic data is submitted and before the response page is loaded. During navigation, we trap the BeforeNavigation event and at that precise moment we capture a snapshot of the current dynamic page DOM and inspect its form elements for dynamic attribute-value pairs.

The logger, a plug-in program to IE, is activated only when scientists go to PubMed and type in keywords to search through the archive. Then, it records the keywords used, the set of articles displayed, and the set of articles that scientists try to read by clicking on the link to its abstract. It also records the time spent on an abstract as well as other relevant information described above. All the archived information is stored in MySQL database for easy access across many clients.

5 Entity-to-Entity Relevance Networks

First, gene-abstract occurrence matrix (GA matrix, **GA**) is constructed for the entries in the log. **GA** matrix is a matrix where its element, ga_{ij}, is 1 if a gene i appears in an abstract j. Otherwise, it is zero. Similarly, we build user-abstract matrix (UA matrix, **UA**). ua_{ij} is 1 if user i read an abstract j. Otherwise, it is zero. Based on these matrices, we find gene-gene, abstract-abstract and user-user relevance networks as follows.

5.1 Gene-Gene Relevance Networks

Once GA matrix is constructed, we then compute gene-gene relevance matrix, GG matrix (**GG**), by multiplying **GA** by the transpose of **GA**, and normalizing it by dividing each row of GG by the number of abstracts. gg_{ij} is 0 if genes i and j never appear in an abstract at the same time. gg_{ij} is 1 if genes i and j appear in all of the abstracts looked at. The value obtained will be in the normalized range of [0,1] , 1 indicating that the two genes co-occur all the time and 0 indicating that the two genes never co-occur together. The idea is to assume if two genes are relevant either positively or negatively, they would tend to appear often in same abstract. Often this assumption may not be true; it is not rare to find an abstract to claim two genes are irrelevant in particular context. However, we found that, even with this crude assumption, some of the genes with high relevance could be identified.

5.2 Abstract-Abstract Relevance Networks

Abstract-abstract relevance, AA_G matrix ($\mathbf{AA_G}$), can be built, by multiplying the transpose of **GA** by **GA**, and normalizing it by dividing each row of $\mathbf{AA_G}$ by the number of genes appeared in either abstracts. aa_{ij} is 0 if abstracts i and j do not have any gene in common. aa_{ij} is 1 if any gene appeared in one abstract appears in the other. This $\mathbf{AA_G}$ matrix corresponds to content-based relevance since the more genes are shared between these two abstracts, the more relevant they are to each other. Another way to define an abstract-abstract relevance matrix is by using the user-abstract access matrix, **UA**. The access matrix **UA** can be multiplied to its transpose to construct another access-based relevance matrix, $\mathbf{AA_U}$. In this preliminary work, we relied on a definition of the abstract-abstract relevance, **AA,** by using a weighted sum of these two different similarity measures $\mathbf{AA_G}$ and $\mathbf{AA_U}$. Similarly User-User relevance matrix can be defined as a weighted sum of commonly accessed gene and abstract based relevance matrices.

6 Biolog Recommendation System

Our hybrid recommendation system utilizes a combination of the above relevance networks and a collaborative filtering based approach.

Content Based Clustering (of Genes and Abstracts): The log gives us information about the abstracts accessed so far by various users. One can extract the list of genes/proteins from these abstracts. The intention here is to find co-occurring genes based on the abstracts they are present in. Similar logic can be used in finding co-occurring abstracts based on their composition of genes in each abstract.

Algorithm (in finding co-occurring genes)
    a.   Build the gene-gene relevance network
    b.   Normalize the obtained GxG matrix using the following formula.

$$S_{uv} = \frac{C_{uv}}{C_{uv} + C_{vv} - C_{uv}}$$. In this equation, Cxy denotes the un-normalized entries of

GxG. Each cell in the matrix is normalized according to the equation shown above. The value obtained will be in the range of [0,1] , 1 indicating that the two genes co-occur all the time and 0 indicating that the two genes never co-occur together.

    c. Perform Hierarchical Agglomerative Clustering (HAC) [16] to reach a fixed number of clusters or some termination condition. Genes that co-occur together fall into one cluster.

This way we can identify similar genes. A similar approach can be done on clustering abstracts. Here we build up a normalized AxA matrix from the AxG matrix. Co-occurring abstracts (based on the composition of their genes) fall into one cluster. Therefore, we could find similar abstracts. In fact, this method was used in the preliminary analysis of archives from our pilot studies.

Collaborative Filtering Based Approach: As contrast to content-based filtering, we can also define the similarity between two abstracts/genes in terms of number of users who have accessed both the abstracts/genes. To recommend similar abstracts, from the log, we build the User by Abstract (UxA) matrix, and compute the AxA normalized co-relational matrix from the UxA matrix. Given any abstract, we could rank the 'k' most similar abstracts based on the correlation similarity measure. Alternatively, User by Abstract (UxA) matrix can be used to find the closest neighbours (similar users), whose preferences can be used to predict the interest/vote on other abstracts. Pearson's correlation co-efficient can be used to find the neighbours, but this strategy would fail if the UxA matrix is sparse.

Hybrid Approach – Combining Content and Collaborative Filtering Based Approach: This approach combines a collaborative filtering and a content based mining in finding similar abstracts. Two abstracts are similar:

i) if they have a good set of genes common in them (Content based perspective) and
ii) if many users view both the abstracts (Collaborative Filtering based perspective). In this way, we consider both the content and the user browsing pattern in associating similarity between abstracts. An approach, using weights to combine two different similarity matrices is detailed Figure 3.

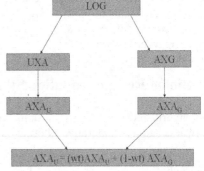

Fig. 3. Similarity matrix computation in the hybrid approach using weights

7 Subspace Clustering for Recommendation with Sparse High Dimensional Data

Finding subspace clusters in the gene-abstract occurrence matrix can reveal relationships between genes and abstracts allowing us to recommend relevant subsets of articles for each query. In search of abstracts with shared genes, we can improve efficiency and accuracy by focusing on clusters of abstracts that share relevant genes. On one hand, the number of genes can be as many as fifty thousand and the number of abstracts can be millions; on the other hand, each abstract usually has a small number of genes (from 1 to 6 genes). That is, although the Abstract-Gene matrix has an extremely high dimensionality, clusters of abstracts can only exist in low dimensional subspaces. By finding these low dimensional subspaces, we can achieve the following: (1) given a new set of genes, subspaces defined by associated genes can be quickly identified; (2) clusters of abstracts in these subspaces can be efficiently located; and (3) similar abstracts can then be ranked and recommended as the number of abstracts in the subspaces is significantly smaller than the total number of available abstracts for search.

Given the Abstract-Gene matrix, abstracts are compared using a similarity measure that considers only the positive (non-zero) values in the matrix. This comparison is done first in low dimensional space, revealing those genes that occur frequently together in abstracts. Searches in the low dimensional space allow us to eliminate genes or gene combinations that are not frequent which helps to reduce the search space. The subspaces represent groups of genes that occur often together in abstracts. The clusters represent abstracts that mention many of the same genes. When analyzed, the smaller data set yields 10 clusters in 2-dimension (using only two words as features), 5 clusters in 3-dimension and 1 cluster in 4-dimension. The size of clusters in 2-D ranges from two to 5 abstracts and the cluster found in 4-dimension is composed of 3

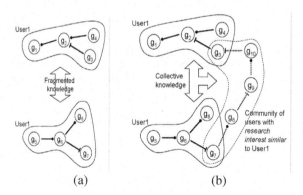

(a)   (b)

Fig. 4. Subspace clustering finds closely related genes based on user's access patterns of articles. Each cluster indicates that the genes grouped together appeared many times in the set of articles accessed by the user. The set of articles in which the clustered genes appear together can be pulled from each cluster as knowledge support. (a) The knowledge of User1 is fragmented due to the lack of relevant knowledge (links) in individual access patterns. (b) Collective knowledge helps User1 realize two pathways are connected

abstracts. For the larger dataset, the cluster with the largest dimension was in 12-dimensions with two abstracts belonging in the cluster. There were 4 clusters in 11-D, 9 clusters in 10-D with at least 2 abstracts. In general, more clusters were found in lower dimensional subspaces.

Adding the Abstract-User matrix further improves the utility of the tools, as illustrated in Fig. 5. As hypothesized, dynamic communication with other colleagues studying similar subjects would help locate relevant information for biologists. Let us consider a user (U1) has accessed many abstracts and accumulated knowledge during his/her previous and current querries. The knowledge acquired through a previous query might often be relevant to the current search based on information that has not been realized by the user. If the proposed approach can identify this information by pulling together and analyzing knowledge (abstracts) utilized by other scientists with a similar research interest, such guidance will speed up adopting new knowledge, such as new pathways.

Also, if two different biologists (U1 & U2) may not have a link (common research interest; same gene or transcription factor) to directly connect them even if they might indeed benefit from talking to each other due to some indirect links, the tool might be able to locate such links by analyzing various links embedded in knowledge access patterns, hence, enable their connection. Synergism resulting from such collaboration would yield much faster knowledge discovery. An illustration similar to Fig. 5, replacing one of User1s with User2 can visualize our approach.

The Figure 4 above exhibits how subspace clustering can be applied effectively to discover implicit knowledge for a researcher. Figure 4 (a) shows that two subspaces exist for User1 alone where a subspace represents a set of genes occurring together. Here, User1 thinks that genes 1,2,3,4 are linked to each other and genes 5,6,7,8 are linked with each other independently with no connection between the subspaces. Figure 4 (b) shows that there exists a subspace generated from all users where the subspace suggests that there is a link between gene3 and gene7. Notice that User1 did not realize or was not aware of the connection between the two genes but by using the knowledge from the community of users, User1 can be given such knowledge. This kind of knowledge could be very useful for User1 because if he was working independently, it might have taken him a longer period of time or in the worst case the user might not have been aware of this knowledge at all. Preliminary experimental results of subspace clustering on large Web logs indicate that such knowledge can be effectively discovered from the data.

8   Pilot Study in Two TGen Labs

Two biology labs at TGen [7] were selected to perform pilot study with BioLog. Both labs are part of the Neurogenomics program at TGen. We set up two central servers to archive their access patterns on PubMed separately.

Since both study brain-related diseases, they could share some commonality. However, they are two different labs studying different specific diseases; therefore, they would differ significantly in accessing literatures in PubMed. We would like to see if the archives show such difference as well as similarity. During one study 25 abstracts accessed, while the other archive returned 253 abstracts accessed.

The gene relevance network from smaller archive is shown in Figure 5. The networks are visualized to emphasize the co-occurrence of two genes; if two genes co-occur more often than others, they were put close to each other in the visualizations. Also, the thickness of edge represents the normalized frequency of co-occurrence of the pair; thicker the edge, more often they co-occur. For example, in Figure 5, genes **smn, sma, smn1, smn2,** and **kinase** are very close to each other, indicating they appear in the same abstract often. We also found it interesting that these genes were found in the second network which is constructed from the archive from the other lab. Therefore, this shows that these two labs sometimes study similar genes. This is critical because it might imply that two lab studying similar subjects, brain-related disease in this case, share the genes of their interests, and we might be able to use this clue to find out other group or people that could study some of the subject common to one's research. However, since they do have many other genes that are not in the other's. This could indicate either that one is studying some other subjects that the other does not (most likely), or that each one is taking a different route to find answers. In the latter case, one might be interested in what other genes the other group is after.

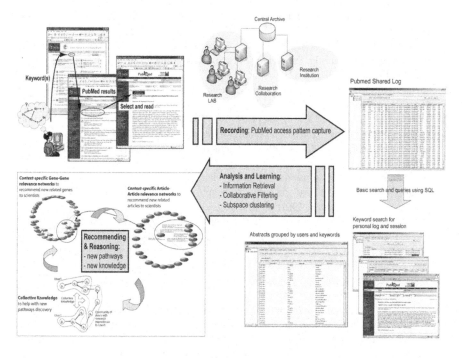

Fig. 5. BioLog: PubMed Recording, Reasoning and Recommending (R3) Navigation Assistant Pilot Study

Figure 5 visualizes abstract-abstract relevance network. Interestingly, we have identified a distinct cluster of abstracts in the relevance network from the smaller archive shown in Figure 5, it was related to the cluster of genes identified in the previous section; all describing smn, sma, smn1, or smn2. Such clusters form the basis of BioLog recommendations.

9 Future Work

The components built as a part of the Biolog system (Figure 5) can also be suitable for domains other than Biology, where a group of people is searching and interacting with a set of entities. Once the recommendation algorithm is embedded into a browser component we plan to perform detailed user evaluations in order to determine the usefulness and validity of BioLog's recommendations as compared to other existing recommender algorithms.

References

[1]   Pubmed by NCBI and National Library of Medicine http://www.ncbi.nlm.nih.gov/entrez/query.fcgi

[2]   Resnick, P., Iacovou, N., Suchak, M., Bergstrom, P., and Riedl, J. Grouplens: An Open Architecture for Collaborative Filtering of Netnews. In Proceedings of the ACM Conference on Computer Supported Cooperative Work, New York, ACM, 1994. (175-186).

[3]   G. Salton and M. J. McGill. Introduction to Modern Retrieval. McGraw-Hill Book Company, 1983.

[4]   Breese, J., Heckerman, D., and Kadie, C. Empirical Analysis of Predictive Algorithms for Collaborative Filtering. Proceedings of the 14 th Conference on Uncertainty in Artificial Intelligence, 1998 (43-52).

[5]   AnHai Doan, Robert McCann. Building Data Integration Systems: A Mass Collaboration Approach. IJCAI 03.

[6]   Browser Helper Objects: The Browser the Way You Want It. Dino Esposito. Microsoft Corporation. http://msdn.microsoft.com/library/en-us/dnwebgen /html/bho.asp

[7]   Legendre, P. & Legendre, L. (1998). Numerical Eoclogy. Second English Edition. Ed. Elsevier.

[8]   Translational Genomics Research Institute (TGen) http://www.tgen.org/

[9]   Bar-Joseph, Z., et al., K-ary clustering with optimal leaf ordering for gene expression data. Bioinformatics, 2003. **19**(9): p. 1070-8.

[10] Getz, G., et al., Coupled two-way clustering analysis of breast cancer and colon cancer gene expression data. Bioinformatics, 2003. **19**(9): p. 1079-89.

[11] Parsons, L., Haque, E., and Liu, H. Subspace clustering for high dimensional data: a review. SIGKDD Explor. Newsl., 1004. Vol. 6, No. 1: p. 90-105.

[12] Devroye, L., Gyorfi, L. and Lugosi, G. A Probabilistic Theory of Pattern Recognition. 1996, Springer-Verlag: New York.

[13] Parsons, L., Haque, E., and Liu, H. Evaluating Subspace Clustering Algorithms. in Workshop on Clustering High Dimensional Data and its Applications, SIAM International Conference on Data Mining (SDM) 2004. 2004.

[14] Friedman, J.H. and Meulman, J.J. Clustering Objects on Subsets of Attributes. Journal of the Royal Statistical Society Series B, Volume 66, Issue 4, page 815, November 2004.

[15] Liu, H. and Yu, L. Toward Integrating Feature Selection Algorithms for Classification and Clustering. IEEE Transaction on Knowledge and Data Engineering, forthcoming.

[16] A.K. Jain and R. C. Dubes. Algorithms for Clustering Data. Prentice Hall, 1988.

[17] Burr Settles. Biomedical Named Entity Recognition Using Conditional Random Fields and Rich Feature Sets. International Joint Workshop on Natural Language Processing in Biomedicine and its Applications (NLPBA), Switzerland. 2004.

Learning Layouts of Biological Datasets Semi-automatically

Kaushik Sinha, Xuan Zhang, Ruoming Jin, and Gagan Agrawal

Department of Computer Science and Engineering,
Ohio State University, Columbus, OH 43210
{sinhak, zhangx, jinr, agrawal}@cse.ohio-state.edu

Abstract. A key challenge associated with the existing approaches for data integration and workflow creation for bioinformatics is the effort required to integrate a new data source. As new data sources emerge, and data formats and contents of existing data sources evolve, wrapper programs need to be written or modified. This can be extremely time consuming, tedious, and error-prone.

This paper describes our semi-automatic approach for learning the layout of a flat-file bioinformatics dataset. Our approach involves three key steps. The first step is to use a number of heuristics to infer the delimiters used in the program. Specifically, we have developed a metric that uses information on the frequency and starting position of sequences. Based on this metric, we are able to find a superset of delimiters, and then we can seek user input to eliminate the incorrect ones. Our second step involves generating a layout descriptor based on the relative order in which the delimiters occur. Our final step is to generate a parser based on the layout descriptor. Our heuristics for finding the delimiters has been evaluated using three popular flat-file biological datasets.

1   Introduction

Bioinformatics research frequently requires accessing data from multiple data sources, and analyzing this data. As the number of data sources is large, and continues to grow, this is becoming an increasingly challenging task. Currently, the number of molecular biology databases is between 500 and 1000 [18]. Even DBcat [5], a *metadatabase* designed to keep track of all biological databases, fails to report all activities in this rapidly evolving field. Biological databases are quite diverse in their goals, structure, and use patterns. A variety of approaches are used for data modeling, storing, and analysis. Out of 111 databases studied by Kroger in 2003 [18], 36% to 40% are implemented as flat files collections, and others use a variety of database technologies.

With increasing amount and heterogeneity of data, biological data management and data integration have become important topics. The Sequence Retrieval System (SRS) [1], K2/BioKleisli [23], TAMBIS [11], DiscoveryLink [13], and Biomediator [22] are some well-known examples of systems for biological data integration. More recently, bioinformatics workflow systems like

B. Ludäscher and L. Raschid (Eds.): DILS 2005, LNBI 3615, pp. 31–45, 2005.
© Springer-Verlag Berlin Heidelberg 2005

Pegasys [21] and IBM BioWBI[1] are being built. These data integration and workflow systems typically use *wrapper* programs to integrate data from multiple sources, or to translate from the format of a data source to the format expected by an analysis program.

A key challenge associated with these existing approaches for data integration and workflow creation is the effort required to integrate a new data source. As new data sources emerge, and data formats and contents of existing data sources evolve, wrapper programs need to be written or modified. This can be extremely time consuming, tedious, and error-prone.

In recent years, the topic of automatic wrapper generation has received much attention in the information integration community [3, 20, 6, 10, 7, 2]. Most of existing approaches in this field are applicable to HTML pages only, and even the other approaches require that a number of pages with identical layout be available. Therefore, these approaches are not directly applicable to flat-file biological datasets.

This paper describes our semi-automatic approach for learning the layout of flat-file bioinformatics datasets. Our approach involves three key steps. The first step is to use a number of heuristics to infer the delimiters used in the dataset. Specifically, we have developed a metric, *delimiter score* or *d_score*, which uses information on the frequency and starting position of sequences. Based on this metric, we are able to find a superset of delimiters, and then we can seek user input to eliminate the incorrect ones. Our second step involves generating a layout descriptor based on the relative order in which the delimiters occur. Our final step is to generate a parser based on the layout descriptor.

We have evaluated our approach for finding the delimiters in a dataset. We have used three popular flat-file biological datasets, Swissprot, GenBank, and Pfam. The effectiveness of our method varied across these datasets. In the case of Swissprot, the 21 delimiters in the dataset were the sequences with the top 21 *d_score* values. For Genbank, all 18 delimiters were among the sequences with the top 37 *d_score* values. Finally, for Pfam, the 31 delimiters were among the sequences with the top 81 *d_score* values. In each of these cases, combining both positional and frequency information turned out to be very important.

The rest of the paper is organized as follows. Our approach for identifying delimiters is described in Section 2. The method for generating a layout descriptor is presented in Section 3. Experimental results from evaluating our techniques for identifying the delimiters are presented in Figure 4. Finally, we compare our work with related research efforts in Section 5 and conclude in Section 6.

2   Finding Delimiters Semi-automatically

Given a flat-file dataset, extracting its underlying structure is a difficult problem. In the domain of biological databases, the data is often written in a file assuming a human will read the data. In order to facilitate human reading, a delimiter

---

[1] See http://www.alphaworks.ibm.com/tech/wsbaw

always precedes a specific data field of interest, specifying what that particular field is. Thus, the first step in our process of wrapper generation is to determine the set of delimiters used in a data file. This section describe our approach for this step. We initially describe some heuristics we tried, and then, describe the approach we finally implemented and evaluated.

2.1 Frequency Counting

As each file contains a set of similar records, we can expect the frequency of delimiters to be quite high. In comparison, across the set of records, we expect several different values for each data field. Thus, if we consider each token (a word separated by a space) in the file, the delimiter tokens are expected to occur more frequently than the other tokens. So, the simplest heuristic we used was to count the frequency of all tokens appearing in the file, and take the most frequent ones. If this set comprises a relatively small superset of the actual set of delimiters, a domain expert could help remove the false positives.

This simple scheme, though intuitive and promising, fails to find the delimiters efficiently. The reason is that many tokens which are not delimiters can also occur very frequently in a dataset. Also, in many datasets, the delimiters could be a sequence of tokens and not just a single token.

2.2 Sequence Mining

As we just mentioned, one reason for the failure of the frequency counting heuristic was that some delimiters could be a sequence of tokens, rather than a single token. So, instead of counting single token frequencies, we could count frequently occurring token sequences.

Sequence mining is a well known data mining problem, and several efficient algorithms exist [15]. However, there were several problems that we had to address in our implementation. The first problem with this approach is that we do not know the delimiter sequence length in advance. This problem can be addressed by finding token sequences in an iterative manner. We can start with finding all sequences of a specified *min_length*, and sorting them in a list S, the set of possible delimiter sequences. In the next iteration, we find all token sequences of length *min_length* +1. Again, domain knowledge or interaction with a domain expert can be used to decide when to terminate the search.

We further use the following two rules to focus on sequences that are most likely to be delimiters. In the following, s_i represents any token sequence of length i, $f(s_i)$ represents the frequency of the sequence s_i, and s_{i-1}^j represents the jth subsequence of s_i with the length $i - 1$. The rules are as follows and are applied when i is 3 or greater.

1. If $f(s_i) = f(s_{i-1}^j) \ \forall j$, then remove $s_{i-1}^j \ \forall j$ from S and insert s_i into S.
2. If $\exists j$ such that $f(s_i)$ is much smaller than $f(s_{i-1}^j)$, then remove all s_{i-1}^j from S except the one having the highest frequency. Do not insert s_i into S.

The following example illustrates the above two rules. Suppose ABC is a token sequence of length 3 found in the current iteration with $f(ABC) = 10$. We need to decide if ABC is to be inserted into S. If AB, BC, and CA are already in S with $f(AB) = 10$, $f(BC) = 10$, and $f(CA) = 10$, then we remove AB, BC, CA from S and insert ABC into S. This is done because the information about AB, BC, and CA is already embedded in ABC.

Consider, alternatively, the following scenario. Again, let $f(ABC) = 10$. If AB, BC, and CA are already in S with $f(AB) = 20$, $f(BC) = 10$, and $f(CA) = 10$, then we remove BC, CA from S and do not insert ABC into S. The intuition for using this rule is that since BC and CA occur less frequently than AB, ABC is not likely to be a delimiter sequence. Also, since AB occurs more frequently that BC and CA, it is more likely that, AB will be a possible delimiter sequence and not BC or CA.

This approach gave better results than just using frequency of tokens, but had several limitations as well. This approach does not work well if token frequencies are distributed in a *skewed* fashion. To illustrate this problem, consider the following. In the Swissprot dataset $/n$,DR, and $EMBL$ are tokens, with $f(/n, DR) > f(DR, EMBL)$ and $f(/n, DR, EMBL) < f(/n, DR)$. Thus, we only keep $/nDR$ as a possible delimiter, which turns out to be correct. However, in the case of Pfam dataset. $/n$, $\#=GF$, and AC are tokens, with $f(/n,\#=GF) \gg f(\#=GF,AC)$ and $f(/n,\#=GF,AC) \ll f(/n,\#=GF)$. Thus, we only keep $/n$ $\#=GF$ as the possible delimiter. But, this is incorrect because $/n$ $\#=GF$ AC is a valid delimiter. This happens when some of the delimiter token subsequences have very high frequency as compared to the other delimiter token subsequences. Moreover, since biological databases are created for humans to read, it is very unlikely that the tokens will be distributed at different positions within the line in the file they appear in. This fact is not exploited by just using the frequency of sequences.

2.3   d_score Based Pruning

We now describe the final approach we implemented. This approach does not completely rely on the frequency of sequences, but does give a high weightage to such frequencies. In addition, it exploits the fact that delimiters are very likely to start only at certain positions within a line. The overall metric we use is referred to as *d_score* or *delimiter score*, and has two components, *positional weight*, and *frequency weight*.

Positional Weight: As we have stated, biological datasets are often written for a human to read. Thus, all the delimiters are expected to appear in a specific position in a file. We capture the positional information in the following way.

Let P be set of different positions within a line where a token can appear. Clearly, this is equal to the maximum number of tokens appearing in any line in the file. For any position $i \in P$, let $tot\_seq_i^j$ represent total number of token sequences of length j starting at position i. Similarly, $tot\_unique\_seq_i^j$ represents the total number of unique token sequences of length j, starting at the position i. We define for any tuple (i, j), denoting sequences of length j starting at the position i in a line, a metric $p\_ratio(i, j)$.

$$p\_ratio(i, j) = \frac{tot\_seq_i^j}{tot\_unique\_seq_i^j}$$

For any specific sequence length j we take the log of $p\_ratio(i, j)$ and normalize it as follows,

$$p\_wt(i, j) = \frac{log(p\_ratio(i, j)) - \min_{i \in P} log(p\_ratio(i, j))}{\max_{i \in P} log(p\_ratio(i, j)) - \min_{i \in P} log(p\_ratio(i, j))}$$

Clearly, $p\_wt(i, j) \in (0, 1)$ with the property that if the delimiters in a file usually start at the position i then, $p\_wt(i, j) > p\_wt(k, j)$, for any $k \neq i$.

Frequency Weight: Let S^j represent the set of all token sequences of length j. For any sequence s_i^j, which has length j and starts at the position i within its file, we can find the log normalized frequency weight as follows:

$$f\_wt(s_i^j) = \frac{log(f(s_i^j)) - \min_{s_i^j \in S^j} log(f(s_i^j))}{\max_{s_i^j \in S^j} log(f(s_i^j)) - \min_{s_i^j \in S^j} log(f(s_i^j))}$$

Similar to the positional weight, $f\_wt(s_i^j) \in (0, 1)$ with the property that if $f(s_i^j) > f(s_k^j)$ then, $f\_wt(s_i^j) > f\_wt(s_k^j)$.

d_score: For any sequence s_i^j, once we have $p\_wt(i, j) \in (0, 1)$ and $f\_wt(s_i^j) \in (0, 1)$, we take a linear combination of these to define d_score:

$$d\_score(s_i^j) = \alpha \times p\_wt(i, j) + (1 - \alpha) \times f\_wt(s_i^j)$$

where $\alpha \in (0, 1)$. The value of α can be chosen to vary the relative weight of positional and frequency weights. Overall, d_score has the property that $d\_score(s_i^j) \in (0, 1)$, and $d\_score(s_i^j) > d\_score(s_k^j)$ implies that s_i^j is more likely to be a delimiter than s_k^j.

2.4 Finding Delimiters Using d_score

Even though the d_score value is closely correlated with the likelihood of a sequence being a delimiter, several issues still need to be addressed. First, a sharp and clear cut-off point, separating delimiters and other sequences, is often not achieved. Second, we still do not know the most number of tokens a delimiter might have.

We have used the following method to address this problem. We proceed in an iterative fashion, trying to find delimiters of length i in the iteration i. Let the cut-off point for d_score values be c_i. This cut-off is determined by finding a substantial difference in the d_score values between two consecutive sequences in a sorted list. Let the number of sequences found with d_score greater than c_i be N_i. We consider this as the set of potential delimiters.

The termination condition used in our algorithm is based on the following heuristic. If the highest length of a delimiter sequence is i, and if we use the same cut-off score, we will expect $N_{i+1} < N_i$. This observation can be explained in the following way. If delimiters are of length i, then any token sequence of length $i + 1$ (delimiter of length i appended by some token etc) will be less frequent, thus having much lower f_wt, whereas, p_wt will remain almost the same. Note that this observation should not be used with a very small value of i, otherwise, we could terminate the process too soon. It should also be noted that in the case of a skewed frequency distribution, this observation may not hold true. However, it serves as a reasonable heuristic for most datasets. Once we have a termination criteria, the delimiters can be found iteratively as shown in the following algorithm.

Algorithm

1. Set initial value of i. Set $S = \varnothing$.
2. Find the potential delimiters (token sequences with $d\_score > c_i$) of length i and store them in S.
3. Set $Last\_N_i = N_i$. Set $i = i + 1$.
4. Find potential delimiters of length i.
5. If $N_i > Last\_N_i$ store the set of potential delimiters in S. Go to the step 3.
6. Sort S in descending order of d_score.

Once the above algorithm stops, the list S contains potential delimiters sorted by their d_score. As we had stated earlier, a domain expert can help in identifying the frequent sequences which are not delimiters. These values are usually frequently occurring values in data fields, and can be identified by using domain knowledge.

3   Towards Generating Wrappers

After the delimiters have been identified, the next step is to generate a wrapper. This involves understanding the structure of the dataset. Once the structure is identified, a parser for the dataset can be generated automatically. In this section, we initially describe the technique we use for determining the structure of the dataset. This technique is based on constructing an Non-deterministic Finite Automata (NFA) from the relative order of occurrence of delimiters. Then, we give a quick overview of our work on generating parsers from such descriptors.

3.1   Generating Layout Descriptors

The set of states of the NFA is the set of delimiters. We insert an edge from a delimiter A to the delimiter B if B is the immediate next delimiter following A in any record in the data file. Because of optional or repeated fields, there could be multiple out-going edges from a node.

To understand the structure, we carry out the following analysis. Initially, a topological sort is done on the nodes in the NFA, breaking cycles by the order of first appearance of a delimiter in the dataset. Based on the topological sort, we classify the edges in the NFA to be in two groups:

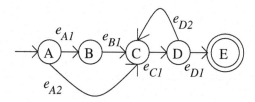

Fig. 1. NFA with Optional and Repeating States

Forward Edge: An edge between the delimiters A and B is a forward edge if A appears before B in the topological sort.

Backward Edge: An edge between the delimiters A and B is a backward edge if B appears before A in the topological sort.

We use the notion of forward and backward edges to define *strict precedence* of appearance. If there exists a forward edge between delimiters A and B, A is said to strictly precede B. The precedence relationship is based on the transitive closure of the strict precedence relationship.

The NFA and the above definitions can be used to determine the structure. In the following, with the help of Fig 1, we describe how we can extract simple structures, as well as optional and repeating structures.

For simple structures without any repeating or optional fields, we can simply use the precedence of appearance to find which delimiter appears before which delimiter. For more complex structures involving optional and repeating patterns, we use the following two rules.

1. *Repeating Fields:* If any node A has a backward edge to the node B, and we have a set of nodes $\{Y_i | i = 1, 2, \ldots, n\}$, such that B strictly precedes Y_1, Y_i strictly precedes Y_{i+1} $(i = 1, 2, \ldots, n - 1)$, and Y_n strictly precedes A, then we say that the set of nodes B, $\{Y_i | i = 1, 2, \ldots, n\}$, and A, in that order, repeat themselves.
2. *Optional Fields:* If there exists a forward edge from A to C, and there exists another state B such that A precedes B and B precedes C, then B is an optional state.

As a simple application of the above two rules, using the NFA in Figure 1, we find that the node B as an optional node and nodes C and D together, in that order, are repeating nodes. The fields following the delimiters which correspond to these nodes are called optional and repeating delimiters, respectively.

To represent the structure, we use annotation which is much similar to regular expressions. If by x, we represent any unknown data field value, then Xx represents unknown data field value x followed by delimiter X. If a delimiter followed by its data field value is optional, we represent it by $(Xx)^{opt}$. Likewise, if a number of delimiters $X_1, \ldots, X_n$ are repeating, in that order, we represent it by $(X_1x\ldots X_nx)^r$. With this notation, structure corresponding to Fig 1 can be represented as

$$Ax(Bx)^{opt}(CxDx)^r Ex$$

3.2   Generating Parsers

After generating the descriptor of the type described above, the next task is to be able to parse the data. As we have discussed, the datasets show a pattern of alternate delimiters and variables, which we will refer to as a DLM-VAR pairs. We have designed a tree data structure to capture the layout. As an example, the tree view for the TRANSFAC data is shown in Figure 2. TRANSFAC [16] is a database on eukaryotic transcription factors, their genomic binding sites, and DNA-binding profiles. In the tree, the leaves are DLM-VAR pairs. The last leaf for TRANSFAC is a generalized DLM-VAR pair with a dummy variable. The internal nodes in the tree, also called the *environment* nodes, indicate how the children are repeated. The advantages of this view are that it is easy to interpret and build. The depth first scan of the tree resembles the data layout. It also simplifies the conversion process by interpreting the data at variable, instead of data field level. Working at the finer level, the wrapper avoids the overhead of reconstructing data fields. This reconstruction would consist of two processes, the process of composing a field by merging variables when reading and the reverse process of partitioning a field into several variables when writing.

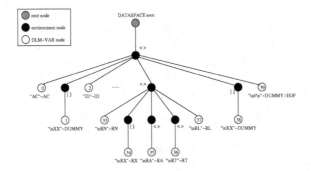

Fig. 2. Logical View of TRANSFAC Data Layout as a Tree

4   Results

Our experiments focused on evaluating the effectiveness of our *d_score* based method for semi-automatically determining the set of delimiters in the dataset. As we had discussed, this method reports the list of potential delimiters, sorted by their *d_score*. The metric we use for determining the effectiveness of the method is as follows. In the sorted list of sequences, we consider the position at which the last delimiter appears, and compare it with the number of potential delimiters. We believe this represents the ease with which a domain expert can prune false positives. Besides evaluating the effectiveness of *d_score* based mechanism, we also considered other heuristics that we had described in Section 2. Results from these heuristics are presented towards the end of this section.

We used three different datasets for our experiments. The first is Swissprot protein dataset from the Uniprot Knowledgebase[2]. The UniProt Knowledgebase is the central database of protein sequences with sequence and functional annotation. Swissprot is a section of Uniprot containing manually-annotated records with information extracted from literature and curator-evaluated computational analysis. The second dataset we used is the Genbank dataset from the National Center for Biotechnology Information (NCBI)[3]. Data is stored in a flat-file format where each data field is preceded by a tag which describes what that data field stands for. The third data set is the Pfam dataset[4]. Pfam is a collection of protein family alignments which were constructed semi-automatically using hidden Markov models (HMMs). The alignment is in Stockholm format. This includes mark-ups of four types:

$$\#=GF < featurename >< freetext >$$
$$\#=GC < featurename >< freetext >$$
$$\#=GS < seqname >< featurename >< freetext >$$
$$\#=GR < seqname >< featurename >< freetext >$$

where $freetext$ means any data field. Introducing mark-ups before the $featurenames$ make this data set different from the other two and difficult to find delimiters from.

We initially report on the effectiveness of the d_score approach. For all the experiments, we gradually vary α value from 0 to 1. Setting $\alpha=0$ implies that we only consider the frequency information, whereas, $\alpha=1$ implies that only positional information is considered. For other values, α linearly combines both positional and frequency information. As we will show in the results, non-extreme α values provide best results.

Swissprot data: Among the three datasets, the Swissprot dataset is the simplest. All the delimiters are two character long, and they appear at the beginning of a line. Table 1 shows the result for the Swissprot dataset. The values in the second column is the metric we stated earlier, which is the position in the sorted list where the last delimiter appears. The results from Table 1 show that if both frequency and positional information is used, i.e., α is neither 0 nor 1, this approach is very effective. The results are the same as the value of α is changed from 0.3 to 0.9. In these case, the 21 actual delimiters appear as the top 21 sequences in the sorted list.

Genbank data: Genbank data is more involved as compared to the Swissprot data. This is because all delimiters are not of the fixed size. Further, certain words appear in the same position in every record, but are not delimiters. Thus, the performance of our approach is not as good. Table 2 shows the results for Genbank dataset. Even though all delimiters were successfully found, unlike

---

[2] http://us.expasy.org/sprot/
[3] http://www.ncbi.nlm.nih.gov/Genbank/index.html
[4] http://www.sanger.ac.uk/Software/Pfam/ftp.shtml

Table 1. Results from Using *d_score* Approach on the Swissprot dataset

α	Position where last delimiter found	All 21 found?
0.0	-	20 found within top 100
0.1	29	yes
0.2	25	yes
0.3	21	yes
0.4	21	yes
0.5	21	yes
0.6	21	yes
0.7	21	yes
0.8	21	yes
0.9	21	yes
1.0	-	10 found within top 50, 20 found within top 100

Table 2. Results from Using *d_score* Approach on Genbank dataset

α	Position where last delimiter found	All 18 found?
0.0	62	yes
0.1	60	yes
0.2	53	yes
0.3	48	yes
0.4	43	yes
0.5	39	yes
0.6	38	yes
0.7	37	yes
0.8	37	yes
0.9	37	yes
1.0	-	Only 1 found within top 50, 14 found within top 100

Swissprot data , the 18 actual delimiters were not necessarily found within the top 18 positions. Best results are obtained when relatively high weightage is given to the positional information. However, completely ignoring frequency information gives poor results.

Pfam dataset: Pfam dataset is the most difficult one to work with. The reason is that we have delimiters that are a combination of words. On one hand, we have delimiters like "//", which stand for the end of a record entry. On the other hand, we also have data fields in between a delimiter sequence, which is not part of the delimiter. For example, "#=GF AC" could be a delimiter, and "#=GS * AC" could be a delimiter, where * represents an actual value. Identifying the latter can be a complex task. Because of this added complexity, the performance for *d_score* method over Pfam data is worse, as shown in Table 3.

Table 3. Results from Using *d_score* Approach on Pfam dataset

α	Position where last delimiter found	All 31 found?
0.0	-	2 found
0.1	-	2 found
0.2	-	28 found
0.3	116	yes
0.4	82	yes
0.5	82	yes
0.6	82	yes
0.7	82	yes
0.8	82	yes
0.9	82	yes
1.0	-	Only13 found within top 100

Table 4. Comparison of different heuristics

	Swissprot (21 delimiters)	Genbank (18 delimiters)	Pfam (31 delimiters)
Frequency	All found within top 41	All found within top 71	Simple frequency count could not find the delimiters
Sequence based pruning	20 found within top 100	14 found within top 100	Dropping lower frequent subsequence could find only 1 delimiter
d_score	All found within top 21	All found within top 37	All found within top 81

While for the other two datasets, *d_score* based method works at least somewhat effectively for α=0.1 and α=0.2, it is not the case for the Pfam data. The reason is that markups in Pfam data, like "#=GF" and "#=GS" have a very high frequency, as compared to the other tokens or words. Because of this, frequency information alone cannot find all delimiters. Thus, for this dataset, *d_score* based method works only for high α values, as we have shown in Table 3.

Comparison with other Heuristics: Finally, we compare the *d_score* based method with two other simple heuristics we had described earlier. These heuristics were, simply using the frequency of tokens, and using frequency of tokens with some pruning. Table 4 compares these approaches with the *d_score* approach.

As we can see, these simple heuristics are not very effective, and the *d_score* based approach has much better results.

5   Related Work

Automatic wrapper generation has been an active research topic. Currently, most of the automatic wrapper generation research has focused on extracting information from tabular structures in HTML files [3, 20, 6, 10]. ROADRUNNER [7] generates record layout structure by comparing HTML pages. Data fields are annotated by the user after this inference process. Heuristic about HTML pages are crucial to ROADRUNNER. For example, it relies on tags to tell field name from field instance, the presence of closing tags to distinguish optional and repeating patterns. These features make it hard to extend the application of this approach to data files other than HTML files. Arasu *et al.* have proposed an approach [2] where no heuristics on HTML were used. However, multiple pages generated from a same template must be collected for template construction. This, although useful for web-service-based applications, is not directly suitable for some bioinformatics applications when all records are listed in only one flat file. The Web extractor developed by Hammer [14] could be used for flat files besides HTML pages. However, it requires a declarative specification which states how to extract information hierarchically.

A number of efforts exist on mediator-based bioinformatics integration, as reviewed in [17]. Our goal is to enable integration of a larger number of sources, and allow data formats to evolve over time, through automatic or semi-automatic wrapper generation. In comparison, the existing mediator-based systems require hand-written wrappers. K2/BioKleisli uses a specialized language called Collection Programming Language (CPL) [4]. It requires source specific wrappers and uses these to map queries to heterogenous sources [23]. TAMBIS [11] also needs external wrappers. The query plan in this system is also written in CPL, which is supplied with a library that has wrapper services. Biomediator [22] relies on wrappers to convert all data from various sources to XML format before further processing. DiscoveryLink[13] allows its users to define their own wrappers and re-configure the system through a registration process at a relatively higher level. Yet, the wrapper still has to be hand-coded. BACIIS [19] is the only federated biological databases that we are aware of that is able to automatically derive extraction rules and store them in the source wrappers. However, the data source schema files used by BACIIS can only describe HTML pages and the individual schema is mapped to a common domain ontology contained by BACIIS.

Besides these mediator-based systems, there are other efforts on biological information integration and query processing. Genomics Unified Schema (GUS) uses datawarehousing [8]. Knowledge-based Integration of Neuroscience Data (KIND) combines wrappers for each source with ontologies [12]. The Sequence Retrieval System (SRS) [1] is a keyword-based retrieval system, which is based on a locally stored index to retrieve entries. Eckman *et al.* have focused on optimizing the execution of queries that access multiple biological databases in a distributed environment [9].

The myGrid project has been developing technologies for integrating a variety of services in the web, through the use of web service composition language [24]. IBM has been developing Bioinformatic Workflow Builder Interface (BioWBI) for creating web service based workflows for biological researchers[5]. These efforts typically require: 1) Use of XML for exchange of data between different sources, which can introduce high overheads, 2) Java wrappers on existing applications, which can also introduce overheads, and 3) users' familiarity with web services. Our proposed system can overcome each of the above three limitations, though it cannot provide as much interoperability as is possible through web services.

6 Conclusion

This paper has described our semi-automatic approach for learning the layout of flat-file bioinformatics datasets. Our approach involves three key steps. The first step is to use a number of heuristics to infer the delimiters used in the dataset. Specifically, we have developed a metric, *delimiter score* or *d_score*, which uses information on the frequency and starting position of sequences. Based on this metric, we are able to find a superset of delimiters, and then we can seek user input to eliminate the incorrect ones. Our second step involves generating a layout descriptor based on the relative order in which the delimiters occur. Our final step is to generate a parser based on the layout descriptor.

We have evaluated our approach for finding the delimiters in a dataset. We have used three popular flat-file biological datasets, Swissprot, GenBank, and Pfam. The effectiveness of our method varied across these datasets. In the case of Swissprot, the 21 delimiters in the dataset were the sequences with the top 21 *d_score* values. For Genbank, all 18 delimiters were among the sequences with the top 37 *d_score* values. Finally, for Pfam, the 31 delimiters were among the sequences with the top 81 *d_score* values. In each of these cases, combining both positional and frequency information turned out to be very important.

References

1. Srs at the european bioinformatics institute. http://srs.ebi.ac.uk/.
2. Arvind Arasu and Hector Garcia-Molina. Extracting structured data from web pages. In *Proceedings of the 2003 ACM SIGMOD international conference on Management of data*, pages 337–348, 2003.
3. Naveen Ashish and Craig A. Knoblock. Semi-automatic wrapper generation for internet information sources. In *Proceedings of the Second IFCIS International Conference on Cooperative Information Systems*. IEEE Computer Society, 1997.
4. P. Buneman, S.B. Davidson, K. Hart, C. Overton, and L. Wong. A data transformation system for biological data sources. In *Proceedings of the Twenty-first International Conference on Very Large Databases*, 1995.
5. Discala C., Benigni X., Barillot E., and Vaysseix G. DBcat: a catalog of 500 biological databases. *Nucleic Acids Res*, 28, Jan 200.

---

[5] See http://www.alphaworks.ibm.com/tech/wsbaw

6. Liangyou Chen, Hasan M. Jamil, and Nan Wang. Automatic wrapper generation for semi-structured biological data based on table structure identification. In *14th International Workshop on Database and Expert Systems Applications*. IEEE Computer Society, 2003.

7. Valter Crescenzi, Giansalvatore Mecca, and Paolo Merialdo. Roadrunner: Towards automatic data extraction from large web sites. In *Proceedings of 27th International Conference on Very Large Data Bases*, pages 109–118, 2001.

8. S. Davidson, J. Crabtree, B. Brunk, J. Schug, V. Tannen, C. Overton, and C. Stoeckert. K2/Kleisli and GUS: Experiments in Integrated Access to Genomic Data Sources. *IBM Systems Journal*, 40(2):512–531, 2001.

9. Barbara A. Eckman, Zoe Lacroix, and Louiqa Raschid. Optimized seamless integration of biomolecular data. In *Proceedings of the 2nd IEEE International Symposium on Bioinformatics and Bioengineering*, 2001.

10. X. Gao and L. Sterling. Autowrapper: automatic wrapper generation for multiple online services. In *Asia Pacic Web Conference '99*, 1999.

11. C. A. Goble, R. Stevens, G. Ng, S. Bechhofer, N. W. Paton, P. G. Baker, M. Peim, and A. Brass. Transparent access to multiple bioinformatics information sources. *IBM Systems Journal*, 40(2), 2001.

12. A. Gupta, B. Ludascher, and M. E. Martone. Knowledge-Based Integration of Neuroscience Data Sources. In *Proceedings of Conference on Scientific and Statistical DataBase Management (SSDBM)*, 2000.

13. L. M. Haas, P. M. Schwarz, P. Kodali, E. Kotlar, J. E. Rice, and W. C. Swope. Discoverylink: A system for integrated access to life sciences data sources. *IBM Systems Journal*, 40(2), 2001.

14. Joachim Hammer, Hector Garcia-Molina, Junghoo Cho, Arturo Crespo, and Rohan Aranha. Extracting semistructured information from the web. In *Proceedings of the Workshop on Management of Semistructured Data*, 1997.

15. Jiawei Han and Micheline Kamber. *Data Mining: Concepts and Techniques*. Morgan Kaufmann Publishers, 2000.

16. T. Heinemeyer, E. Wingender, I. Reuter, H. Hermjakob, A. E. Kel, O. V. Kel, E. V. Ignatieva, E. A. Ananko, O. A. Podkolodnaya, F. A Kolpakov, N. L. Podkolodny, and N. A. Kolchanov. Databases on transcriptional regulation: Transfac, trrd, and compel. *Nucleic Acids Research*, 26:364–370, 1998.

17. Thomas Hernandez and Subbarao Kambhampati. Integration of biological sources: Current systems and challenges ahead. *SIGMOD Record*, 33(3):51–60, 2004.

18. Peer Kroger and Francois Bry. A computational biology database digest: Data, data analysis, and data management. *Distributed and Parallel Databases*, 13:7–42, 2003.

19. Zina Ben Miled, Nianhua Li, Yang Liu, Yue He, Eric Lynch, and Omran A. Bukhres. On the integration of a large number of life science web databases. In *Proceedings of the First International Workshop on Data Integration in the Life Sciences*, 2004.

20. Daniel Rocco and Terence Critchlow. Automatic discovery and classification of bioinformatics web sources. *Bioinformatics*, 19(15), 2003.

21. Sohrab P Shah, David YM He, Jessica C Druce, Gerald Quon, Drew Lett, Grace XY Zheng, Tao Xu, and BF Francis Ouellette. Pegasys: Software for Integrating and Executing Analyses of Biological Sequences. *BMC Bioinformatics*, 40(5), 2004.

22. Ron Shaker, Peter Mork, J Scott Brockenbrough, Loren Donelson, and Peter Tarczy-Hornoch. The biomediator system as a tool for integrating biologic databases on the web. In *Proceedings of the Workshop on Information Integration on the Web*, 2004.

23. Limsoon Wong. Kleisli, a functional query system. *Journal of Functional Programming*, 10(1), 2000.

24. C. Wroe, R. Stevens, C. Goble, A. Roberts, and M. Greenwood. A Suite of DAML+OIL Ontologies to Describe Bioinformatics Web Services and Data. *Journal of Cooperative Information Science*, 2003.

Factors Affecting Ontology Development in Ecology

C. Maria Keet

KRDB Research Centre, Faculty of Computer Science,
Free University of Bozen-Bolzano Piazza Domenicani 3,
39100 Bozen-Bolzano, Italy
Tel: +39 04710 16128
keet@inf.unibz.it

Abstract. Few ontologies in the ecological domain exist, but their development can take advantage of gained experience in other domains and from existing modeling practices in ecology. Taxonomies do not suffice because more expressive modeling techniques are already available in ecology, and the perspective of flow with its centrality of events and processes cannot be represented adequately in a taxonomy. Therefore, formal ontologies are required for sufficient expressivity and to be of benefit to ecologists, which also enables future reuse. We have created a formal mapping between the software-supported ecological modeling method and software tool STELLA and ontology elements, which simplifies bottom-up ontology development considerably and has excellent potential for semi-automated ontology development. However, the conducted experiments also revealed that ontology development for ecology is close to being part of ecological research that through the formalized representation of the knowledge more clearly points to lacunas and suggestions for further research in ecology.

1 Introduction

It is well-known that ontologies can be a valuable artifact for data(base) integration. However, for ontologies to be useful, one first needs to develop a good ontology that covers the domain accurately and precisely and has the right balance between utility and ontological correctness (the ontological trade-off). Although multiple engineering artifacts exist, from structured controlled vocabularies to formalized foundational ontologies, ontologies in the domain of ecology do not exist to the extent as, for instance, in cell biology. We can take advantage of lessons learned from developing ontologies in other biology disciplines, most notably in molecular biology and anatomy, and from suggestions made by philosophical ontologists. The former includes experiences with GO[1], OBO[2], and FMA[3], the latter comprises the use of foundational ontological aspects like the nature of entities/concepts and (primitive) relations [1] and OntoClean [2] which provides a methodology for removing incorrect ontological

---

[1] Gene Ontology: http://www.geneontology.org.
[2] Open Biological Ontologies: http://obo.sourceforge.net.
[3] Foundational Model of Anatomy: http://sig.biostr.washington.edu/projects/fm/index.html.

B. Ludäscher and L. Raschid (Eds.): DILS 2005, LNBI 3615, pp. 46–62, 2005.

decisions made in a taxonomy by relying on types of properties (characterising, sortal, phased sortal. etc.) and metaproperties (rigidity, identity, etc.). However, whether we can use a similar approach as taken by the Gene Ontology Consortium depends on the result of a comparative analysis between molecular biology and ecology (§2). One of the differences is that there is an established practice of modeling in ecology, that, albeit different from computer science and ontology research, can be advantageous to enhance ontology development. A widely used, software-supported, ecological modeling technique is STELLA[4], which we have exploited in formulating formal correspondences between STELLA model elements and ontology elements (§4). This was identified and put to the test with formalizing ecological knowledge contained in a STELLA pollution example and the Microbial Loop (ML) model [3], reported in §5 that also contains several ontology development considerations. Apart from simplifying and speeding up ontology development by using the formalization, related facets benefiting ontology development for ecology are discussed in §6 and the potential for semi-automatic bottom-up ontology development based on STELLA models is assessed. We finalize with some conclusions in §7.

2   Some Salient Features of the Ecology Subject Domain

An important factor in ecological and biogeochemical models is the *flow* of components in a eco(sub)system[5], i.e. the path components take or sequence of processes it is involved in. A component can be a specific nutrient, such as nitrogen- or carbon-containing substances, pollutant, energy, and so forth, hence the centrality of endurants (entities that are wholly present in time) and instances thereof. However, the 'component of concern' is firmly embedded in the flow. For example, the nitrogen cycle from nitrate in soil to bacterium (nitrogen fixation by e.g. a *Rhizobium* sp.), transfer to a leguminous plant (like clover) with which the bacteria live in symbiosis with, transport within the plant, consumption by a ruminant, metabolism of the animal, excretion by animal, return of (some of) the nitrogen-containing molecules back to soil. One also can consider such cycles as a process of nested processes, i.e. from a perspective of a specific combination or sequence of distinct perdurants (entities that are partially present and happen in time). Thus the relation between 'stuff' (a substance, amount of matter etc.) and what happens to it are inextricably linked to one another. Conversely, molecular biologists do distinguish more clearly a separation between structural components, their functions and the processes in which they can be involved. GO consists of three distinct ontologies: Molecular Function (MF, describing activities), Biological Process (BP, with biological goals), and Cellular Component (CC, for locations) [4]. This approach treats perdurants *as if* they are endurants, but this objectification does not solve the connection between endurants and perdurants. For example, if one wants to couple some biological process with a cellular component, new relationships between the two ontologies need to be created (e.g. [5]). Thus, adding new knowledge about the combination that may result in a separate

---

[4] ISEE Systems: http://www.iseesystems.com; *ithink* is the same tool but used for business modeling.

[5] For the remainder of the article, 'ecological model' comprises both types – a biogeochemical model is element-conserving, but this aspect is irrelevant for ontology development.

new 'situation ontology', or a mapping ontology that is positioned between BP and CC. However, that ecological modelers use tightly coupled endurants and perdurants does not necessarily prevent an ontologist to create artificial divisions between the two.

Perdurants include types of entities such as processes, events and states, in contrast with modeling paradigms in informatics and most ontological investigations, where the center of attention is the entity of the thing-quality paradigm. Philosophically, there are arguments for and against such emphasis: processes can only exist when there are endurants that are the 'carriers' of the process [5]. On the other hand, objects only come into existence through a process (refer to [6] and [7] for a wider scope of arguments). Few agreed-upon ontological categorizations exist, as can be observed in *Figure 1* or the Process Specification Language[6], Business Process Management Initiative[7], and Petri-nets. From an ontology engineering perspective, the approaches vary. One tactic is to separate perdurants from endurants linked by a participation relation as in DOLCE (Descriptive Ontology for Linguistic and Cognitive Engineering) [8], [1]. The Basic Formal Ontology (BFO[8]) consists of SNAP and SPAN ontologies where the latter includes a time perspective. However, none addresses the thing-process aspect as fully *interdependent*, which poses a potential problem when representing ecological knowledge in an ontology. Bittner *et al* [9] go to some length in formalizing the difference between endurant and perdurant, but this does not solve the nature of the relation when viewed from different perspectives. The Standard Upper Ontology has set up a 4D Ontology Working group[9], without useable results as of yet.

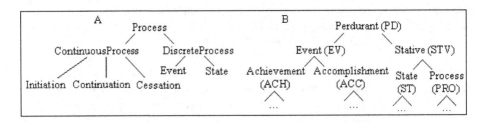

Fig. 1. Two examples of process-related categorizations. A: [6]; B: DOLCE [1]

A second difference lies in the level of granularity and demarcation of the discipline. The boundary of molecular biology lies at the cell-level and smaller entities, whereas in ecology 'ecosystem' and 'the environment' have fiat boundaries. On the one hand, earth is one ecosystem, but also the Amazon basin or the micro-environment in tree tops. Other methods of differentiation include trophic levels and 'grand processes' such as eutrophication and El Niño. While there are different ways of partitioning the domain at the molecular biology level, such scope in ecology is wider, thus when developing an ontology it requires involvement of a broader range of (sub-)disciplines that are less reductionist than molecular biology.

---

[6] http://www.mel.nist.gov/psl/index.html.
[7] http://www.bpmi.org/.
[8] http://ontology.buffalo.edu/bfo/, http://ontology.buffalo.edu/smith/articles/SNAP_SPAN.pdf.
[9] http://suo.ieee.org/SUO/SUO-4D/index.html.

Third, ecologists have a different starting position compared to molecular biologists when modeling domain knowledge. Whereas in molecular biology no established modeling tradition existed, ecologists do have multiple established standards such as Odum's conventions [10] and STELLA that are, depending on the subdiscipline in ecology, more or less often used. STELLA is relatively widely used and has software support comprising graphical elements and automatic generation of equations. STELLA is used in research and education for systems analysis and simulations of, for example, predator-prey interaction, effects of contamination, and food chains. Ontologists should take advantage of these models for bottom-up ontology development. However, this also means that one cannot begin with a structured controlled vocabulary: whatever ontology is developed has to surpass semantic expressivity of existing ecological models for it to be perceived to be of benefit to ecologists.

A preliminary experiment in ontology development for ecology was carried out with a simplified pollution example, which confirmed that an informal ontology limited to isA and partOf relationships could not capture the expressivity of its corresponding STELLA model. The "extended semantic representation of equations" via "placeholder objects" [11] did not represent the (partially implicit) semantics of the STELLA model fully either (results omitted). An additional advantage of using STELLA as a starting point for ontology development is that, with the mapping between STELLA and ontology elements, the STELLA representation serves as an intermediate representation. Thereby it bridges the two disciplines with a common ground for communication. This will speed up bottom-up ontology development, which will facilitate data integration sooner rather than later.

3 Methodology

The first experiment was carried out with STELLA v8 for Windows from ISEE Corporation and the demonstration model *Amalgamated Industries*. The abstraction of this model, including the STELLA terminology and modeling elements, was matched to ontology jargon. Protégé v2.1.1 with OWL Plugin v1.2 Beta (Build 139) was used to improve the level of formalization and test the translation. Racer v1.7.21 was used for the 'classify taxonomy' and 'check consistency' features; Graphviz v1.12 plug-in to activate OWLViz, and ezOWL plug-in (v20040412) were added for ontology visualization of the developed ontology. To test the translation between STELLA and ontology elements, we used the larger ML model (*Figure 4*), converted it into a list of candidate entities and relations, which was structured into a formal ontology, also in Protégé. Development of both the *Pollution* and *MicrobialLoop* ontologies was aided by structuring the candidate entities and relationships adhering to the formalized DOLCE foundational ontology, which is intended for making already formed conceptualizations explicit (refer to [1] for explanation and categories).

4 Abstractions and Matching

Before addressing the formalization, a small STELLA model (*Figure 2*) of the simplified pollution scenario is outlined for illustration. This model captures a scenario where a factory disposes toxic waste in the river that flows into the pond downstream, in turn killing organisms living in the pond depending on the pollutant concentration. The ecological 'concept of concern' is the concentration of the pollutant in the pond, which has the related influencing factors modeled 'around' it, such as the released amount of pollutant by the chemical plant. There are three main aspects: water and pollutant in/outflow of the bound system, the combination of water volume and amount of pollutant determining the pollutant concentration in the pond, and the combination of water outflow and pollutant concentration determines the amount of pollutant outflow. There are two factors of interest in comparing this type of model with its variants in computing, such as UML class diagrams, (E)ER and ontologies:

1. The ecological model is *event centered*, hence contains the representation of time, diagrammatically represented with the horizontal thick arrows with an open shaft, or phrased as the *route* taken by an element.
2. Key aspects in the ecological model are Flow, Stock, Converter, and Action Connector. A Stock correspond to a noun, being it particulars or universals, Flow to verb, Converter to attribute or property related to Flow or Stock, and Action Connector relates the former. *Figure 3* contains the comparison with computing verbiage (top half). *Object* is a candidate for an entity, *event_or_activity* in OO terms a candidate for a method and in an ontology categorised under a perdurant hierarchy and converter maps to *attribute_or_property*, which says something about the object, such as the outflow rate. The Action Connector (thin line with arrow) may be candidate for binary (ternary?) relationship between any two of Flow, Stock and Converter.

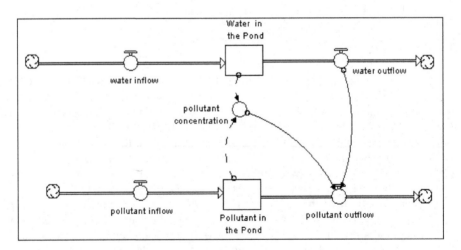

Fig. 2. Abstraction of the pollution example

Following from 2), the formalization for the translation is:

$$\forall x\,((\mathrm{Stock}(x) \leftrightarrow \mathrm{Entity}\,(x)) \rightarrow \mathrm{ED}(x)) \tag{1}$$
$$\forall x\,((\mathrm{Flow}(x) \leftrightarrow \mathrm{Entity}\,(x)) \rightarrow \mathrm{PD}(x)) \tag{2}$$
$$\forall x\,((\mathrm{Converter}(x) \leftrightarrow \mathrm{Entity}\,(x)) \rightarrow (\mathrm{Q}(x) \vee \mathrm{ST}(x))) \tag{3}$$
$$\forall x\,(\mathrm{ActionConnector}(x) \leftrightarrow \mathrm{Relationship}\,(x)) \tag{4}$$

where ED means endurant, PD perdurant, Q quality ('attribute' belonging to an entity), and ST state. Axiom (3) is open to experimentation: for example, the Converter *Pollutant concentration* in *Figure 2* can be a quality of the pond, liquid mixture, or detritus – anything that can be polluted – but also a state as in 'the pollutant concentration of the pond where the river enters' or 'the pollutant concentration of the pond on 20-7-2004'. Further, *Pollutant concentration* may be subsumed by *Concentration* that can be applicable to a wider range of endurants and as such is not necessarily an *essential* property (a pond is still a pond with or without some concentration of a pollutant) but a situational one, or having concentration as a non-rigid property. Therefore, (pollutant) *Concentration* is better modeled as an ST, but at this stage of the investigation, the mapping of Converter to Q cannot be excluded with certainty. This ambiguity will be resolved by applying the proposed formalization to a formal pollution ontology and the larger ML model, which will clarify if the mappings are correct, shed light on the distribution of Q and ST from a Converter, and might be solved by adding additional axioms taking into account the context of the STELLA elements, such as how the converters are related to the other elements.

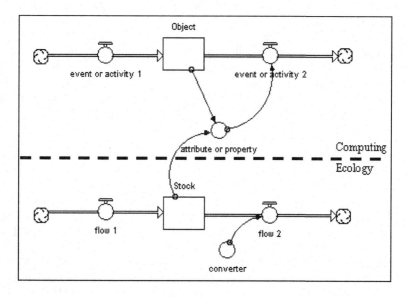

Fig. 3. Comparing the ecological model components with its analogue in a computing model

The consequences of translating an ecological model into an ontology based on the provided mapping is that temporality and the movement of energy or nutrients is not yet represented exactly as is captured in the ecological model apart from incorporating the fact that types of processes exist; however these extras in STELLA are epistemological aspects. The original ecological model now can be remodeled into an ontology consisting of three features: endurants, 'attributes', and perdurants; what remains to be solved are the relations between them, i.e. the Action Connectors. With further testing of larger STELLA models such as LEEDS (Lake Eutrophication, Effect, Dose,

Sensitivity model) and ML, and the provided formalization, it is possible to generate an ontology by 'loading' several of the STELLA ecological models into one of the ontology development tools.

5   Ontology Development

5.1   A Pollution Ontology

5.1.1   Motivating Example: Some Issues with an Informal Ontology

An informal ontology restricted to concepts and the isA or partOf relation does not suffice. For example, an isA relationship between *Water* and *Molecule*: although *Water* is indeed a molecule, *Water in the context of some ecological site* is not pure H_2O, but water containing dissolved molecules and suspended particles, i.e. water as a mixture (an amount of matter). The methodology of ontology base & commitment layers of DOGMA [12] may be more advantageous, because *Water isA Molecule* can be included in the ontology base and omitted from a commitment layer for an ecological site, whereas it would be included in a commitment layer of a chemicals ontology that omits *Water isA LiquidMixture*. In a simple taxonomy these options are unavailable.

Volume, *Rate* and *Concentration* capture a characteristic of their respective entity they are attached to, alike an attribute. *Molecule hasA Concentration*, but it can only have a concentration dissolved or suspended *in something* and not *of itself*; of itself are properties like melting temperature and structure of a molecule. However, to conclude it is an attribute or state of only water in the pond is premature: if modeled as such, the model will be unable to accommodate pollutants in sea, air, soil etc.

5.1.2   Upgrading to a Pollution Ontology

It is sub-optimal if one has to use different types of representation models (multiple taxonomies and placeholder objects) to capture the semantics. For a formalization to be exploited in full, one needs a formal ontology and a tool that is based on logic. Two widely used ontology development and editing tools are DAG-Edit and Protégé; the former provides functionality for structured controlled vocabularies (and taxonomies), whereas the latter is frame-based with Description Logic (DL) and OWL support. The DL version supports "maximum expressiveness without losing computational completeness ... and decidability ... OWL DL includes all OWL language constructs with restrictions such as type separation (a class can not also be an individual or property, a property can not also be an individual or class)"[10]. DAG-Edit is insufficient for the task, because relationship types are limited to isA and underspecified partOf (sometimes also developedFrom). Protégé, on the other hand, supports reasoning and allows higher expressivity by allowing specification of e.g. properties, range restrictions, and disjointness of entities. However, it also supports multiple inheritance, which complicates inferencing over the ontology and multiple inheritance may indicate bad modeling decisions, but this is not necessarily so.

Multiple Inheritance. In the initial categorization, *Molecule* directly subsumed *Pollutant*, *Nutrient*, *OrganicMolecule*, and *AnorganicMolecule*, where some molecules,

---

[10] http://www.w3.org/TR/2004/REC-owl-guide-20040210/.

like *PHB* and *Phosphate*, were subsumed by *OrganicMolecule* and *Pollutant*, and *Nutrient* respectively. Biologically, these are correct statements, but it would be better to specify (not possible in Protégé, but implementable in a DL knowledge base):

> *if* (concentration of AnorganicMolecule *x*) *in* (WaterBasin or AmountOfMatter) *is higher than* [some number] *then* (*x* isA Pollutant and *x* isA AnorganicMolecule) *else* (*x* isA Nutrient and *x* isA AnorganicMolecule) *for species y.*

This statement indicates that the difference between pollutant and nutrient is problematic: pollutants are harmful and nutrients beneficial to organisms[11], thus a functional categorization. However, there are two obstacles: first, a pollutant for species *x* can be a nutrient to species *y*; this information may be used for *in situ* bioremediation[12], hence lies within the UoD. Secondly, low molecule concentration can be a nutrient but excess concentration pollutant; but when is 'excess' concentration high enough to classify it as pollutant? Narrowing down *Nutrient* to *NutrientBioremediation* prevents confusion with generic nutrients that never function as nutrients for bioremediation. That *Molecule* subsumes *Pollutant* and *Nutrient* together with *OrganicMolecule* and *AnorganicMolecule* is incorrect, because the former are *functions* assigned to the molecules, whereas the distinction for *OrganicMolecule* and *AnorganicMolecule* is based on the *composition* of molecules. Using the DOLCE categories for guidance, *Pollutant* and *Nutrient* still are physical objects (POB), but classified according to other criteria. For brevity of this experiment, they are separated into structure and the function of molecules. *H₂O* and *Amylose* each had only one isA relation hence were removed, but *PHB* and *3-chlorobenzoate* can be used for bioremediation as each one has 3 isA relationships: being an *OrganicMolecule*, *Pollutant*, and *Nutrient*. Although multiple inheritance has not been eliminated, due to having structure and function in one ontology, there is a major advantage in maintaining this sort of multiple inheritance: when one adds a new entity under *MoleculeStructure*, *Pollutant*, and *Nutrient*, i.e. the new entity has three isA relationships, then one can deduce there is potential for *in situ* bioremediation (as is the case with *3-chlorobenzoate*). On the other hand, if the new entity has two isA relationships, one to *MoleculeStructure* and the other to *Pollutant*, an ecosystem disruptive method to clean up the contaminated site is required; if there is only one isA relationship, then there is no need for the molecule to be in the classification because it does not serve any particular purpose for the UoD, or still needs to be grouped under *NutrientBioremediation* or *Pollutant*, thereby missing essential knowledge in the ontology. Thus from that perspective, multiple inheritance is not a 'bad thing' and can be used to derive additional information from querying the ontology.

Other modeling considerations and limitations include 1) Protégé prohibits creating an entity or individual starting with a number, in this case *3-chlorobenzoate*, which is problematic because names of many chemicals start with a number and appending the number at the end is not an option with more complex chemical struc-

---

[11] Pollutant = "waste matter that contaminates the water, air or soil" (Wordnet) although nature may cause pollution as well; nutrient = "any substance that can be metabolized by an organism to give energy and build tissue" (Wordnet).

[12] *In situ* bioremediation: removing a pollutant from a contaminated site without disrupting the ecosystem by using organisms instead of soil excavation and chemical decontamination.

tures. 2) The STELLA pollution model assumes Pollutant_concentration_in = Pollutant_concentration_dumped_by_plant, thereby ignoring adsorption and absorption to particulates in the river and sedimentation, consumption by organisms, and assuming that the pollutant concentration is uniform throughout the pond. Adsorption and absorption can be added to the ontology, but this extension is omitted from *Pollution* because its purpose is explorative with relation to the axioms. 3) Protégé tolerates unconstrained property creation, which can become prohibitive if one desires to develop an ontology with possible future ontology integration while another ontology has been restricted to a few relationship types. 4) In order to create a sound ontological basis of the categorization of entities, the structure of the DOLCE top-level categorization was used, most notably the amount of matter (M), physical object (POB), and process (PRO) versus ST, resulting in 56 entities for the *Pollution* ontology. Whereas relating entities other than isA or partOf is not possible in a taxonomy, this is possible with a formal ontology and therefore included in *Pollution* via 9 properties and corresponding constraining axioms. The plug-in module OWLViz for Protégé only shows the isA relationships in the graphical representation; for additional expressivity, ezOWL is required. With a very small ontology, this creates a manageable view of the semantics, but even with only 56 entities, the diagram is already too large to be manageable (the OWL file of *Pollution* is available online as supplementary material). Comparing this ontology with the original STELLA pollution model, the 11 elements are 'translated' into 56 entities and 9 properties. This may seem excessive, but the ontology captures more semantics than its STELLA counterpart does, hence has a higher likeliness of being useful for more pollution models than STELLA's dump-river-pond scenario. From the perspective of semi-automated bottom-up development of ontologies based on STELLA models, this poses a challenge: how labor-intensive is the additional structure one needs to add to adhere to sound ontological principles? Is it sensible to develop semi-automatic translation software if a considerable amount of ontology development effort may have to be carried out manually anyway? Instead of generating a structure of the ontology, a viable option is to translate STELLA elements into a list of entities and relationships that one needs to include in the ontology. This reduces the manual analysis because it is possible to develop a backbone domain ontology, 'hang in' the entities generated from the STELLA model, and augment this with the relationships and properties that resulted from the translation.

5.2 The Microbial Loop Model

The formalization was applied to the ML model (*Figure 4*) to examine if the axioms still hold in a real and larger STELLA model, to shed light on the distribution of Q and ST from a Converter, and to investigate if additional axioms are required when taking into account the context of the STELLA elements, such as how converters are related to other elements. ML's initial mapping to ontological categories contain 38 STELLA elements, of which 11 Stock/ED, 21 Flow/PD, two Converters that map to ST, and four Action Connectors/Relationships that are modeled as properties in Protégé (mappings included as supplementary material). All Stock elements can be further categorized as Non-Agentive Physical Object (NAPO) leaf categories. Further, to accommodate these NAPOs in an ontology, extra entities related to the NAPOs were added, such as *Phytoplankton* (which is an Agentive Physical Object APO), and *De-*

tritus (an amount of matter M). Note that "Phyto C" is the organic carbon *component* of phytoplankton, not the phytoplankton as a whole. To accommodate for this in the ontology, adding phytoplankton only as an APO is insufficient. Apart from the phytoplankton carbon and nitrogen, the NPK parameters (Nitrogen, Phosphor, Potassium) are relevant for agriculture and soil science in particular. Should one include other molecules to be more comprehensive? From an ontological viewpoint probably yes, but one might argue a utilitarian restriction "it'll do" for the intended purpose.

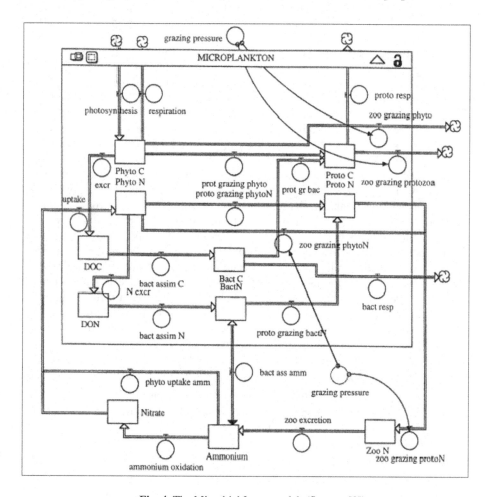

Fig. 4. The Microbial Loop model. (Source: [3])

A separate issue concerning categorization of organisms in the ML as APOs emerged during a conversation with one of the creators of the ML model, Professor Paul Tett. The distinction between individual organisms, their class and a population, are called (phyto)plank*ter* for the individual and (phyto)plank*ton* for the "class with the characteristics of the population". From an ontological perspective, a class is different from a population: a population is a group of individual organisms belonging to the same species living in a given geographic region. The *assumption* in ML emerged that both the *–er* and *–on* have the same properties but have only differing numerical values (quales), i.e. entering the realm of the statistical properties of the *population* of organisms categorized as being of the same type, hence 'average organism classified as belonging to species *x*'. At present, there is an empirical problem differentiating between the characteristics of the individuals in the population, which is a challenge at the epistemological level. One may expect that within several years, ecologists *will* be able to distinguish between properties and their values of population, individual, and class, which may have a knock-on effect on the design decisions made with the *MicrobialLoop* ontology.

Seventeen of the 21 Flow elements are processes (PRO) and the other three accomplishment (ACC) (2x the entity *Uptake* and *Excretion*) and achievement (ACH) (*Oxidation*). There is no particular contextual aspect why these three have to be categorized under a different type of perdurant. The two Converters, both called "grazing pressure", each is a parameter of the process *Grazing*. Hence, it might be a quality of the process of grazing, because it is specifically constantly dependent on the entity it inheres in (grazing): at any time, a quality cannot be present unless the entity it inheres in, in this case a PD, is also present, and a PD is present if its ED bearer is present. However, if there is no plankton, the grazer (ED) may be grazing on something else. If there are no grazers, then grazing (PD) is not there and its grazing pressure as Q is also absent. Alternatively, the grazing pressure simply reaches zero, resulting in "grazing pressure" as a ST: the notion of "grazing pressure" is there, thus a ST and not a Q, which adheres to the ecology interpretation. Thus, this does not imply that Converters automatically always can be translated into states. The Action Connectors map well to properties (relationships between entities). There are 59 entities and 10 properties in the *MicrobialLoop* ontology (the OWL file is available as supplementary material), of which a summarized section is shown in *Figure 5*. For instance, the left-, most rectangle is a visual representation of *Protozoa* as subtype of *Microorganisms*, with (5) inherited from *Microorganisms*, (6) and (7) as necessary properties of *Protozoa*, and (8) a necessary property of *PhytoPlankton*. *MicroAlgae* and *MacroAlgae* are disjoint subtypes of *Algae*, and *Phytoplankton* and *ZooPlankton* are disjoint subtypes of *Plankton*, and so forth.

$$\forall x \exists y\ (\text{Protozoa}(x) \rightarrow \text{hasProcess}(x, y) \wedge \text{Respiration}(y)) \tag{5}$$
$$\forall x \exists y\ (\text{Protozoa}(x) \rightarrow \text{grazesOn}(x, y) \wedge \text{PhytoPlankton}(y)) \tag{6}$$
$$\forall x \exists y\ (\text{Protozoa}(x) \rightarrow \text{grazesOn}(x, y) \wedge \text{Bacteria}(y)) \tag{7}$$
$$\forall x \exists y\ (\text{Phytoplankton}\ (x) \rightarrow \text{accomplishes}(x, y) \wedge \text{Uptake}(y)) \tag{8}$$

The translation axioms provide an accurate high-level mapping for both the *MicrobialLoop* and *Pollution*, although the amount of Converters and Action Connectors in the models may be to be too few to statistically conclusively confirm correctness of the mapping.

6 Discussion

Additional entities had to be added to the ontology compared to its STELLA variant. In spite of this, several factors ameliorate this issue. Ecology already divides concepts into three types: natural, functional and integrative concepts. The first two types of ecological concepts can be identified in the ontology: the functional concepts are categorized under *Perdurant* and the natural concepts subsumed by *PhysicalEndurant*. Imposing a separation and categorization may actually benefit ecology. Ford [13] presents the interdependencies between the three types of concepts indicating that "[n]ew functional concepts arise to describe newly understood structures or interactions in natural concepts and research into functional concepts is constantly used to refine the definition of existing natural concepts and their classifications" and "[d]evelopments in measurement lead to refinements of functional concepts". Hence, by defining the concepts more clearly with the aid of formal ontology, the discipline of ecology itself may advance at a faster pace. However, realize that the *change in definition of concepts* and *how they may be classified* is the very essence of scientific advance [13]. Consequentially, software for development of an ecological ontology *must* contain extensive features for ontology maintenance, such as described by Klein and Noy [15]. Using the DOLCE concept CN [14] or GO Guidelines, it means that a change in the definition of a concept implies creation of a new concept, because $\neg(CN_{old} = CN_{new})$ even though the domain expert may perceive that the meaning is 'updated'. Remains the challenge of representing the integrative concepts of ecological models, which are sometimes established and captured in axioms, but also may be conjectures or in the process of being refined, where the second and third include alternative views of some ecological theory. This indicates that the chosen ontology development process should be capable of representing alternative views. This is not possible in Protégé, but DOGMA features [12] do allow this in the ontology commitment layers.

A generated translation list from STELLA to entities and relationships as outlined in the previous section may be used as ontology base where each commitment layer represents a different view. An alternative can be to develop software that allows ontology browsing from different perspectives exploiting a theory of granularity applied to the subject domain. Aside from diverging ideas on theories, the 'windows on reality' differ depending on what the scientist is looking for. In ecology, it is common to start with flows as opposed to starting with the object where processes act upon. From a formal ontological perspective, this is not necessarily problematic: creating the ontology starting with perdurants and subsequently axiomatizing their influence on endurants is possible. In addition, two distinct methodological approaches in ecological research exist. In theoretical ecology, one devises a theory that is tested on its applicability in nature afterwards. On the other hand, ecological modeling via the empirical approach involves tweaking the model until it fits the observed data, where only a limited set of parameters of the subject matter is used [16]. The former approach indicates that one starts with a framework that will be filled up inwardly and more densely, where the latter starts small and gradually will evolve by spreading outward once more research has been conducted. If one methodology is better than the other is irrelevant here, however, it is important to realize that both approaches are used in ecology, and, at least initially, might not result in the same output due to divergent design decisions. Also, cooperating with domain experts of either type involves a different approach commencing ontology development.

Although engineering challenges of ecology ontology development can be solved, the philosophical formal ontology development entertains itself with *what* and *how* to represent what is known, where an ontologist for ecology will have to cooperate in the process that otherwise logically occurs before ontology creation, i.e. the *why* in the semantics as part of regular science. The output must not only be usable for computer science (the ontology), but also of use to ecologists, who will be pushing the boundaries of their discipline by clarifying relevant concepts, thereby better formulating research questions, and later their theories. Provided alternative views of integrative concepts and theories can be accommodated for, it will aid the advance in ecological science. Apart from the difficulty on class/population (§5.2), a second aspect emerged during discussions with Tett: to compare and contrast more precisely a to-be-developed ontology of the STELLA model for the 'model organism' ERSEM[13] with his *MicrobialLoop*. In addition, this ERSEM-ontology or *MicrobialLoop* might function as template or backbone for other ecological models in marine science. Another suggestion how ecology can benefit from the ontological approach is during the "fitting stage" of simulation models to match empirical data, where, according to Tett, parameters are added and removed "arbitrarily" and their values changed to fit observations. Adding the reasoning power of ontologies can ensure consistency. Inconsistencies introduced during the fitting stage provide a focus for (re)assessment and investigation of (a section of) the domain.

Considering some practical aspects of ontology development, the mapping between STELLA and ontology elements do not imply these correspondences will always be applicable, although the devised correspondences were confirmed to be sufficient for the *MicrobialLoop* development experiment. Accommodating the Converters remains less straightforward, because decisions have to be made to translate it to a quality or state. The use and meaning of the Action Connectors aided in determining the properties and relations in the *Pollution* and *MicrobialLoop*. The relative absence of serious difficulties during the modeling of characteristics of the biological entities may be due to the size of the (randomly) chosen models and/or the author's domain knowledge. Initial challenges during the development of a taxonomy were absent during the 'upgrade' to *Pollution*, because the expressivity and flexibility of DL is much greater than the limited hierarchies in a taxonomy. Our experience confirms observations made by many other researchers that more expressive modeling languages *do* capture a richer semantics. This is not only because it compels the user to do so but also because one has the possibility to 'squeeze in' more knowledge, which in turn enforces closer inspection of the domain, resulting in ontologies with less errors and higher precision, hence are more stable. This is in contrast with e.g. DAG-Edit or standard UML class diagrams when one can gloss over such details. Moreover, where the flow dynamics cannot be addressed in a taxonomy, this is dealt with in the developed ontologies by first categorizing the relevant perdurants under *Process* and *State* and using properties to create the relationships between these entities and the endurants they act with/upon, all captured within *one* ontology instead of different representations. An alternative considered was BFO. However, developing two ontologies (SNAP and SPAN) that need to be 'connected' to capture the ecological semantics is prohibitive. The DOLCE top-level categories intuitively make sense and aids understanding of how distant or close biological semantics is from ontologies with a cognitive bias.

---

[13] European Regional Seas Ecosystem Model; refer to [17] for the structure and methodology of ERSEM and [18] for the microbial food web in marine systems.

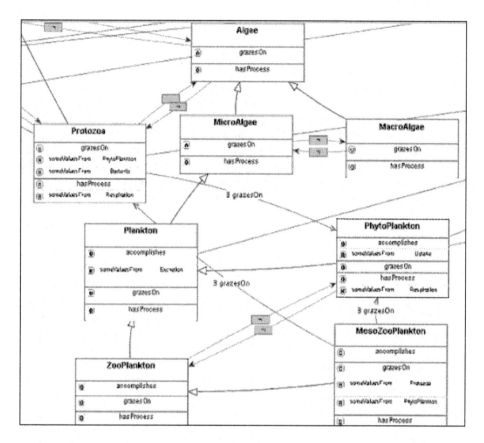

Fig. 5. Section of the *MicrobialLoop* ontology graphical representation with ezOWL

The two developed ontologies include more entities and relations than their STELLA counterpart and easily can be further extended to increase reusability. The latter can step up efforts to resolve ambiguities and assumptions; a very large ontology may be practically difficult to work with or requires full support of many sub-disciplines within ecology, analogous to the GO project [4]. Alternatively, one can take advantage of the extant modularization of ecological models: ML and e.g. Sea-Weed are composed of smaller sub-models, where the former contains Riley+, MicroPlankton and Autotroph-Heterotroph [3], and the latter Vollenweider[14] and a tide & light simulation. Thus it is conceivable create 'mini-ontologies' based on the same foundational ontology principles for each ecological model separately, then develop a library where the user can choose the desired sections to create larger models, supported by a backbone ontology where for each situation one or a few ontologies can be attached to it. This approach also facilitates representation of diverging views of integrative ecological concepts. Challenges are the development of a backbone ontol-

---

[14] http://tejo.dcea.fct.unl.pt/resources.asp. Vollenweider models form the basis for eutrophication control, which in turn is used in the LEEDS model and semantically related to ML.

ogy and prospects of integrating ontologies. Analysis of differences between 'simple' and 'complex' versions of the Vollenweider models revealed that the more complex models contain both *additional* sections as well as filling the existing structure with more *detail*, thus both coverage and granularity changes. Tett and Wilson [3] indicate that this may be the case with multiple ecological models, because there is a desire to keep the amount of Stock elements to a minimum for reasons of computational power and practical as well as theoretical challenges of estimating parameters. Smith [19] claims that, in ecology terms, good *simulations* should include as much detail as possible, whereas good *models* should include as little as possible to capture the most important factors. This will affect ontology development for ecology, depending on with which kind of model/simulation one starts ontology development. Perceptions and knowledge about the subject domain change over time, is not always consensual, and have the potential destabilizing effect of cascading uncertainties to larger modular simulations and models, which are, according to [20], neither possible nor desirable to include in one model. A design decision about one larger ontology versus multiple mini-ontologies will have to be made.

Concerning the *MicrobialLoop*, one may argue that the author's knowledge of the subject domain prevented the need for making excessive amounts of assumptions, such as knowing what "Phyto C" is, and microbiology in general. The outcome likely would have been different without such prior domain knowledge. Temporal factors such as accommodating changes in the rate of in/outflow are not addressed fully, because they are in the realm of instances. The richer expressiveness of the formal ontology approach using Protégé and DOLCE categories proved to be flexible enough for the task as it allowed correct representation of entities from taxonomies, entities that emerged from the semantic representation of equations, and other implicit knowledge of the STELLA models. The (untrained) ecologist indicated that the richer ezOWL graphical representation (*Figure 5*) that includes properties and constraints was preferred over a simplified taxonomic tree generated with OWLViz. Further, and more importantly, the ecologist judged the logic made the knowledge captured in the formal ontology become *clearer* than both STELLA and natural language, and considered to have useful potential to disambiguate the semantics to advance ecological research. The translation of the STELLA models into ontologies did introduce many new concepts, especially with the simple pollution experiment, but this was much less the case with the larger ML model. This indicates that with larger ecological models the issue of manual intervention during ontology development decreases. The translation axioms simplified ontology development from the ML considerably (a first version including initial mappings including comments was created within three hours and required only a few minor changes afterwards); therefore, utilizing other STELLA models with the provided formalization of the translation will also speed-up the overall development process of ontologies in ecology.

7 Conclusions

Although few ontologies in the ecological domain exist, their development can take advantage from existing modeling practices in ecology in particular. Taxonomies are insufficiently expressive compared to existing ecological modeling techniques and the perspective of flow in ecological models cannot be represented adequately in a taxonomy. We have created a formal mapping between the software-supported ecological modeling method STELLA and ontology elements, which simplifies bottom-up ontology development and has excellent potential for semi-automated ontology development. We developed two formal ontologies, *Pollution* and *MicrobialLoop*, in Protégé, exploiting the expressivity of OWL DL to capture the semantics of 'flow' in salient in ecology models. STELLA serves as an intermediate representation, widely used by ecologists and is translatable to a representation usable for ontologists. In addition, the more comprehensive semantics of the ontologies have not only a higher level of reusability within the domain, but also for future ontology integration as both *Pollution* and *MicrobialLoop* were developed with the same ontological foundational principles which shall facilitate reuse of ontologies. However, the experiments also revealed that ontology development for ecology is close to being part of ecological research that through the formalized representation of the knowledge more clearly points to lacunas and suggestions for further research in ecology, thereby aiding hypothesis generation. We are currently extending this research with ontology development and management aspects such as modularization and ontology integration.

Acknowledgements

We wish to thank Paul Tett for his contribution to discuss the *MicrobialLoop* & ML. Some of the research was carried out while CMK was at Napier University, UK.

Supplementary Material: The MicrobialLoop *and* Pollution *ontologies, a color version of* Figure 5, *and the initial mapping between ML and ontological categories are available online at http://www.meteck.org/supplDILS.html.*

References

1. Masolo, C., Borgo, S., Gangemi, A., Guarino, N., Oltramari, A.: WonderWeb Deliverable D18 – Ontology library. WonderWeb. http://wonderweb.semanticweb.org/ (2003)
2. Guarino, N., Welty, C.A.: An overview of OntoClean. In: Handbook on ontologies. Staab, S., Studer, R. (eds.). Springer Verlag, Berlin (2004) 151-159
3. Tett, P., Wilson, H.: From biogeochemical to ecological models of marine microplankton. J. Mar. Sys. 25 (2000) 431-446
4. Gene Ontology Consortium: The Gene Ontology (GO) database and informatics resource. Nucl. Acids Res. 32 (2004) D258-D261
5. Fielding, J.M., Simon, J., Ceusters, W., Smith, B.: Ontological theory for ontological engineering: biomedical systems information integration. 9th International Conference on the Principles of Knowledge Representation and Reasoning (KR2004), Whistler, Canada (2004)
6. Sowa, J.F.: Signs, Processes, and Language Games – Foundations for Ontology. http://www.jfsowa.com/pubs/signproc.htm. Date accessed: 16-6-2004. (2003)
7. Rescher, N.: The revolt against process. J. Philosophy 59 (1962) 410-417
8. Gangemi, A., Guarino, N., Masolo, C., Oltramari, A., Schneider, L.: Sweetening Ontologies with DOLCE. Proceedings of EKAW 2002, Siguenza, Spain (2002)

9. Bittner, T., Donnelly, M., Smith, B.: Endurants and Perdurants in Directly Depicting Ontologies. AI Communications. IOS Press, Amsterdam (2004)
10. Odum, H.T.: Systems ecology. John Wiley and Sons, New York (1983)
11. Keller, R.M., Dungan, J.L.: Meta-modeling: a knowledge-based approach to facilitating process model construction and reuse. Eco. Mod. 119 (1999) 89-116
12. Jarrar, M., Demy, J., Meersman, R.: On Using Conceptual Data Modeling for Ontology Engineering. J. on Data Semantics 1(1) (2003) 185-207
13. Ford, E.D.: Scientific methods for ecological research. Cambridge University Press, Cambridge (2000)
14. Masolo, C., Vieu, L., Bottazzi, E., Catenacci, C., Ferrario, R., Gangemi, A., Guarino, N.: Social Roles and their Descriptions. Ninth International Conference on the Principles of Knowledge Representation and Reasoning (KR2004), Whistler, Canada (2004)
15. Klein, M., Noy, N.F.: A Component-Based Framework for Ontology Evolution. Workshop on Ontologies and Distributed Systems at IJCAI-2003, Acapulco, Mexico (2003)
16. Todorovski, L., Dzeroski, S.: Using domain knowledge on population dynamics modeling for equation discovery. Proceedings of the Twelfth European Conference on Machine Learning. Lecture Notes in Computer Science, Springer Verlag, Berlin (2001) 478-490
17. Blackford, J.C., Radford, P.J.: A structure and methodology for marine ecosystem modelling. Neth. J. Sea Res. 33(3/4) (1995) 247-260
18. Baretta-Bekker, J.G., Baretta, J.W., Koch Rasmussen E.: The microbial food web in the European Regional Seas Ecosystem Model. Neth. J. Sea Res. 33(3/4) (1995) 363-379
19. Smith, J.M.: Models in ecology. Cambridge University Press, Cambridge (1979 ed.) (1974)
20. Nihoul, J.C.J.: Modelling marine ecosystems as a discipline in Earth Science. Earth-Science Reviews 44(1) (1998) 1-13

Querying Ontologies in Relational Database Systems

Silke Trißl and Ulf Leser

Humboldt-Universität zu Berlin, Institute of Computer Sciences,
D-10099 Berlin, Germany
{trissl, leser}@informatik.hu-berlin.de

Abstract. In many areas of life science, such as biology and medicine, ontologies are nowadays commonly used to annotate objects of interest, such as biological samples, clinical pictures, or species in a standardized way. In these applications, an ontology is merely a structured vocabulary in the form of a tree or a directed acyclic graph of concepts. Typically, ontologies are stored together with the data they annotate in relational databases. Querying such annotations must obey the special semantics encoded in the structure of the ontology, i.e. relationships between terms, which is not possible using standard SQL alone.

In this paper, we develop a new method for querying DAGs using a pre-computed index structure. Our new indexing method extends the pre-/ postorder ranking scheme, which has been studied intensively for trees, to DAGs. Using typical queries on ontologies, we compare our approach to two other commonly used methods, i.e., a recursive database function and the pre-computation of the transitive closure of a DAG.

We show that pre-computed indexes are an order of magnitude faster than recursive methods. Clearly, our new scheme is slower than usage of the transitive closure, but requires only a fraction of the space and is therefore applicable even for very large ontologies with more than 200,000 concepts.

1 Introduction

Ontologies play an important role in biology, medicine, and environmental science. The probably oldest ontology in biology is the taxonomic classification of flora and fauna. The NCBI taxonomy [1] is represented as rooted, directed tree, where nodes represent organisms or families, while edges represent an evolutionary relationship between two nodes.

In the area of medicine and molecular biology several ontologies were introduced in the last years, including the Gene Ontology (GO) [2]. The project aims at providing a structured, precisely defined, commonly used, and controlled vocabulary for describing the roles of genes and gene products in any organism. In contrast to the NCBI taxonomy, which resembles a tree, the Gene Ontology is structured in the form of a rooted directed acyclic graph (DAG). Each GO term represents a labeled node in the graph, while an edge represents a direct relationship between two terms.

Ontologies as those mentioned before are used to annotate biological and environmental samples, or to define functional characteristic of genes and gene products. Both, the annotated data and the ontologies are stored in information systems, usually in

B. Ludäscher and L. Raschid (Eds.): DILS 2005, LNBI 3615, pp. 63–79, 2005.

relational database systems. Clearly, these data are not just stored, but also queried to answer biologically interesting questions and to find correlations between data items.

The main advantage of ontologies lies in their hierarchical structure. When a query asks for all samples annotated with a certain concept, not only the term itself needs to be considered, but also all its child, grand-child, etc. concepts. Consider the question "Is the concept *transcription factor activity* defined as a kind of *nucleic acid binding* in the Gene Ontology?".

1.1   Motivation

Graph structures are usually stored using two tables, one for nodes and one for edges. Each edge represents a binary relationship between two nodes, i.e., a father and a child concept. Using this model, it is easy to get parents or children of a node, but not ancestors or successors as these are in arbitrary distance of the start node. Answering this simple question above using standard SQL alone is therefore impossible.

Generally, there are two different approaches for answering the question. The simplest method is to program a recursive function – either as stored procedure or using a host language – that traverses the ontology at run time to compute the answer to the query. However, a recursive functions requires time proportional to the number of traversed nodes in the tree or the DAG, leading to bad runtime performance. The second possibility is to index the graph in some way. For instance, one could compute and store the transitive closure of a tree or DAG before queries are posed. Then, a question as the one above can be answered in almost constant time by a simple table lookup. But index structures require time for computation and space for being stored, rendering them inapplicable for very large ontologies.

In this paper we present a new index structure for DAG-formed ontologies that is an order of magnitude faster than recursive functions and in most situations consumes an order of magnitude less space than a pre-computed transitive closure.

The rest of the paper is organized as follows. Section 2 describes our model of storing ontologies, defines typical queries for ontologies, and describes how these queries can be answered using recursive functions. Section 3 describes two well-known indexing schemes for tree structures, i.e., pre-/ postorder ranks and transitive closure. Section 4 describes how these indexing structures can be extended to index DAGs. The extension of the pre-/ postorder ranking to DAGs is the main contribution of the paper. Section 5 shows our results on implementing and benchmarking the different methods. Finally, Section 6 concludes the paper.

2   Storing and Querying Ontologies

In this section we first describe our model of storing graphs in relational database systems and we then introduce and specify common questions on ontologies. We demonstrate how these data can be queried using recursive database functions. In the next section we then present index structures and how to query them.

2.1   Data Model

We consider ontologies that are rooted, directed trees or DAGs. In both structures, a path is a sequence of nodes that are connected by directed edges. The length of a path is the number of nodes it contains. The length of the shortest path between two nodes is called the distance between the nodes. In a tree each node can be reached on exactly one path from the root node. The same is true for any other two nodes in a directed tree, if a path between the two nodes exists. DAGs are a simple generalization of trees, as nodes may have more than one parent. Therefore, nodes may be connected by more than one path.

In any directed graph successors of a node v are all nodes w for which a path from v to w exists. The successor set of v are all nodes w that can be reached from v. In analogy ancestors of node v are all nodes u where a path from u to v exists. The ancestor set of v are all nodes u from which v can be reached.

Graphs are stored as a collection of nodes and edges. The information on nodes includes a unique identifier and possibly additional textual annotation. Information on edges is stored as binary relationship between two nodes. Additional attributes on edges can be stored as well. In a relational database system both collections are stored in separate tables. The NODE-table contains all node information including the unique identifier, node_name. The second table is called EDGE, where the binary relationship between two nodes is stored in the attributes from_node and to_node.

2.2   Typical Queries on Ontologies

The main questions on taxonomies and ontologies can be grouped into three categories, namely reachability, ancestor- or successor set, and least common ancestor of two or more nodes.

Q1: Reachability is concerned with questions like 'Does the species *Nostoc linckia* belong to the phylum *Cyanobacteria*?'. To answer the question, one has to find out, if the node labeled 'Nostoc linckia' has an ancestor node labeled 'Cyanobacteria' in the NCBI taxonomy. The length of the path between the two nodes does not matter.

Q2: Ancestor-/ Successor set of a given node contains all ancestor and successor nodes, respectively. Given a set of proteins, annotated by Gene Ontology terms, a researcher may want to find all proteins that are involved in nucleic acid binding. Of course, not only the proteins directly annotated by the term 'nucleic acid binding' are of interest, but also all proteins that have a successor term of the original term as annotation. The first step in answering the question is to retrieve all successor nodes of the given start term – in short the successor set.

Q3: Least common ancestor is of interest when a common origin of a set of nodes should be computed. For instance, microarray experiments produce expression levels of thousands of different genes within a single experiment. A typical analysis is the clustering of genes by the expression levels. A biologist now wants to find commonalities among genes in a cluster. In this situation, GO annotations of genes are helpful, as the least common ancestors of the annotated GO terms defines the most specific common description of the genes in the cluster. Note that for computing the least common ancestor of a set of nodes, the lengths of the paths between nodes is crucial. Ancestor sets of nodes may have several nodes in common, and one has to decide which of these is the closest to all given nodes. Obviously, for answering this question it suffices to know the distance between the nodes.

2.3   Querying Ontologies

The conventional way is to use recursion to traverse a tree or graph on query time. Algorithm 1. performs a depth-first search over a tree and returns the successor set for a node v. The function first looks for children of the start node v and appends each child, m to the successor set. It then searches for successors of m by calling itself with node m as the new start node. Doing so, it also holds a counter for the length of the path v and the current node. As in trees only one path between any two nodes exists, this is equivalent to the distance. As soon as no more child nodes are found the by then accumulated successor set is returned.

Algorithm 1. Recursive Algorithm to retrieve the successor set of a node v

FUNCTION successorSet(v, dist) **RETURNS** succcessors
BEGIN
  FOR EACH m $\in$ $\sigma_{from\_node=\ v} EDGE$ **DO**
    append (m,1) to *successors*;
    successorList(m) := successorSet(m, dist+1);
    append successorList(m) to *successors*;
  END FOR;
  return *successors*;
END;

To compute the ancestor set of a node a second function has to be created, called `ancestorSet()`. This function takes the same parameters as the one presented in Algorithm 1., but instead of looking for child nodes the algorithm will look for all parent nodes and append them to an ancestor set, which will be returned at the end.

Using these stored procedures, it is possible to query tree and DAG structures. However, for DAGs the function is not optimal. Using the functions the exemplary questions presented in Section 2.2 can be answered with the following SQL statements:

– **Q1: Reachability**
```
SELECT 1
FROM successorSet(v, 0)
WHERE suc = w;
```

– **Q2: Ancestor/Successor set**
```
SELECT suc
FROM successorSet(v, 0);
```

– **Q3: Least common ancestor**
```
SELECT A.anc,
    A.dist+B.dist AS dist
FROM (SELECT anc, dist
    FROM ancestorSet(s)) A
INNER JOIN (SELECT anc, dist
    FROM ancestorSet(t)) B
ON A.anc = B.anc
ORDER BY dist;
```

3   Indexing and Querying Tree Structures

We now show how to index and query tree structures using the pre- and postorder ranking scheme as well as the transitive closure.

3.1   Pre- and Postorder Ranks

The pre- and postorder rank indexing is well studied for trees [3]. Several systems suggested to use it for indexing XML documents in relational databases [4]. The advantage of pre- and postorder rank indexing for an XML document is that the document order is maintained, i.e., the user is able to query for descendant nodes as well as for following. Note that in our case only descending and ascending nodes are of interest, as ontologies usually do not contain any order among children of a node. In chapter 4, we will extend the pre-/ postorder ranking scheme to DAGs. Therefore, we describe the method in detail in the following.

Algorithm 2. shows the function for assigning pre- and postorder ranks to a node in a tree. Ranks are assigned during a depth-first traversal starting at the root node. The preorder rank for a node is assigned as soon as this node is encountered during the traversal. The postorder rank of a node is assigned before any of the ancestor nodes and after all successor nodes have received a postorder rank. We store pre- and postorder ranks together with the node ID in a separate table forming the index. Clearly, the space requirement of the ranks is proportional to the number of nodes in the tree.

Algorithm 2. Pre-/postorder rank assingments of nodes, starting with root node r

---

```
var pr:=0; var post:=0;
FUNCTION prePostOrder(r)
BEGIN
   FOR EACH child, m ∈ σ_{from_node= r} EDGE DO
      pre:=pr; pr:=pr+1;
      prePostOrder(m);
      INSERT m, pre, post, pr-pre INTO prePostOrder;
      post:=post+1;
   END FOR;
END;
```

To illustrate the steps of the algorithm consider the tree in Figure 1(a). Starting at the root node A, we traverse the tree in depth-first order. Node B gets the preorder rank of 1, while E gets 2. As node E has no further child nodes it is the first node to get a postorder rank and is stored with both ranks in table prePostOrder. This way the rest of the tree is traversed. The pre- and postorder rank of root node A is assigned separately.

In addition to the ranks, we also store the number of descendants, s for each node, which we will use later for improving queries. This number can be computed as the difference between the current preorder rank and the preorder rank of the node to be inserted next. To clarify this, consider node C in Figure 1(a). This node is inserted with

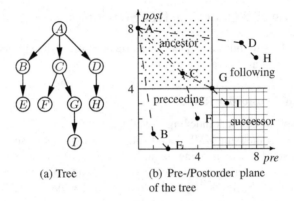

(a) Tree

(b) Pre-/Postorder plane
of the tree

Fig. 1. Pre-/postorder rank assignment of a tree

the preorder rank of 3. The current preorder rank is 6 as the last successor node of C, I has this preorder rank. The difference between the two preorder ranks is 3, which is exactly the number of successor nodes of C.

Pre- and postorder ranking becomes clearer when it is plotted in a two dimensional co-ordinate plane, with the preorder rank on the x-axis and postorder rank on the y-axis as shown in Figure 1(b).

Querying Pre-/Postorder Indexed Trees. As indicated for node G in Figure 1(b) the pre-post plane can be partitioned into four disjoint regions for each node v. The upper-left partition contains all ancestors of v, while the successors can be found in the lower-right area. The remaining two areas hold the preceeding and following nodes of v.

As ontological structures are usually order-independent, only the ancestor and successor sector are of interest. Using the preassigned ranks, nodes in these two partitions can be retrieved without recursion, since any successor of node v must have a preorder rank that is higher and a postorder rank that is lower than that of v. The location of the successors of a node v within the lower-right partition can be further restricted. Let node v have preorder rank pre_v. If v has s successor nodes, then each successor w of node v will have a preorder rank pre_w with $pre_v < pre_w \leq pre_v + s$.

To find the least common ancestor of two nodes the ancestor sets of both nodes have to be joined on the attribute node_name and the common ancestor with the highest preorder rank is least common ancestor of both nodes.

Using the refinement on the location of the successors the queries for answering questions Q1, Q2, and Q3 are the following:

- **Q1: Reachability** (is w successor of v)

```
SELECT 1
FROM prePostOrder p1,            AND p2.node_name = v
     prePostOrder p2            AND p1.pre > p2.pre
WHERE p1.node_name = w          AND p1.pre ≤ p2.pre+p2.s;
```

- **Q2: Ancestor set**
  ```
  SELECT p1.node_name AS u
  FROM prePostOrder p1,
       prePostOrder p2
  WHERE p2.node_name = v
    AND p1.pre < p2.pre
    AND p1.post > p2.post;
  ```
- **Q2: Successor set**
  ```
  SELECT p1.node_name AS w
  FROM prePostOrder p1
       prePostOrder p2
  WHERE p2.node_name = v
    ANDp1.pre > p2.pre
    AND p1.pre ≤ p2.pre+p2.s;
  ```
- **Q3: Least common ancestor**
  ```
  SELECT A.node_name, A.pre
  ```

```
FROM (
  SELECT p1.node_name, p1.pre
  FROM prePostOrder p1,
       prePostOrder p2
  WHERE p2.node_name = s
    AND p1.pre < p2.pre
    AND p1.post > p2.post) A
INNER JOIN (
  SELECT p1.node_name, p1.pre
  FROM prePostOrder p1,
       prePostOrder p2
  WHERE p2.node_name = t
    AND p1.pre < p2.pre
    AND p1.post > p2.post) B
ON A.node_name = B.node_name
ORDER BY A.pre desc;
```

3.2 Transitive Closure

The transitive closure of a graph is a set of edges. Edge (v, w) is inserted into the transitive closure if either (v, w) is an edge in the graph or if there exists a path between node v and w. Using the transitive closure, queries on reachability and queries for ancestor and successor sets can be answered very efficiently. Finding the least common ancestor of two or more nodes requires to store the length of the shortest path between two nodes.

In the past, several algorithms have been developed to compute the transitive closure within a relational database system [5]. We found that the so called 'Logarithmic algorithm' [6] performed best for trees as well as DAGs. The function is presented in Algorithm 3..

Algorithm 3. Computing the transitive closure

```
FUNCTION transtiveClosure()
BEGIN
  INSERT INTO TC SELECT from_node, to_node, 1 FROM EDGE;
  max_dist:=1;
  REPEAT
    INSERT INTO TC SELECT TC1.anc, TC2.suc, min(TC1.dist+TC2.dist)
    FROM TC TC1, TC TC2 WHERE TC1.suc=TC2.anc AND TC1.dist=max_dist;
    max_dist:= SELECT max(dist) FROM TC;
  UNTIL INSERT = ∅
END;
```

This algorithm first inserts all tuples of the initial edge relation with the distance 1 to the transitive closure table TC. In the next step the tuples from TC with a distance equal to the maximum distance are self-joined with TC. The join condition is that the successor node of one relation must be equal to the ancestor node of the other. The ancestor nodes of the first relation, the successor node of the second and the minimal distance between the two nodes is stored in TC. This step is repeated until no further tuples can be inserted into TC.

Note that the transitive closure requires space that is in worst case $O(|V|^2)$. Clearly, the real space requirements are much smaller for trees, as they are for DAGs. In Section 5, we will measure space consumption of transitive closures in more detail.

Querying the Transitive Closure. The transitive closure essentially contains one tuple for each pair of ancestor - successor nodes. Accordingly, queries answering our three problems may look as follows:

- **Q1: Reachability**
  ```
  SELECT 1
  FROM TC
  WHERE anc = v
     AND suc = w;
  ```

- **Q2: Ancestor/Successor set**
  ```
  SELECT suc
  FROM TC
  WHERE anc = v;
  ```

- **Q3: Least common ancestor**
  ```
  SELECT A.anc, A.dist+B.dist
  AS distance
  FROM (SELECT anc, dist
    FROM TC WHERE suc = s) A
  INNER JOIN (SELECT anc, dist
    FROM TC WHERE suc = t) B
  ON A.anc = B.anc
  ORDER BY distance;
  ```

4   Extending Index Structures to DAGs

So far, we only considered trees for querying. In this section we extend the indexing schemes to work on DAGs, as ontologies often have the form of directed acyclic graphs. Specifically, we present how the pre- and postorder ranking scheme can be used for DAGs and how this structure can be queried.

4.1   Pre- and Postorder Ranks for DAGs

The pre-/ postorder ranking scheme we described in the previous chapter is restricted to trees. The reason is that in DAGs, where nodes may be reached on more than one path from root, neither the pre- nor the postorder rank is unique for a single node. If multiple paths exist, a node is reached more than once during the traversal.

Obviously, it is no option to simply take any one of the ranks, e.g., the first to be assigned, because then the relationships between the ranks of ancestors and successors do not hold any more. Consequently, we would loose successors or ancestors during querying.

In the following, we describe a new and simple extension of the ranking scheme that is also capable of indexing DAGs. We will show in Section 5 that our method can be seen as a compromise between recursive query methods, which are slow for queries but need no further storage and the transitive closure, which allows for very fast queries, but also requires considerable storage space. We will also show that the advantages of our method depend on the "tree-likeness" of a DAG. For DAGs that are almost trees, our method has considerable advantages when compared to the transitive closure, however, these advantages are lost the less tree-like a DAG is.

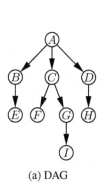

(a) DAG

node	pre	post	s
A	0	12	12
B	1	1	1
C	3	5	3
D	7	11	5
E	2	0	0
F	4	2	1
G	5	4	1
H	8	10	4
I	6	3	0
C	9	9	3
F	10	6	0
G	11	8	1
I	12	7	0

(b) Table with pre-/ postorder ranks and number of successors

Fig. 2. Pre-/postorder rank assignment of a directed acyclic graph

The basic idea of our extension is very simple. Instead of assigning only one pair of ranks to a node, we allow for multiple rank pairs. More specifically we assign an additional pre- and postorder rank to a node each time this node is encountered during the depth-first traversal. Actually, Algorithm 2. already performs this computation, as it inserts a new node-rank combination each time a node is encountered. After running the function on a DAG, each node will have as many pre- and postorder ranks as this node occurs in a path from the root node.

As an example, we add one more edge (the dotted edge) to the tree from figure 1(a). Table 2(b) shows the resulting pre- and postorder ranks for each node in the DAG. As one can see node C and all descendants of C get two different rank pairs, because these nodes are encountered on two different paths, one directly from A to C and one from A over D and H to C.

Clearly, the number of node-rank pairs is higher for DAGs than for trees, leading to an increase in space consumption for the index. The degree of increase depends on the number of additional non-tree edges and the location of such an edge in the graph. Clearly, additional edges in the upper levels of the tree will lead to an addition of rank pairs for a large number of nodes, while additional edges close to the leaves of a DAG only have marginal impact. Potentially there is an exponential growth of the index structure in the number of edges added. However, we observed that in practice the increase in size is not critical. The reason for this is that in ontologies concepts on the upper level usually only have one parent concept. For instance, in the Gene Ontology the first level where a node has two or more parents is on level four. In Section 5, we will show the impact in size more precisely both on real ontologies and on randomly generated trees and DAGs.

Like for trees, all rank pairs in the DAG can be plotted on a two-dimensional coordinate plane (see Figure 3). Nodes appear as many times in the plane as they have rank

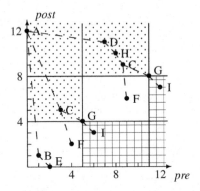

Fig. 3. Pre-/postorder plane of a DAG

pairs. This shows that, intuitively, our method multiplies all subtrees of nodes that have more than one parent.

To query our new indexing scheme, we need to adapt the methods for querying pre-post order indexes for trees. As an example, consider node G in Figure 3. This node as well as its successor set appears twice in the coordinate plane as it can be reached on two different paths from the root node A. However, the successor sets are identical for each instance of G, because this set is independent of the number of paths G can be reached from root. Thus, for successor queries it suffices to select any instance of a node and query for all its children using the conditions on pre- and post order rank used for trees. This is reflected by limiting the number of returned preorder ranks in the query to 1. As for trees the search space can be reduced by using the information on the number of descending nodes. However, caution must be taken to filter the result for duplicates.

The situation is more complex for ancestor queries, e.g., the ancestor set of a node v. Computing this set requires to merge all nodes in the upper-left partition of any instances of v, as the set of one instance only contains nodes for one possible path from root to v. Again, duplicates must be removed from the result.

– **Successor set:**
```
SELECT DISTINCT
    p1.node_name AS w
FROM prePostOrder p1
WHERE p1.pre > (
    SELECT p1.pre
    FROM prePostOrder p2
    WHERE p2.node_name = v
    LIMIT 1)
AND p1.pre ≤ (
    SELECT p2.pre+p2.s
    FROM prePostOrder p2
    WHERE p2.node_name = v
    LIMIT 1);
```

– **Ancestor set for DAGs:**
```
SELECT DISTINCT
    p2.node_name AS u
FROM prePostOrder p1,
    prePostOrder p2
WHERE p1.node_name = v
AND p2.pre < p1.pre
AND p2.post > p1.post;
```

We only gave the code for computing the successor and the ancestor set. Reachability can be computed in the same way as for trees. Least common ancestor requires to compute the ancestor sets of all nodes, intersect them, and find the node with the minimal sum of the differences between the preorder ranks of the two nodes and the common ancestor node.

4.2   Transitive Closure on DAGs

Algorithm 3. can be applied without changes to index DAGs. The space complexity will not change if only the minimal distance between any two nodes is stored. If all possible path lengths between two nodes are needed, the situation would be different and the upper bound would be exceeded.

Querying the transitve closure of DAGs is the same as for trees.

5   Results

In this section we compare both indexing methods and the recursive algorithm. We measure in detail run time of queries, space consumption of the index structures, and time necessary for building the indexes. We give results on generated tree and DAG structures and on real data, i.e., queries against the Gene Ontology.

We have implemented both indexing algorithms and the recursive algorithm as stored procedures in ORACLE 9i. Tests were performed on a DELL dual Xeon machine with 4 GB RAM. Queries were run without rebooting the database. Given the relative small size of the data being studied (in the range of a couple of megabytes), we expected that all computation is very likely performed solely in main memory, as both data and index blocks can be cached completely. Thus, secondary memory issues were not considered.

5.1   Time and Space Consumption of Graph Indexing Algorithms

To systematically measure the construction time and space consumption of the two index algorithms we generated trees with a given number of nodes and a given average degree of 8.0. The average degree is the average number of incoming and outgoing edges of a node, therefore in our trees each node has on average 7 children. DAGs were created by randomly adding additional edges to the tree, independent of the depth of the newly connected nodes. Added edges had to fulfill two conditions: First, it was not allowed to introduce parallel edges, and second, no edge between node v and an ancestor node of v was allowed, as this would introduce a cycle. The index structures of the generated trees and DAGs were created using Algorithms 2. and 3..

·   Figure 4 shows the size of the index structures given as the number of tuples inserted in the index relation. The starting point of a curve always stands for the tree with the indicated number of edges. To create DAGs we have iteratively added additional 10 % of the number of edges from the corresponding tree. For instance, starting with a tree of 10,000 edges, the second measurement contains 11,000, the third 12,000 edges, ect. Thus, in each line all but the first point represent DAGs. Altogether, we used 11 start points of trees with 1,000 to 200,000 nodes, performing 1 to 5 rounds of edge additions.

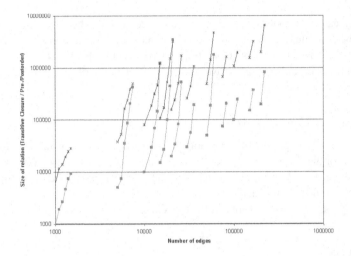

Fig. 4. Size of index depending on the size of the tree or DAG, respectively. Black stars give the size for `TC`, gray boxes for `prePostOrder`. Note that both axes use logarithmic scale

For a tree, the number of tuples inserted into `prePostOrder` equals the number of nodes. For most measured cases, the size of the index using our method is an order of magnitude smaller than the size of the transitive closure. However, we see that sizes of `TC` and `prePostOrder` are converging as the number of non-tree edges in a DAG increases. Adding up to 30 % more edges still leads to more than 50 % less tuples in the pre-/postorder index than for transitive closure in any of the examined sets.

However, adding 40 % more edges reverses the situation in two of the shown sets. The reason for this behavior is that, when adding additional edges to the tree, the end node of the added edge plays an important role for the pre- and postorder ranking, but not so much for the transitive closure. Imagine, you have already added a certain amount of additional edges to the tree, and now you add a new edge. The pre-/ postorder ranking now has to traverse another sub-structures more than once, and the nodes within that structure will get an additional rank pair. The transitive closure will also increase, as new connections are established. But the number of newly found connections decreases the more edges already exist in the DAG, as many new edges only introduce new paths between already connected nodes, thus not increasing the size of the transitive closure.

We can conclude that our method uses considerably less space than the transitive closure for DAGs that are tree-like. Note that the measurements on a real ontology are even more favorable for our method (see below).

The time required to construct the pre-/ postorder index for trees is always 3 to 10 times higher than for the transitive closure (data not shown). However, the actual time difference is marginal, as both structures can be computed very fast even for large trees. Computing the transitive closure for a tree of 200,000 nodes takes 58 seconds, while the pre-/postorder ranking index needs 3:45 minutes.

The time difference increases quickly with the number of edges added. For up to 20 % more edges, the difference remains within the order of the differences for trees. Adding more edges leads to a dramatic increase in the time necessary for computing the pre-/ postorder index. For a DAG with 10,000 nodes and 13,999 edges, it already takes 71 times more time to compute the pre-/ postorder than the transitive closure, although the number of inserted tuples in both tables is nearly equal. The reason for this difference is that pre-/ postorder ranks require extensive graph traversal, while the transitive closure can be efficiently computed using dynamic programming - style algorithms over increasing path lengths.

5.2   Querying Ontologies

We measured query times for three exemplary questions described in Section 2.2, based on real ontologies. We used real ontologies and not generated ones to obtain more realistic results, as in human curated ontologies concepts on higher levels usually do not have more than one parent. This specific edge distribution is not included in our DAG generator.

Table 1. Number of tuples inserted in each relation and time (in min:sec) required for computing the index structures

	NCBI Taxonomy		Gene Ontology	
	Tuples	Time	Tuples	Time
Pre-/Postorder ranking	230.559	5:26	76.734	1:24
Transitive Closure	3.583.760	1:44	178.033	0:04

Table 1 shows for the two ontologies, i.e. the NCBI Taxonomy and the Gene Ontology the size of the index structure and the time required for computing both indexes. As the NCBI Taxonomy is a tree, the pre-/ postorder index is much smaller than the transitive closure. The figures are more interestingly for the Gene Ontology. We used a version with 16.859 nodes and 23.526 edges. Although the number of edges exceeds the number of nodes by approximately 40 %, the size of the pre-/ postorder index is still considerably smaller than the transitive closure, confirming our observation about the edge distribution in real ontologies.

In the following, due to space restrictions, we only give query times for Gene Ontology. For each of the queries, Q1, Q2, and Q3, 25 % of the nodes of the Gene Ontology were randomly selected. The query for each node was issued 20 times. The following figures give average query execution times.

Reachability.   We computed times for answering the query 'Is w a successor node of v?' for randomly selected w and v. Figure 5(a) shows the times for 4,300 single queries using either of the two index structures. As one can see, querying the transitive closure is faster than querying the pre-/ postorder index, but only by a small and almost constant factor. The recursive function, whose running time depends on the number of nodes traversed, is not displayed, as it required between 6 and 11,000 times more time than querying the indexing schemes.

Successor Set.   The successor set for 25 % randomly selected nodes from the Gene Ontology was retrieved using the queries presented in the former sections. Results can

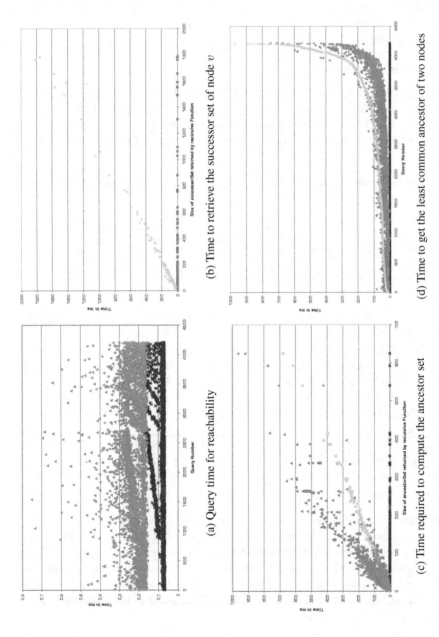

(a) Query time for reachability

(b) Time to retrieve the successor set of node v

(c) Time required to compute the ancestor set

(d) Time to get the least common ancestor of two nodes

Fig. 5. Query times for the three presented methods for the different queries. The light-gray squares are the query times for the recursive function, dark-gray for pre-postorder ranking, and black stars for transitive closure

be found in Figure 5(b). Note that the successor set returned from the recursive function and from querying the pre-/ postorder index can contain successor nodes several times. The successor set from the transitive closure will contain any node only once.

Query times using a recursive function is linearly dependent on the number of tuples returned. Times for both index structures remain fairly constant over the number of tuples. Times for querying using the pre-/ postorder index are on average 1.5 times higher than using the transitive closure.

Ancestor Set. Figure 5(c) shows the time needed to retrieve ancestor sets. In this case, the indexing methods differ considerably. While query times using transitive closure are similar to the times for the successor set, times for querying the pre-/ postorder index is even more costly than using a recursive function. The reason that the pre-/ postorder index is slow is that the ancestor set has to be calculated for every instance of the start node leading to an extremely redundant ancestor set.

Least Common Ancestor. Computing the least common ancestor of two nodes first requires to compute the ancestor sets of each node, second to find common nodes in both sets, and third to select the node with the minimal distance to both original nodes. Figure 5(d) shows the time necessary to compute the least common ancestor of 4,300 randomly selected pairs of nodes, sorted by the time required for computing the answer using the recursive function. The figure shows that querying the pre-/ postorder index structure is better than using a recursive database function and worse than using the transitive closure. The results resemble the one shown in Figure 5(c), as the cost-dominating operation is the computation of the ancestor sets. The steep rise in time for some "pathological" node pairs, i.e., queries where both sets have extremely large ancestor sets, is somewhat surprising and deserves further study.

6   Discussion

Indexing tree and graph structures is a lively research area. In the XML community the pre-/ postorder ranking scheme is widely used as it preserves information about the document order and allows very fast queries at four axis of the XQuery model. To further optimize access to tree data in relational databases, Mayer et al. [7] have created the so called 'Staircase Join', a special join operator for queries against pre-/ postorder ranking schemes. It is unclear of this method could also be extended to DAGs.

Vagena et al. [8] presented a different numbering scheme for DAGs. This scheme also conserves the document order, but it is restricted to planar DAGs. As we can not guarantee that every ontology has such a structure, the algorithm is not universally applicable. Another numeric indexing structure for DAGs was presented in [9], where they label spanning trees with numeric intervals. In DAGs not the nodes with several parent nodes get more than one interval, but all ancestor nodes get the first interval of that node. They proposed a reduction, but as intervals are propagated upwards in real ontologies this would probably lead to an index size in the same order of magnitude.

A different indexing method for trees and graphs was proposed by Schenkel et al. [10]. Their method uses the 2-hop cover [11] of a graph, which is more space efficient than the transitive closure and allows to answer reachability queries with a single join.

Since computing optimal 2-hop covers is NP hard, they use an approximation optimized for very large XML documents with XPointers. However, 2-hop covers do not allow for least common ancestor queries, as no distance information can be preserved.

[12, 13] are examples of attempts to index graph structures, one by finding and indexing all frequent subgraphs, and one by exploiting properties of the network structure. However, both methods are for full graphs, and we would expect them to perform rather poor on DAGs. In the ontology community, we are not aware of any work on optimized indexing and querying of large ontologies.

We have presented a novel structure for indexing and querying large ontologies, extending the well known pre- and postorder ranking scheme to DAGs. Our method has favorable properties for ontologies that are tree-like, which is true for most ontologies we are aware of. In those cases, most queries for successors are almost as fast as using the transitive closure, while space consumption is an order of magnitude lower. One drawback of our method is the time for creating the index. Our current research is geared towards reducing this time and speed up ancestor queries.

Acknowledgement

This work is supported by BMBF grant no. 0312705B (Berlin Center for Genome-Based Bioinformatics).

References

1. DL Wheeler, C Chappey, AE Lash, DD Leipe, TL Madden, GD Schuler, TA Tatusova, and BA Rapp. Database resources of the National Center for Biotechnology Information. *Nucleic Acids Research*, 28(1):10 – 14, Jan 2000.
2. Gene Ontoloy Consortium. The Gene Ontology (GO) database and inforamtics resource. *Nucleic Acids Research*, 32:D258 – D261, 2004. Database issue.
3. P. Dietz and D. Sleator. Two algorithms for maintaining order in a list. In *Proceedings of the nineteenth annual ACM conference on Theory of computing*, pages 365–372. ACM Press, 1987.
4. Torsten Grust. Accelerating XPath location steps. In *Proceedings of the 2002 ACM SIGMOD international conference on Management of data*, pages 109–120. ACM Press, 2002.
5. Hongjun Lu. New strategies for computing the transitive closure of a database relation. In *Proceedings of the 13th International Conference on Very Large Data Bases*, pages 267–274. Morgan Kaufmann Publishers Inc., 1987.
6. P. Valduriez and H. Boral. Evaluation of recursive queries using join indices. In L. Kerschberg, editor, *First International Conference on Expert Database Systems*, pages 271–293, Redwood City, CA, 1986. Addison-Wesley.
7. Sabine Mayer, Torsten Grust, Maurice van Keulen, and Jens Teubner. An injection of tree awareness: Adding staircase join to postgresql. In Mario A. Nascimento, M. Tamer Özsu, Donald Kossmann, Renée J. Miller, José A. Blakeley, and K. Bernhard Schiefer, editors, *VLDB*, pages 1305–1308. Morgan Kaufmann, 2004.
8. Zografoula Vagena, Mirella Moura Moro, and Vassilis J. Tsotras. Twig query processing over graph-structured xml data. In Sihem Amer-Yahia and Luis Gravano, editors, *WebDB*, pages 43–48, 2004.

9. Rakesh Agrawal, Alexander Borgida, and H. V. Jagadish. Efficient management of transitive relationships in large data and knowledge bases. In James Clifford, Bruce G. Lindsay, and David Maier, editors, *SIGMOD Conference*, pages 253–262. ACM Press, 1989.
10. Ralf Schenkel, Anja Theobald, and Gerhard Weikum. Efficient creation and incremental maintenance of the hopi index for complex xml document collections. In *ICDE*, 2005.
11. Edith Cohen, Eran Halperin, Haim Kaplan, and Uri Zwick. Reachability and distance queries via 2-hop labels. *SIAM J. Comput.*, 32(5):1338–1355, 2003.
12. Andrew Y. Wu, Michael Garland, and Jiawei Han. Mining scale-free networks using geodesic clustering. In Won Kim, Ron Kohavi, Johannes Gehrke, and William DuMouchel, editors, *KDD*, pages 719–724. ACM, 2004.
13. Xifeng Yan, Philip S. Yu, and Jiawei Han. Graph indexing: A frequent structure-based approach. In Gerhard Weikum, Arnd Christian König, and Stefan Deßloch, editors, *SIGMOD Conference*, pages 335–346. ACM, 2004.

Scientific Names Are Ambiguous as Identifiers for Biological Taxa: Their Context and Definition Are Required for Accurate Data Integration

Jessie B. Kennedy, Robert Kukla, and Trevor Paterson

School of Computing, Napier University, Edinburgh, EH10 5DT, UK
{j.kennedy, r.kukla, t.paterson}@napier.ac.uk

Abstract. Biologists use scientific names to label the organisms described in their data; however, these names are not unique identifiers for taxonomic entities. Alternative taxonomic classifications may apply the same name, associated with alternative definition or circumscription. Consequently, labelling data with scientific names alone does not unambiguously distinguish between taxon concepts. Accurate integration and comparison of biological data is required on taxon concepts, as defined in alternative taxonomic classifications. We have derived an abstract, inclusive model for the diverse representations of taxonomic concepts used by taxonomists and in taxonomic databases. This model has been implemented as a proposed standard XML schema for the exchange and comparison of taxonomic concepts between data providers and users. The representation and exchange of taxon definitions conformant with this schema will facilitate the development of taxonomic name/concept resolution services, allowing the meaningful integration and comparison of biological datasets, with greater accuracy than on the basis of name alone.

1 Introduction

Scientific names are inherently poor identifiers for organisms, because although names are formalized and validated according to strict codes of nomenclature, the same name can be applied by taxonomists to alternative taxonomic views of the extent or definition of a taxon (e.g. a species, genus *etc.*). Biologists (i.e. the 'users' of taxonomic classifications) identify and label their data with scientific names, by identifying their organisms according to a particular taxonomic classification, as found for example in field guides, but without recognizing and recording that taxonomic context. As a consequence datasets cannot be reliably integrated on the basis of the scientific names because the context or meaning of the name is not captured.

Taxonomic identification is emerging as a significant problem for the integration and comparison of diverse datasets across all fields of biology from genomics to ecology. For example, annotations of Genbank DNA sequences typically label the source species according to the NCBI Taxonomy (www.ncbi.nlm.nih.gov/Taxonomy). Whilst specifically disclaiming any 'taxonomic authority' NCBI attempts to provide a single consensus view on taxonomy and represent name alterations and 'corrections'

B. Ludäscher and L. Raschid (Eds.): DILS 2005, LNBI 3615, pp. 80–95, 2005.

by encoding synonym relationships for use by their search engines (for example the genus *Fugu* has recently been 'renamed' *Takifugu*). Such an approach cannot deal with complex, changing and unrecorded relationships between names as used according to alternative taxonomic views. For example, the alternate classification and reclassification of Orangutans into separate species or subspecies means that sequence data might be labelled according to a variety of alternative classifications. (Currently over 50,000 nucleotide sequences are ascribed to *Pongo pygmaeus,* with fewer than 100 for each 'subspecies' *abelii* and *pygmaeus*). It is not clear how the NCBI Taxonomy might handle the alternative reclassification of these sub-species as species or whether the 50, 000 *P. pygmaeus* sequences include data that some taxonomists would ascribe to *abelii* (species or subspecies). These problems impact on other areas of biology and beyond. For example, the increase between 1996 and 2000 in the number of officially endangered primate species is partly attributable to the decision in 2000 to accept the reclassification of some subspecies (including Orangutan) at the species level [1]. Clearly consideration of species names in isolation, without the appropriate classificatory context, makes it difficult to interpret biodiversity data such as the distribution of Orangutans, when collected at different times, and labelled according to different (unrecorded) classification contexts.

1.1 Taxonomy and Nomenclature

Taxonomists classify organisms into hierarchically ranked taxa according to their evolutionary relatedness, based on any of a variety of types of biological evidence (morphology, genetics, palaeontology *etc.*). Alternative classifications (taxonomic revisions) arise over time reflecting new or alternative taxonomic opinion following more detailed study, the discovery of new taxonomic information such as evidence about relationships between taxa, description of new species, and increasingly molecular phylogenies based on DNA sequence comparison. Therefore taxonomy is itself an investigative science, and taxonomic classifications represent partial and evolving hypotheses rather than static identifications of absolute taxa. Any recorded taxonomic classification represents an opinion, according to one authority, at a given time. Relationships may be expressed or inferred between successive or alternative taxonomies, relating the concepts (taxa) in one classification to concepts in another, but without knowing the total genetic history of all life on earth it is not possible to derive a final, 'true' classification of existing (and extinct) organisms.

Taxonomists use scientific names in order to label and communicate about the taxonomic concepts that they create. Names are applied to the taxa in a given classification according to the codified rules of nomenclature, based on 'typification' (i.e. by reference to archived 'type' specimens) and following the principle of 'priority' where names are dependent on the oldest type specimen included in the circumscription of a taxon. This system provides stability to scientific names over time, as they are preserved in relation to their original use and type specimen. However, as a direct consequence of the application of these rules the same valid scientific name will apply to different views of a taxon according to different postulated taxonomic classifications. Indeed it is also true that very similar taxonomic concepts may have different names according to different classifications.

Names therefore are a *part* of a 'taxon concept', and cannot be used to unambiguously identify a concept. The identifiers used by experimental biologists to label organisms as a member or instance of a particular taxonomic concept should unambiguously refer to the taxon concept itself: true integration therefore requires unique identifiers for taxon concepts. We propose these concept identifiers should minimally include the *scientific name applied* and the *classification context*. This context is represented by the authorship of the concept, i.e. an 'According To' or *secundum* (*sec.*) citation. Assigning identifiers for concepts allows simple resolution of taxon concepts based on identity, particularly if GUIDs were to be adopted for concepts.

Taxonomic concepts are created and defined (or revised) in taxonomic publications. These publications may include various levels of detail defining each taxon, which might include: character descriptions (i.e. a list of structure, attribute, value triples), lists of archived specimens which are included in the taxon (specimen circumscription), relationships to other concepts in the same classification (including parent-child relationships between a taxon and its subordinate taxa), relationships with concepts in earlier alternative classifications, assignment of rank (family, genus, species *etc.*) and application of a scientific name for this taxon. Individual taxonomists have different perceptions or models for what constitutes and defines a taxon. This makes comparison of alternative taxon concepts problematic, even if the full rationale for the classification is available. However, comparing components of concept definitions might allow experienced Taxonomists to establish and record relationships between concepts with different GUIDs (e.g. two concepts can be considered equivalent for some particular purpose).

1.2   The Users of Taxonomic Classifications

The complex issues of ambiguity surrounding taxonomic classification and naming are well understood by expert taxonomists, but their importance and consequences are probably not considered relevant by experimental biologists who wish to use the names as static identifiers for the organisms described in their data. The explosion in biological data makes the accurate identification of source organisms critical. For example a researcher will frequently wish to identify which available datasets contain information on a particular organism of interest. Typically datasets are annotated by scientific name. However, correct identification of these datasets requires matching the taxonomic concepts as used in the source datasets, with the taxonomic concept of interest to the researcher (as defined by their reference classification). This requires either the use of identifiers for concepts, or comparison of the actual definitions of the concepts of interest with the definitions used by the authors of each dataset. A corollary of this is that datasets should be marked up with unambiguous taxonomic concept identifiers, for example they should reference the identification guide or classification system used by the researcher: identification by scientific name alone is insufficient.

By way of example a researcher wishing to access data on a fictitious species *Aus bus* from globally distributed databases might minimally want to recover data about any species that had ever been known as species *Aus bus*, or they might want to extend this query to recover information about all named species asserted to be synonymous with *Aus bus* at some level. Alternatively, if they have precise knowledge of the underlying concept described as species *Aus bus* they may only want to retrieve information about concepts closely related to their own concept of

information about concepts closely related to their own concept of *Aus bus*, regardless of their identifying names. Such detailed exploration of all species that overlap or are equivalent with *Aus bus* is only possible if 'names' are resolved according to the concepts to which they have been attached, so that data is retrieved on the basis of concept comparison, regardless of nomenclatural issues. Firstly however we require a common exchange schema to facilitate the representation, exchange and query of concepts.

In the following section we describe the current use of biological nomenclature and present an example to illustrate the problems associated with relying upon scientific names as identifiers for organisms. In section 3 we discuss the variety of approaches taken by biologists when describing taxonomic concepts and in section 4 argue the case for a standard schema to allow the exchange of this data to permit potential comparison and resolution of taxonomic concepts. In section 5 we present our work in defining the Taxonomic Concept Schema, an XML exchange standard for taxonomic concepts and names and compare this to other models in section 6. Finally some conclusions are drawn in section 7.

2 Using Names as Identifiers of Concepts

The formulation and application of valid scientific names for taxonomic groups is governed by separate codes of nomenclature for botany, zoology, bacteria and viruses (ICBN [2]; ICZN [3]; ICSP [4], ICTV [5]). According to these rules the name of a taxon is usually determined by the oldest type specimen included in its circumscription. The history of the fictitious genus *Aus* detailed in Figure 1 (and described more fully online [6]) illustrates how the rules of nomenclature provide stability for names throughout the history of taxonomic revisions, but automatically mean that names cannot be used as unique, non-ambiguous identifiers of taxon concepts. In fact the use of species names can never be truly separated from a taxonomic classification because the rules of binomial nomenclature obscure the boundary between classification and nomenclature for taxon names below the level of 'genus' (see for example [7]).

Where a full scientific name is used with attribution to the authors of the name and of the taxonomic revision, this represents a clear identifier for a concept. However, this level of detail is rare outwith specialist taxonomy. Most users and creators of biological data are not expert in taxonomy, and the names or labels that they use to refer to specimens and organisms include ad-hoc labels, common names or the (sometimes approximate or inaccurate) scientific name for a species or higher taxonomic group. Published and electronically deposited data might therefore be labelled with a variety of names, of varying precision and specificity. For example data about a particular species of 'daisy' can be found labelled as: lawn daisy, English lawn daisy, european lawn daisy, USDA code BEPE2, APNI code 163507-3, ITIS TSN 36862, *Bellis perennis*, *Bellis perennis* L., *Bellis perennis* L. Sp.Pl. 886, *Bellis perennis* L. Species Plantarium 2 1753, *Bellis perennnis* L. Species Plantarium (1753): 886, *Erigeron perennis* (L.) Sessé & Moc., *Conyzopsis bellis* EHL Krause. Integration and resolution between such diverse and semantically distinct names is clearly non-trivial, where even a 'single' name might be recorded with minor variations due to errors and corrections in spelling, or there may be variation in the abbreviations used.

A growing number of taxonomic resources and databases are available online, which seek to provide an integrated record of the names and taxonomic relationships for a particular narrow or wide taxonomic range (e.g. FishBase, www.fishbase.org; ITIS, www.itis.usda.gov). These taxonomic databases require quite complex models of taxonomic names in order to represent their data and to account for the needs of their users. Historically such databases only represented single, aggregated views of taxonomy, but it is now recognized that the issue of multiple classifications should be addressed. This requires consideration both of the synonymies between names as used in alternative classifications, and the application of the same name to different concepts in alternative classifications. Current representations of synonymy between names fail to capture the full complexity of these relationships which imply differences between concept definitions not simply between names.

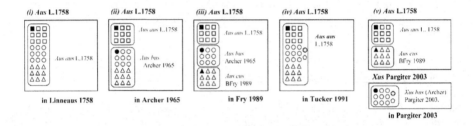

Fig. 1. Taxonomic history of the imaginary genus *Aus* L. 1758 (i) through four subsequent revisions (ii – v). Individual specimen organisms are represented by the symbols ○, □, △ *etc.*, with nomenclatural type specimens infilled: ▲, ■, ●. In 1965 Archer split *Aus bus* Archer 1965 from *Aus aus* L.1758 (ii), which was in turn 'split' creating *Aus cus* Fry 1989 (iii). Discovery of new specimens in 1991 caused Tucker to re-'lump' taxa in a single species *Aus aus* L.1758 (iv), but according to Pargiter these new specimens indicated that *bus* (Archer) in fact belonged in a separate new genus as *Xus bus* (Archer) Pargiter 2003 (v). Comparing the specimen circumscription of the various views on the taxa it is clear that the underlying concepts referred to by the various names change over time. For example compare *Aus aus* L.1758 in (i) versus (ii); or *Aus bus* Archer 1965 in (ii) and (iii); or the relationship of *Aus bus* with *Xus bus*

3   Defining Taxonomic Concepts

A taxonomic concept is one view of what constitutes a taxonomic entity, be it a species, genus or taxon of higher rank. Typically this would be represented as a published opinion or hypothesis according to a given author team, and include a valid scientific name as controlled by the rules of nomenclature. Care should be taken to distinguish between published taxonomic concepts, representing taxonomists' classification hypotheses, and the publication of data by biologists who are only identifying organisms according to some preexisting taxonomic concept, i.e. name usage [8].

A minimal representation of a taxon concept is therefore a scientific name plus citation of definition (i.e. an attribution). In this respect any first usage of a scientific name represents an original taxon concept, as published by the author of the name. As the rules of nomenclature require the original author to be included as part of the name, e.g. *Aus aus* L. 1758, this combination does not uniquely distinguish the original concept in a taxonomic database, as the same name might be valid for subsequent revision concepts, which should be distinguished by recording the originator of the concept, in addition to the author of the name (as part of the full scientific name), e.g. *Aus aus* L. 1758 *sec.* Fry 1989. Recording the originating (*sec.*) authorship for a concept therefore distinguishes between concepts, but does not provide any information with which to *compare* different concepts. The meaningful comparison of defined concepts would require the user to consult and interpret the original citations, where available. Any computer-assisted automatic comparison and resolution of concepts will require that the elements of the concept definition are stored as part of the electronic representation of the concept in the taxonomic database sources.

We have modelled how taxon concepts can be represented with varying complexity by a range of creators and users of concepts (including taxonomists, database providers and experimental biologists). Detailed analysis of the components that are used by taxonomic databases or found in taxonomic publications to define their taxon concepts includes (i) specimen and taxon circumscriptions, (ii) character descriptions or circumscriptions and (iii) relationships with other taxon concepts.

There are a wide variety of relationships that might be expressed between taxon concepts, which have been considered in detail by others (e.g. [9]; see online documentation, section 2.3 [10]). These relationships may implicitly or explicitly represent set-based relationships defining the extent of overlap with or inclusion of other concepts, or they may capture 'nomenclatural' relationships. However, the description of types of relationships is complicated by the interdependence of nomenclature and classification. A strict interpretation of terms such as synonymy, homonymy *etc.* implies relationships between the definitions of names, and it is questionable whether a relationship between names can be asserted in the absence of the context or usage of those names. Any relationship between taxon names at least minimally considers relationships between the type specimens determining the names. In the Taxonomic Concept Schema (TCS) model presented in this paper a 'nomenclatural' relationship is expressed as a relationship between two concepts, implying between the names of the concepts.

4   The Requirement for Data Exchange Standards

Given that there are an increasing number of important database providers of taxonomic information, and a large potential user base amongst biologists and non-scientists, it is necessary to facilitate data exchange between the providers and the users, so that data can be integrated from multiple sources, without losing or misrepresenting the semantics of the data according to the providers' information models. This is necessary both from the perspective of database providers who wish to aggregate information from multiple data sources into a single representation of taxonomy without duplication of concepts, as well as for taxonomically naive users who wish to integrate data from multiple database providers. If no exchange standard is globally adopted, it will be necessary for any application or service that seeks to query multiple taxonomic databases to implement bespoke query and exchange protocols for each provider. It would then be impossible to develop standard mechanisms to match or resolve concepts between different sources, and no guarantee of any protocol's stability or longevity.

The need for data exchange standards across the domains of biology, particularly in the context of biodiversity studies, has been identified by GBIF [11] and SEEK [12] amongst others. The common approach being taken to provide these standards is the development of XML Schemas that define the data transfer structure as an XML document, including the structure of the metadata associated with the actual data. This approach mirrors that already taken to provide Data Description, or ' Mark-up' Languages such as EML (EcologicalML [13]), CML (ChemicalML [14]) and GML (GeographyML [15]). The necessary information exchange standards for taxonomy might include those for taxon concepts, Specimen Records, Collection Details, Publications, Observation Data, Geographical Location and People (i.e. Authors *etc.*). Standards and protocols for some of these facets are already available or under development, including: DIGIR [16] and ABCD [17] for detailing and exchanging information regarding biological specimens; TaxMLit allowing the complete mark-up of the content of taxonomic work [18], and a number of standards for publication information (MODS [19]; XOBIS [20]; XMLMARC [21]; *etc.*).

In order to achieve global data exchange standards it is necessary that the standards process should be open and inclusive, and it is desirable that proposed standards should be consistent, and well documented. TDWG (International Taxonomic Databases Working Group, www.tdwg.org) has taken a lead in providing an international forum for the development of standards for biological data exchange. Current standards being developed (as XML schema) include: the ABCD Task Group On Access to Biological Data (providing standards for transfer and discovery of biological collection data sets); the SDD Task Group on Structure of Descriptive Data (developing a standard for storing and transferring detailed, character-based, descriptions of specimens or taxa) and the Taxonomic Names Task Group on Taxonomic Concept Standards (developing a standard for storing and transferring information about taxon concepts and names, the work we present in section 5). Because of the overlap between these three proposed schemas (for example in their use of taxonomic names and concepts and their referral to specimens and collections) it is proposed to modularize their implementation to allow reuse of each other's data structures. Furthermore, because each type of document will need to provide similar metadata elements describing the data transferred in a document (for example the source, ownership, version *etc.*) it is proposed that documents conforming to each of these three schemata are wrapped in a common format descriptor document.

5   The TDWG Taxon Concept Schema (TCS)

Considered in abstraction, models for both a taxon name and a taxon concept consist of a *label* plus *definition* plus *author*. Therefore, as demonstrated by Pyle [22], a taxon concept can be represented as a taxon name (protonym) plus definition plus author. Taxonomic definitions of names include the type specimen for that name and application of the rules of nomenclature, whereas the taxonomic definition of a concept might take several explicit (or implicit) forms. A model for names that includes relationships between names might be considered as incorporating elements of a concept model as the relationships between names actually refers to both the usage context and typification of that name.

Because of the structural similarity between elements of names and concepts, and to encourage a more rigorous representation of taxonomic identifiers (as defined concepts rather than somewhat ambiguous names), an XML schema is proposed for the representation and exchange of information regarding taxon concepts. Because the schema includes a representation of names it will be possible to use this schema to represent names as being concepts that lack definitions (i.e. as nominal concepts).

By making explicit the differences in composition between various types of taxon concept definition, the schema will allow users to be aware of the variable accuracy or quality of resolution, whether based solely upon names or upon more richly defined taxon concepts. Various service providers, such as uBio (www.ubio.org) and Species2000 (www.sp2000.org), are providing rich mechanisms for resolving names across distributed taxonomic databases. However, resolution services based on taxon concepts as represented by the TCS should provide more meaningful comparison of taxonomic identifiers.

The TCS schema was derived by composing an abstract model of taxonomic concepts as discussed above, which seeks to account for all the facets that different data providers and users might wish to include in their definition of a taxon concept. This was facilitated by detailed consultation with representatives of several taxonomic databases and researchers with an active interest in modelling and implementing taxonomic information systems (see acknowledgments). The abstract model has been represented as an XML schema that defines the structure of XML documents for the exchange of information about taxonomic concepts. This exchange schema aims to capture data as understood by the data owners without distortion, and facilitate the query of different data resources according to the common schema model. The full schema and documentation can be found at tdwg.napier.ac.uk. The TDWG review process is open and inclusive, giving the opportunity to any interested party to comment and suggest amendments to the proposal.

An overview detailing some of the elements of the transfer schema is shown in Figure 2. Each Dataset will carry MetaData detailing the source of the transferred document. To allow cross-referencing within the document, Vouchers (Specimen records), Publications and TaxonConcepts are given local identifiers (IDs) that could be substituted with global IDs (GUIDs) if these are available (see below). As well as recording the details of TaxonConcepts (which can include Relationships with other TaxonConcepts, see Figure 3), the transfer document may also be used to detail third party RelationshipAssertions between existing TaxonConcepts.

Because the model represented by the schema aims to be inclusive no 'components' of a taxon concept definition are required by the schema, but are optional constituents of a concept as represented by a given provider. However, in order to be useful, a minimal representation would generally include both a Name and details of the concept authorship (i.e. AccordingTo, or *sec.*). The representation of a full scientific name (NameDetailed) that conforms to the requirements of all existing codes of Nomenclature has been developed outside the project (by the Linnean Core interest group [23]) and integrated into the schema.

The various elements of the schema materialize information defining the concept according to the original authors of the concept. This might include details of the concept's Relationships to other pre-existing concepts, including its circumscription by (inclusion of) other (lower rank) taxon concepts, or its membership of higher rank concepts. Further Relationships may detail similarity or overlap with concepts created by other authors. These latter relationships can be considered 'horizontal' in the sense that they can relate concepts defined according to different taxonomic classifications, whilst the hierarchical relationships between concepts within a classification are 'vertical'. A full list of the types of relationships that may be expressed between two concepts is provided online [10].

The manner in which a concept may be circumscribed by 'Character' data is as yet undefined in the schema, and would require a formal model for representing character descriptions. Various structured models for character data have been proposed (see for example [24]), and the SDD working group of TDWG is developing a schema for specimen or taxon descriptions that could be included or referenced within a TCS CharacterCircumscription. The TCS schema does however provide the mechanism for circumscribing concepts by reference to identifiers of specimen records (Vouchers in the schema). Individual specimens that circumscribe a taxon can be labelled according to whether they are accepted holotypes, isotypes, neotypes *etc.* for that taxon, according to the codes of nomenclature.

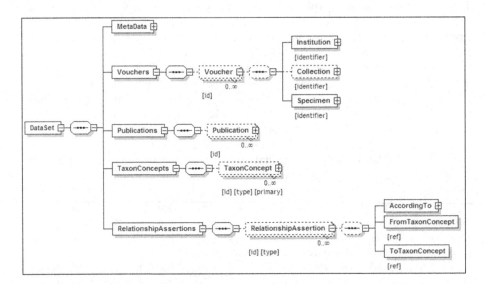

Fig. 2. Overview of the Proposed TDWG TCS XML Schema. The major components of the schema for transferring taxonomic concepts are shown diagrammatically (XML Elements are shown in boxes, with XML attributes listed [below]; generated with XMLSpy.com software). Each document would carry *MetaData* recording source and creation details of the *DataSet*, together with the details of the taxonomic concept information represented. To allow cross-referencing within the document *Vouchers* (Specimen records), *Publications* and *TaxonConcepts* are given local identifiers (ids), which could be substituted with global IDs (GUIDs) if these are available. As well as recording the details of *TaxonConcepts* (which can include *Relationships* with other *TaxonConcepts*, see Figure 3), the transfer document may also be used to detail third party *RelationshipAssertions* between existing *TaxonConcepts*

The structure of the TCS schema allows internal reference and reuse of 'top-level' elements (i.e. TaxonConcepts, RelationshipAssertions, Voucher and Publication records). Indeed it is hoped to standardize the representation of Publications and Vouchers (including Specimens) across the TDWG schemas (see above). Where any of these reusable elements are globally defined and resolvable via Globally Unique Identifiers (GUIDs) it will be possible to represent them in transfer documents simply by reference to this GUID (see below).

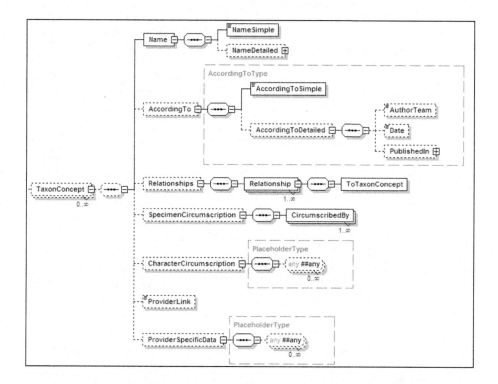

Fig. 3. XML Schema Diagram for a Taxon Concept. A portion of the proposed TDWG TCS schema for transferring Taxonomic Concepts is shown diagrammatically (generated with XMLSpy.com software). Any combination of the optional component elements would be used to detail *TaxonConcept* definitions according to the data model of the data provider, but typically at least *Name* and *AccordingTo* would be required ('Nomenclatural Concepts' may only provide *Name*). For these two components the detail recorded in different data sources will vary, so a simple string representation will always be provided, whether or not detailed decomposition is possible. The *Relationship* element allows the *TaxonConcept* to be defined in relation to existing *TaxonConcepts*. This can include hierarchical relationships to parent or child taxa in the same classification, or synonymy and set based relationships with *TaxonConcepts* defined in alternative classifications, based on the extent to which two concepts are congruent or overlap. *SpecimenCircumscriptions* list the specimen details (*Vouchers* in Figure 2) that the *TaxonConcept* is *CircumscribedBy*, but the nature of *CharacterCircumscriptions* is as yet undefined. The *PlaceholderType* allows standards developed as other schemas to be incorporated; provision of the *ProviderSpecificData* element allows application specific extensions to the representation of a Taxon Concept

Some taxonomic work is concerned with re-using existing taxonomic concepts. For example a taxonomist creating a revision of a large taxon may accept various included taxa according to the work of various other published taxonomists, but wish to record opinions about the relationships between these concepts. Where these relationships are not created as part of a new concept definition they are treated as 'third party' in the schema, and stored as RelationshipAssertions with an AccordingTo authority.

5.1 Globally Identified Taxonomic Concepts

At present each taxonomic database has its own internal (and sometimes external) identifiers for taxon names or concepts (e.g. TSN numbers used by ITIS *etc.*). These are not represented in the core TCS transfer schema, as there is no guarantee that any given database ID would map uniquely to a TCS concept nor remain stable over time.

The TCS schema was devised to allow exchange of concepts together with their definitions, and could be used to represent concepts stored in any global repository or local cache. To provide a stable and resolvable identifier for these concepts it would be highly desirable if GUIDs for taxon concepts were adopted. These could be assigned and maintained locally (by data owners) or globally according to agreed international policies, and would provide a stable reference to a taxon concept as represented according to TCS (i.e. minimally Name plus AccordingTo). Once implemented concept GUIDs would simplify the mark-up of any biological data, according to available defined concepts, and could assist data retrieval based on concept identity. Provision of GUIDs would also help reduce the redundancy and overlap between different data providers who currently reproduce alternative representations of the 'same' concept. Discussion within TDWG, SEEK, GBIF and the wider biological community is investigating the feasibility of providing GUIDs not only for taxon concepts, but also for other stable concepts such as Publications and Specimens.

The availability of stable GUIDs with which any biologist can annotate their data to unambiguously record the organisms described in their work will greatly facilitate the interpretation, integration and accurate reuse of data across the whole of biology and beyond. Furthermore, eventually it should be possible for a given researcher to chose to recognize and use concepts as provided and defined by a preferred taxonomic resource (e.g. ITIS) or even to capture uncertainty by using less well-defined concepts, or collections of possible concepts were identifications are uncertain.

5.2 Resolving Taxon Names and Concepts

The proposed schema was initially conceived in the context of SEEK's requirement for a taxonomic concept/name resolution service with which to resolve taxonomic names as recorded in ecological data sets, following the realisation that resolution by name alone is insufficient, and in the absence of identification through GUID referenced taxon concepts [12]. Typical scenarios would involve the matching of names as provided by users querying the system with the names as found in the metadata of global data repositories, by resolution through the defined concepts provided by taxonomic name providers and servers.

By capturing the individual components of concept definitions, according to any data model, the schema will allow matching to be performed on any combination of the individual components. The type and accuracy of the comparison performed may vary according to the requirements of the user, i.e. concept matching should be 'fit for the purpose'. For example, a match on the abbreviated scientific name *Aus bus*, will be of lower quality (or precision) than matches specifically to the full, attributed name *Aus bus* L. 1758 *sec.* Fry 1989. For some experimental purposes the loose match to *Aus bus* will be sufficient, but for others greater precision is necessary. A related notion is that comparison matches may be of higher or lower quality, and a 'reliability' score might be provided for different concepts returned by the resolution service.

Where the concepts are fully defined in terms of the components of the TCS model, matching on the actual definitions might be possible. When possible this will allow very high quality matches, for example, where resolution is on the basis of comparing full specimen circumscriptions. Alternatively, resolution only on the basis of name-bearing type specimens would provide a less precise, lower quality resolution, which might still be 'fit for purpose'. Whilst it might be possible to assign 'quality scores' to different components of the concept definition model, in practice it might be necessary to weight these scores to reflect the particular taxon model favoured by a user, or the purposes for which they wish to represent a taxon concept. This would allow users to differentially value the alternative components of a concept definition, and recognize higher value in matches according to their favoured criteria. Implementation of a name/concept resolution service would therefore need to include its own quality model for matching, but allow users flexibility in weighting the comparison algorithms or interpreting the results.

6 TCS in Comparison to Other Models for Taxonomy

As stressed earlier the TCS schema and underlying model aims to be inclusive of all other models of taxonomy, and allow data from any data source to be accurately represented. A strength of the TCS schema is that it supports many recent innovative models and implementations of taxonomic information as well as dealing with legacy data. Several of these models have been developed specifically to allow the representation of multiple, alternative taxonomic views (HICLAS [25,26]; PROMETHEUS [27]; BERLIN/IOPI [7-9]; TAXONOMER [22]; NOMENCURATOR [28]; uBIO www.ubio.org), rather than the standardized single view represented by many global taxonomic checklists (e.g. ITIS www.itis.usda.org; Species2000 www.sp2000.org).

In the TCS model the taxon concept is the core object, which includes name, attribution and definition elements. Whilst many database models also represent a central notion of a taxon object, typically the name is used as an identifier for this object. The Nomencurator database model [28] tracks nomenclatural history using a dual name and publication based model to represent potential taxa by 'name usage'. 'Annotations' are used to record relationships between these name usages, providing an implicit notion of taxon concepts. As such Nomencurator was designed to reflect the manner in which taxonomists work in recording revisions, tracking the development of taxonomic theories by changes in name usage. However, as there is no representation of a taxon concept it is not possible to use the model to define taxa, nor does it readily provide identifiable and exchangeable concepts that can be shared

provide identifiable and exchangeable concepts that can be shared amongst the various users of taxonomy. It should be possible to map each Nomencurator 'name usage' (i.e. name plus publication) to a unique TCS taxon concept, using Name and AccordingTo elements.

The Potential Taxon notion, i.e. the representation of subjective views of a taxon, forms the basis of the Berlin IOPI model for botanical databases [8,9]. In this rich and complex model botanical information can be linked to potential taxa (i.e. name plus circumscription reference) rather than to name alone. Such information can include nomenclatural and systematic relationships as well as linked specimen determinations and character descriptions. Alternate taxonomic classifications are related to potential taxa rather than names, closely corresponding to the TCS model. As with Nomencurator it is envisaged that it will be desirable to present a 'Preferred View' of taxonomy to users, by filtering according to preferred reference authorities. A number of databases implement the Berlin model, including the MoReTax database [29, 9] which defines fundamental, set-based relationships which can be expressed between potential taxa. These relationships are included in the types of relation representable in the TCS [10].

The Taxonomer database model [22] also represents potential taxa, by the intersection of a Name and a Reference, called an Assertion. Assertions of the first usage of that name are treated as a special case, as the name (or Protonym) provides the label for the taxon concept. Protonyms form the name for later revised opinions on a taxon concept as implicitly or explicitly circumscribed in a subsequent publication, represented in the model by an Assertion. Protonyms therefore provide common handle for both the name and any taxon concepts or Potential Taxa that use this name. TCS represents protonyms as the Name components of Original taxon concepts, and TCS Revision taxon concepts may express various synonymy relationships to the Original Concepts sharing a taxonomic name. As with TCS taxon concepts, Assertions may be linked by concept relationships (such as those defined by Geoffroy and Berendsohn [31]), and can have attached specimen determinations and character descriptions (as text based 'Excerpts'). In the Taxonomer model, however, common names are represented not as individual concepts (or assertions) but as an attribute of an Assertion (which must be or include a Protonym).

The uBio model of taxonomic information underlying their Taxonomic Name Service (www.ubio.org) seeks to separate 'objective' nomenclatural information into a consensual reference model (NameBank), whilst representing classification information in a separate but linked model of subjective opinions (ClassificationBank). uBio assert that this separation whilst providing a rich representation of taxon concepts through classification relationships will allow nomenclaturists to work with bare names and represent relationships between them, without referring to concepts. The justification being that whereas many aspects of nomenclature are not disputed, taxonomic classifications are inherently unstable, disputed hypotheses. On the other hand the TCS does not represent names independently, and relationships must be expressed through a concept that bears a particular name. This reflects our opinion that it is difficult to find any instances where names are used for identification and communication of taxa without at least an implied notion of the concept to which they apply. Datasets containing only name information, are represented by 'nominal concepts' which capture all concepts that share the same name.

As with the Berlin and uBio models, the Prometheus taxonomic database model, which is based on specimen circumscription, clearly distinguishes nomenclatural from classification information [27] and was built to support the working practices of taxonomists performing botanical revisions. In this model naming is an automatic feature of typification in the specimen circumscription. Alternative classification views, based on specimen circumscription, can readily be compared on the basis of set-based relationships (such as those defined in the MoReTax/Berlin model [9]).

The requirements for simple data discovery and exchange between database providers has favoured the development and implementation of simple generic data query and retrieval protocols, which use simple models for the underlying data structure (for example, the successful DIGIR [16] protocol with the underlying Darwin Core data representation [30]). Whilst such flat, unstructured representations of taxonomic information are certainly simple, they may not be adequate for representing semantically complex information. Species2000 (www.sp2000.org) has developed a Standard Dataset model for exchanging name-based species information according to a single aggregated view of taxonomy, derived from various database sources. Although there is no explicit statement on what 'defines' a named species concept in this model, each species can be recognized as a 'concept' according to the originating source database, or a recorded taxonomic scrutinizer, and could therefore be represented in TCS as a (not well defined) Taxon Concept. The synonymy relationships captured in Species2000 are purely nomenclatural, as the synonyms do not belong to any alternative conceptual hierarchy. Representing such synonymies in TCS would require that each name be represented by a nominal concept.

Whilst the details captured in each of these theoretical and implementation models of taxonomy vary greatly, they tend to converge on a central representation of a potential taxon or taxon concept. TCS can therefore accommodate the salient features of these models, as well as representing database models that use a more traditional representation of taxonomic names as the identifiers.

7 Conclusion

The computerized systems and databases used by biologists and the bioinformatics community are largely blind to the problems inherent in the (ambiguous) identification of organisms by scientific name alone. As we have discussed, accurate integration of biological data sets is problematic due to many reasons including errors in documenting taxonomic names; the lack of standards for capturing the definition of taxonomic concepts; the inherent ambiguity the taxon definitions associated with taxonomic names; the lack of understanding of this ambiguity by users of biological names; and finally the lack of a global repository for taxonomic concepts with GUIDs which can be used to refer to and aid matching concepts for data annotation and integration. Solutions to these problems require ensuring that references to biological taxa in data sets cite the scientific name in the context of a particular classification, which we have modelled as the defining attributes of a Taxon Concept. Data integration can then be achieved either on Concept identity, or on individual components of a defined concept. Where it is not possible to ascribe defined concepts to datasets (such as with legacy data) poorly defined nominal concepts can be used (i.e. concepts with a name but no definition), thus making explicit the deficient quality of the taxon identification. The schema has been used to map data from a variety of sources and is currently being used as the basis for a taxonomic name/concept resolution service in the SEEK project.

Acknowledgments

This work was carried out under the auspices of TDWG and jointly funded by GBIF [12] and SEEK [11], supported by the US National Science Foundation. We are most grateful for detailed and helpful discussions on aspects of individual taxonomic models with representatives of the Berlin Model, GBIF, IPNI, ITIS, Nomencurator, Species2000, Taxonomer, Vegbank, and colleagues within SEEK and TDWG.

References

1. International Union for Conservation of Nature. 2004. IUCN Red List of Endangered Species http://www.iucnredlist.org
2. Greuter W, McNeill J, Barrie FR, Burdet HM, Demoulin V, Filgueiras TS, Nicolson DH, Silva PC, Skog JE, Trehane P, Turland NJ, Hawksworth DL. 2000. (Editors & Compilers): International Code of Botanical Nomenclature. 16th International Botanical Congress St. Louis, Missouri, 1999. (Regnum Vegetabile, 138). Königstein: Koeltz Scientific Books.
3. ICZN 1999. (International Commission on Zoological Nomenclature). International Code of Zoological Nomenclature 4th ed. London : ICZN.
4. ICSP. (International Committee on Systematics of Prokaryotes). International Code of Nomenclature of Bacteria, 1990. Washington: American Society for Microbiology Press.
5. ICTV 2000. (International Committee on Taxonomy of Viruses). International Code of Virus Classification and Nomenclature. http://www.ncbi.nlm.nih.gov/ICTV/rules.html
6. http://www.soc.napier.ac.uk/tdwg/index.php?pagename=TCSAndTheLinneanCore
7. Berendsohn WG. 1995. The concept of "potential taxa" in databases. Taxon 22:207-212.
8. Berendsohn WG. 1997. A taxonomic information model for botanical databases: the IOPI model. Taxon 46:283-309.
9. Berendsohn W, Döring M, Geoffroy M, Glück K, Güntsch A, Hahn A, Jahn R, Kusper W-H, Li J, Röpert D, Specht F. 2003. MoReTax: Handling factual information linked to taxonomic concepts in biology. (Schrift. Veget. 39). Bonn: Bundesamt für Naturshutz.
10. Taxonomic Concept Schema Complementary Documentation for Draft Standard http://tdwg.napier.ac.uk/doc/tdwg_tcs.doc section 2.3
11. GBIF.The Global Biodiversity Information Facility, www.gbif.org
12. SEEK 2004. The Science Environment for Ecological Knowledge. http://seek.ecoinformatics.org
13. EML 2004. Ecological Metadata Language. http://knb.ecoinformatics.org/software/eml
14. CML 2004. Chemical Markup Language. http://wwmm.ch.cam.ac.uk/moin/ ChemicalMarkupLanguage
15. GML 2004. Geography Markup Language. http://opengis.net/gml
16. DIGIR 2004. Distributed Generic Information Retrieval. http://digir.net
17. ABCD 2004. Access to Biological Collection Data http://www.bgbm.org/TDWG/CODATA

18. Weitzman AL, Lyal CHC (2004) An XML schema for taxonomic literature – taXMLit. available at http://web4.si.edu/sil/bca/status.cfm
19. MODS 2004. Metadata Object Description Schema. http://www.loc.gov/standards/mods
20. XOBIS 2004. XML Organic Bibliographic Information Schema. http://laneweb.stanford. edu: 2380/wiki/medlane/schema
21. XMLMARC 2004. XML Machine Readable Cataloging.http://laneweb.stanford. edu:2380/wiki/medlane/xmlmarc
22. Pyle RL (2004) Taxonomer: a relational data model for managing information relevant to taxonomic research. Phyloinformatics 1: 1
23. 23.TDWG Linnean Core Group: http://wiki.cs.umb.edu/twiki/bin/view/UBIF/Linnean-Core
24. Paterson T, Kennedy JB, Pullan MR, Cannon A, Armstrong K, Watson MF, Raguenaud C, McDonald SM, Russell G. A universal character model and ontology of defined terms for taxonomic description. Pages 63-78 in Data integration in the life sciences (DILS 2004) Lecture Notes in Bioinformatics 2994 edited by E. Rahm. Berlin: Springer-Verlag.
25. Zhong Y, Jung S, Pramanik S, Beaman JH. 1996. Data model and comparison query methods for interacting classifications in taxonomic databases. Taxon 45: 223-241.
26. Zhong Y, Luo Y, Pramanik S, Beaman JH. 1999. HICLAS: a taxonomic database system for displaying and comparing biological classification and phylogenetic trees. Bioinformatics 15(2):149-156
27. Pullan MR, Watson MF, Kennedy JB, Raguenaud C, Hyam, R. 2000. The Prometheus taxonomic model: a practical approach to representing multiple taxonomies. Taxon 49(1): 55-75.
28. Ytow N, Morse DR, Roberts DMcL. 2001. Nomencurator: a nomenclatural history model to handle multiple taxonomic view. Biological Journal of the Linnaean Society 73: 81-98.
29. Koperski M, SauerM, Braun W, Gradstein SR. 2000. Referenzeliste der Moose Deutschlands. (Schriftenreihe Vegetationskunde 34). Bonn: Bundesamt für Naturshutz.
30. DWC 2004. The Darwin Core. http://speciesanalyst.net/docs/dwc
31. Geoffroy M, Berendsohn W. 2003. The concept problem in taxonomy: importance, components, approaches. Pages 5-14 in Berendsohn et al. 2003 [9].

The Multiple Roles of Ontologies in the BioMediator Data Integration System

Peter Mork[1, 2], Ron Shaker[3], and Peter Tarczy-Hornoch[2, 3]

[1] Computer Science & Engineering
[2] Biomedical & Health Informatics
[3] Pediatrics, University of Washington,
Seattle, WA 98011, USA
pmork@cs.washington.edu
{rshaker, pth}@u.washington.edu

Abstract. BioMediator is a data integration system that provides a common interface to multiple Internet-accessible databases containing information about genetics and molecular biology. Ontologies play several important roles in the BioMediator system: First, ontologies of genetics and molecular biology can serve as *data sources*. In this role concepts from the ontologies are returned as results of queries. Second, queries are posed against a *mediated schema*, which is an ontology describing the domain of discourse. User queries are expressed using the concepts in the mediated schema to indicate which results to retrieve. Third, each data source is an instance of the *system ontology*. This ontology describes information about the data sources including how often the source is updated and by whom. Finally, we are exploring the use of ontologies as a mechanism for *mapping* data sources to the mediated schema. This will facilitate extending BioMediator from a centralized integration platform to a distributed network of peers.

1 Introduction

Biologists seeking to understand the molecular basis of human health and disease are struggling with large volumes of diverse data (mutation, expression array, proteomic) that need to be integrated and analyzed in order to develop and test hypotheses about disease mechanisms and normal physiology. These data reside in multiple public and private databases maintained by biologists in their laboratories. For example, a set of experiments may generate both gene and protein expression data, which are queried in aggregate to find a set of expression products of potential interest. Each of these products is, in turn, queried against public domain databases such as Entrez [1], SwissProt [2], and the Gene Ontology [3]. Given the dynamic nature of the datasets federated database approaches provide advantages over warehousing approaches in terms of data currency. Federated approaches with flexible mediated schemata representing the entities of interest and their mappings to particular sources are well-suited to handle the diverse schemata necessary, particularly for the laboratory specific private data sets. The BioMediator data integration system [4, 5] takes an ontology driven federated approach to data integration for these reasons.

B. Ludäscher and L. Raschid (Eds.): DILS 2005, LNBI 3615, pp. 96–104, 2005.

In this paper we present an overview of the BioMediator system emphasizing the various roles that ontologies (a term we use loosely to refer to vocabularies such as the Gene Ontology, a database schema, or a terminology expressed in a description logic such as OWL) play in the system. At the source level, the schemata of sources focused on data (*e.g.*, Entrez) and those focused on concepts (*e.g.*, the Gene Ontology) are treated identically by our system and knowledge about the structure and organization of both types of sources can be represented as ontologies (3.1). At mediated level, the schemata used to query across these sources are also represented as ontologies (3.2). We permit multiple mediated schemata customized to different users/query tasks, pieces of which can be shared and reused. At a meta-level the BioMediator system uses a system ontology (3.3) to describe meta-information about the sources (such as information about validation and curation). Finally, we are developing techniques for translating data from specific source schemata into a mediated schema using knowledge stored in a mapping ontology (3.4).

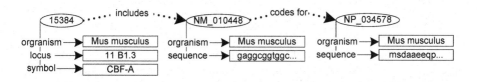

Fig. 1. Sample data viewed as a network of resources and properties; solid lines indicate datatype properties (DTP) and dotted lines, object properties (OP)

2 Background

In BioMediator, the data contained in online public databases are viewed as a network of interconnected records. For example, Online Mendelian Inheritance in Man (OMIM) [6] contains records describing genes and genetic diseases. Entrez publishes records that describe proteins and nucleotide sequences. Entrez also cross-references its protein records with related OMIM records.

2.1 Semantic Web Data Model

The data sources thus constitute a semantic web for the life sciences. In this web, each record corresponds to a node with a collection of attribute/value pairs. This is illustrated in Figure 1. The Entrez node NM_010448 has two solid edges leading from it: the organism edge indicates it pertains to the house mouse, and the sequence edge indicates the nucleic acid sequence. Expressed using RDF [7] terminology, this record is a *resource* with two *datatype properties* that link the resource to values.

LocusLink (LL) [8] provides other information related to this nucleotide sequence. LL resource 15384 describes the CDF-A gene. Also, LL publishes an *object property* that links the LL resource to the Entrez resource. This establishes that one possible sequence for the CDF-A gene is described by the indicated Entrez resource.

We distinguish between datatype properties (DTP) and object properties (OP) for two reasons. First, DTPs indicate the actual content of a resource. DTPss capture what can be thought of as the information represented by the resource. OP are correspondences between resources; they are typically displayed (in a web browser) as hyperlinks. The second distinction pertains to ownership. In BioMediator, each resource is owned by a single data source and only that source can provide DTPs for that resource. OPs, on the other hand, can be provided by any data source. For example, not only does LL provide a property linking CDF-A to a sample sequence (NM_010448), but LL also links this nucleotide sequence to a corresponding protein (Entrez record NP_034578) using RefSeq [9].

Viewing the data sources as a semantic network distinguishes BioMediator from other data integration projects (such as Kleisli [10] or OPM [11]). The semantic network paradigm facilitates organizing the resources with an ontology. This approach was pioneered (for biologic domains) by TAMBIS [12] and, as we describe in this paper, extended by BioMediator. In this context, the ontology organizes the resources (and properties) into a hierarchy of concepts, against which users can query.

2.2 System Interface

BioMediator allows client programs to interact with this semantic web in a number of ways. The most basic interaction, *seed*, retrieves a specific resource and its associated DTPs. The client program provides the resource's accession number, and the database in which the resource can be found. For example, a program can request resource NM_010448 from Entrez, and BioMediator will retrieve the associated attribute/value pairs (*e.g.*, organism/Mus musculus). Microarray researchers with chips annotated using accession numbers use this operation extensively [13].

Resources can also be retrieved using a *query*. In this case, the client program selects one of the classes in the mediated schema (see below) and one or more attribute/value restrictions. BioMediator retrieves all of the resources that are instances of the given class and that include all of the indicated attribute/value pairs. For example, a program can request all phenotype resources whose name is narcolepsy, or genes whose locus is 11 B1.3 and whose organism is the house mouse.

These first two interactions produce DTPs only. OPs can be retrieved using *expand*. Given a resource (or set of resources), this operation retrieves all related OPs (either leading from or pointing to the indicated resource). Both the mediated schema and the system ontology (see below) can be used to restrict which OPs will be retrieved. For example, a client program might be interested in the 'codes-for' property for a sequence, but not the more general 'related-to' property.

Finally, BioMediator can recursively *grow* the network, which expands each new resource it encounters. In this case, it is often useful to limit the OPs using the system ontology (e.g. limiting the growth to include only externally validated properties).

2.3 Architectural Overview

To support these operations, BioMediator relies on a series of components as illustrated in Figure 2. The system relies heavily on the source knowledge base (SKB), which is represented using Protégé-2000 [14], and accessed via the Protégé API. The *SKB* (Fig. 2A) contains the mediated schema and the system ontology, both of which are described in the following section.

The *query processor* (Fig. 2B) provides an API for launching and managing queries posed using the mediated schema. The *metawrapper* (Fig. 2C) translates these mediated schema queries into source specific queries [15]. *Wrappers* (Fig. 2E) pass the remapped queries through to the *data sources* (Fig. 2F). Data sources return results in native format (e.g., HTML, ASN1), which are converted to XML syntax with native semantics by the wrappers. The metawrapper applies mapping rules in translating the XML results from native semantics to mediated schema semantics.

The query processor then retrieves data from the metawrapper, organizes that data and generates events that can be used to synthesize a navigable representation of the result set. Once a result set has been constructed, it may be repeatedly queried, expanded or grown using the query processor's API.

3   Multiple Roles of Ontologies

As described in the previous section, BioMediator uses ontologies in several roles. The SKB contains two ontologies: The mediated schema provides a hierarchical vocabulary for organizing resources published by the underlying data sources and the system ontology describes how the data sources are maintained. In addition, BioMediator can access external ontologies as data sources.

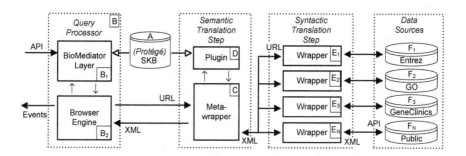

Fig. 2. Architecture pipeline of the BioMediator system

3.1   Data Source

In many cases, an ontology can be represented in the semantic web data model. In this case, resources represent named concepts and properties represent relationships among the concepts. For example, the Gene Ontology (GO) [3] includes two inter-concept properties ('is-a' and 'part-of') and one property relating external resources to concepts ('classified-as').

Properties provided by an ontology are treated no differently than other properties. This means that, for better or worse, we do not attribute any meaning to these properties. For example, given that the nuclear membrane is part of the nucleus, and the nucleus is part of the cell, we should be able to infer that the nuclear membrane is part of the cell. Instead of making this inference, BioMediator returns only those properties explicitly present in the sources.

This simplicity is advantageous because properties relating data resources and properties relating concept resources are treated uniformly. For example, given a collection of nucleotide sequences up-regulated in an experimental group (relative to a control), BioMediator can first identify the corresponding proteins (using the expand operation) and then organize these proteins based on functional classification (using the expand operation a second time). This helps a researcher answer the question, "What do these experimental results mean?"

When a simplistic view of the data is not sufficient (*e.g.*, a user needs to answer a very precise question), more machinery is needed. In this case, the mediated schema provides a common vocabulary for expressing more precise interactions (such as "A mutation of what gene results in dysprothrombinemia, haemophilia caused by an inactive protein?").

3.2 Mediated Schema

At the heart of a data integration system is a mediated schema. The simplest mediated schema is the union of the source schemata which has two key limitations. First, application developers are must understand all of the source schemata to author queries. Second, when a new source is added, each application needs to be modified to reference the new source. For example, both SwissProt [2] and Entrez [1] contain information about proteins. In the absence of a mediated schema, the only way to capture this similarity is by requiring all applications to query for the union of these sources. When another source containing information about proteins (*e.g.*, GeneTests [16]) is identified, every application program must be updated.

Given this limitation, database research has focused on formalisms for expressing the mediated schema in terms of the source schemata. In TAMBIS [12], the mediated schema is an ontology expressed using the GRAIL description logic [17]. The mediated schema is described independently of the underlying sources. The contents of the sources are then described in terms of the mediated schema, and an inference engine is used to determine how the source schemata relate to the mediated schema. For example, an OMIM record can be defined to be the union of genes and phenotypes for which the value of the organism attribute is human.

When a new source is added to the system, neither the existing definitions need to be updated, nor do existing applications. As a result, new sources can more transparently be introduced into the system. However, if the mediated schema is changed, then it becomes necessary to revisit every definition.

BioMediator uses a strategy similar to TAMBIS, but with greater emphasis placed on modularity. Instead of a single mediated schema, one of our goals is to support multiple mediated schemata simultaneously. In Figure 2, each user group can have its own SKB, independent of all other user groups.

Thus, even though the users see the same sources, they may organize these sources differently. One sample mediated schema is shown in Figure 3. This schema was developed for a statistician performing analyses on microarray data (*i.e.*, it is not intended to represent everything about microarray experiments, let alone all of molecular biology). Several concepts in Figure 3 are common to a variety of user groups: Genes are an abstract unit of inheritance. Each gene can include a number of closely related sequences as examples of the gene. These sequences code for proteins, which produce (cause) the manifestation of a phenotype.

Some additional concepts are needed to support microarray analyses. First, we added several classes that describe microarrays. An experiment is performed using a specific chip. That chip contains several spots. Each spot is associated with a specific sequence. The statistical analyses also required functional information (from GO), which was one of the motivations for treating ontologies as data sources. Here GO is modeled as a hierarchical vocabulary, which differs from a controlled vocabulary in that inter-concept properties are allowed (as described above).

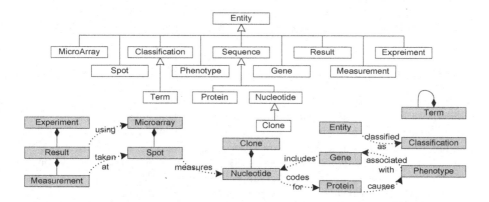

Fig. 3. Sample mediated schema for annotating microarray experiments. The top half displays the inheritance hierarchy; the bottom half displays containment relationships (diamonds) and other object properties

Once the mediated schema has been designed, rules must be written so that the metawrapper can transform source data into the mediated namespace. When multiple groups agree on portions of the mediated schema, they can also share these transformation rules. In the case of disagreement, transformation rules must be modified or removed. Finally, each source must be added into the system ontology.

3.3 System Ontology

Within the system ontology, each data source is represented as an instance of the database class. A database is a collection of resource tables and property tables. Of the resource tables, one is designated as the primary table (references into a database that do not specify a type are assumed to index into the primary table).

A resource table stores the metadata needed to retrieve a collection of resources. Each *resource table* is associated with a class from the mediated schema; all resources in the table are instances of that class. Likewise, each *property table* is associated with a property from the mediated schema. The domain and range of each property table must also be specified (*i.e.*, the resource tables connected by the property table).

Moreover, for each property table, we also record metadata describing how the property table is maintained. These metadata include descriptions of: a.) population, b.) validation, c.) update, and d.) causality (i.e., whether the correspondance indicates a causal mechanism, such as gene coding for protein vs. merely observed correlation).

Metadata can be used to constrain the property tables that will be considered when using the expand or grow operations. For example, a clinician might be interested in browsing only those property tables 'validated' by an external review process, whereas a researcher might choose to browse only 'causal' relationships (even if the relationship has not yet been proven experimentally).

Each table is also associated with rules used by the metawrapper to convert source data into the BioMediator data model. For example, a rule is used to indicate that when OMIM returns a disease record it should be converted to a resource that is an instance of the mediated class phenotype. The value of the title attribute is mapped to a name datatype property.

3.4 Mappings

We have begun exploring OWL [18] as an alternative to the current rule language for expressing relationships between the source schemata and the mediated schema. The hope is that OWL constructs will allow us greater flexibility. Not only will it be possible to translate from a source namespace to the mediated, but the inverse will also be possible. This will allow us to distribute our system in a peer-to-peer fashion.

For example, we can declare that an OMIM record describes a gene or a phenotype, *i.e.*, an OMIM record is defined to be the union of these two classes. A GeneTests record for a gene is equivalent to the class, Gene, in the mediated schema. A query requesting information about a specific gene can be *rewritten* as a query against GeneTests (because Gene ≡ GeneTests Gene Record).

More sophisticated rewritings are also possible. At first, it does not seem that a gene query can use OMIM because an OMIM record is more general than gene (Phenotype ⊆ Gene ∪ Phenotype ≡ OMIM Record). However, assume the mediated schema asserts that the domain of the property, AssociatedWith, is NucleotideSequence ∪ Gene, we can rewrite the query to request OMIM records that participate in the AssociatedWith property (OMIM Record ∩ ≥1 AssociatedWith). We are exploring algorithms for efficiently generating all valid rewritings.

4 Conclusions

BioMediator is a data integration system that uses ontologies in several roles. The network-based data model allows us to use an ontology such as the Gene Ontology as a data source. This is particularly useful for organizing experimental results into functional groups. To support more precise interactions, users can formulate queries in terms of a mediated schema. The role of this mediated schema is to provide a common nomenclature applicable to multiple local or remote data sources. The mediated schema also defines the object properties that can link data instances. These properties are further annotated using the system ontology, which describes how the underlying data sources are maintained. This approach provides several benefits. First, the results returned by BioMediator are as current as the underlying sources. Second, each user group can customize its mediated schema, and the mappings that relate the data sources to that common namespace. Finally, our architecture supports both precise queries (the database standard) and more generic browsing. These advantages make BioMediator an excellent platform for supporting a variety of biomedical data needs.

Acknowledgments

We would like to thank Hao Mei (microarray mediated schema) and Scott Brockenbrough (editorial comments). Funding: R01HG02288 & T15LM07442.

References

[1]   Entrez search and retrieval system. National Center for Biotechnology Information, National Library of Medicine [Online]. Available: http://www.ncbi.nlm.nih.gov/Entrez/

[2]   Swiss-Prot Protein knowledgebase. Swiss Institute of Bioinformatics [Online]. Available: http://us.expasy.org/sprot/

[3]   Gene Ontology™ Consortium (GO). Gene Ontology (GO).[Online]. Available: http://www.geneontology.org/

[4]   Mork, P., Halevy, A. Y., Tarczy-Hornoch, P.: A Model for Data Integration Systems of Biomedical Data Applied to Online Genetic Databases. In: Proc. Proceedings of the American Medical Informatics Association (AMIA) Annual Symposium (2001) 473–477

[5]   Tarczy-Hornoch, P., Halevy, A. Y., Rossini, A. J., Mork, P., Shaker, R., Donelson, L. BioMediator: A Data Integration System for Biomedical Databases. University of Washington [Online]. Available: http://www.biomediator.org

[6]   Online Mendelian Inheritance in Man, OMIM™. McKusick-Nathans Institute for Genetic Medicine, Johns Hopkins University (Baltimore, MD) and National Center for Biotechnology Information, National Library of Medicine [Online]. Available: http://www.ncbi.nlm.nih.gov/entrez/query.fcgi?db=OMIM

[7]   Brickley, D., Guha, R. RDF Vocabulary Description Language 1.0: RDF Schema. World Wide Web Consortium (W3C®) [Online]. Available: http://www.w3.org/TR/rdf-schema/

[8]   LocusLink. National Center for Biotechnology Information, National Library of Medicine [Online]. Available: http://www.ncbi.nlm.nih.gov/LocusLink/

[9]   Pruitt, K. D., Maglott, D. R.: RefSeq and LocusLink: NCBI gene-centered resources. Nucleic Acids Research 29 (2001) 137–140

[10] Chung, S. Y., Wong, L.: Kleisli: a new tool for data integration in biology. Trends in Biotechnology 17 (1999) 351–355

[11] Chen, I.-M. A., Markowitz, V.: An Overview of the Object Protocol Model (OPM) and the OPM Data Management Tools. Information Systems 20 (1995) 393–418

[12] Baker, P. G., Brass, A., Bechhofer, S., Goble, C. A., Paton, N., Stevens, R.: TAMBIS: Transparent Access to Multiple Bioinformatics Information Sources. In: Proc. Proceedings of the Sixth International Conference on Intelligent Systems for Molecular Biology (1998) 25–34

[13] Mei, H., Tarczy-Hornoch, P., Mork, P., Rossini, A. J., Shaker, R., Donelson, L.: Expression array annotation using the BioMediator biologic data integration system and the Bioconductor analytic platform. In: Proc. Proceedings of the American Medical Informatics Association (AMIA) Annual Symposium (2003) 445–449

[14] Musen, M., Crubézy, M., Fergerson, R., Noy, N. F., Tu, S., Vendetti, J. The Protégé Ontology Editor and Knowledge Acquisition System. Stanford Medical Informatics [Online]. Available: http://protege.stanford.edu/

[15] Shaker, R., Mork, P., Barclay, M., Tarczy-Hornoch, P.: A Rule Driven Bi-Directional Tranlslation System Remapping Queries and Result Sets Between a Mediated Schema and Heterogeneous Data Sources. In: Proc. Proceedings of the American Medical Informatics Association (AMIA) Annual Symposium (2002) 692–696

[16] GeneTests. University of Washington [Online]. Available: http://www.genetests.org/

[17] Zanstra, P. E., van der Haring, E. J., Flier, F., Rogers, J. E., Solomon, W. D.: Using the GRAIL language for Classification Management. In: Proc. Fifteenth International Congress of the European Federation for Medical Informatics (1997) 897–901

[18] Web-Ontology (WebOnt) Working Group. OWL Web Ontology Language Overview. World Wide Web Consortium (W3C®) [Online]. Available: http://www.w3.org/ TR/owl-features/

Integrating Heterogeneous Microarray Data Sources Using Correlation Signatures

Jaewoo Kang[1], Jiong Yang[2], Wanhong Xu[1], and Pankaj Chopra[1]

[1] NC State University, Raleigh NC 27695, USA
[2] Case Western Reserve University, Cleveland OH 44106, USA

Abstract. Microarrays are one of the latest breakthroughs in experimental molecular biology. Thousands of different research groups generate tens of thousands of microarray gene expression profiles based on different tissues, species, and conditions. Combining such vast amount of microarray data sets is an important and yet challenging problem. In this paper, we introduce a "correlation signature" method that allows the coherent interpretation and integration of microarray data across disparate sources. The proposed algorithm first builds, for each gene (row) in a table, a correlation signature that captures the system-wide dependencies existing between the gene and the other genes within the table, and then compares the signatures across the tables for further analysis. We validate our framework with an experimental study using real microarray data sets, the result of which suggests that such an approach can be a viable solution for the microarray data integration and analysis problems.

1   Introduction

Microarrays are one of the latest breakthroughs in experimental molecular biology. It provides a powerful tool by which the expression patterns of thousands of genes can be monitored simultaneously and are already producing huge amount of valuable data. Analysis of such data is becoming one of the major bottlenecks in the utilization of the technology. The gene expression data are organized as matrices — tables where rows represent genes, columns represent various samples such as tissues or experimental conditions, and numbers in each cell characterize the expression level of the particular gene in the particular sample. Application of microarray technology to biological problems, ranges from understanding of metabolic responses of microbes, to cancer in humans. The main challenge of analyzing microarray is the virtual explosion in the volume and complexity of gene expression data. Thousands of different research groups generate tens of thousands of microarray gene expression profiles. Different experiments utilize different tissue types, examine different treatment strategies, and consider different stages of disease development. This, along with differences in microarray platform, technology and protocols used in different labs, leads to difficulties in integrating microarray data across experiments.

B. Ludäscher and L. Raschid (Eds.): DILS 2005, LNBI 3615, pp. 105–120, 2005.

How to combine the data (gene expression levels) in different microarrays is a challenging problem since these gene expression levels are not necessarily directly comparable. The same gene may exhibit different bias at different data sets. For instance, a gene in the liver tissue may have higher expression level (higher values in a microarray) than that in the skin tissue (lower values in another microarray) by the nature. As a result, directly integrating the microarrays according to the gene ids would result in inconsistency. In addition, microarrays may contain different (overlapping) sets of genes. This increases the difficulties in the integration of the microarray data sets.

In this paper, we consider the problem of integrating heterogeneous gene expression data sets. We try to tackle this problem by employing a novel **correlation signature** method. The correlation signature captures the data set-wise characteristics of a gene in terms of its correlations to a set of landmark genes. Various methods can be used to choose the landmarks, e.g., genes from a particular pathway or deemed important by domain experts, etc. The expression level of a gene at a microarray table can be converted into the similarity (or correlation) to the set of landmark genes. For example, if there were 10 landmark genes, then at each microarray table, a gene will have 10 correlation values each of which corresponds to a landmark. We call these correlation values as the *gene signature vector*. The signature vector removes the bias in the expression values and can be used to compare genes across heterogeneous experiments.

The amount of data for all signature vectors could be very large, $O(|G| \times |L| \times |S|)$ where $|G|$, $|L|$, and $|S|$ are the average number of genes in a study, the number of landmark genes, and the number of studies, respectively. This could range to tens or hundreds of Gigabytes. How to organize and represent the entire set of signature vectors is a challenging problem. A novel multi-dimensional data model, *gene signature cube*, is proposed in this paper to represent the set of gene signature vectors. The entire cube may not be fully instantiated because of its size. We introduce two virtual signature cube organizations that materialize parts of the cube on the fly upon request in the query time, and present the result of the performance comparison of the two models. In summary, we make the following contributions in this paper.

- We introduce a statistical model, *correlation signature*, that captures system-wide dependency relations among data instances. The correlation signature projects semantically non-conforming data instances from disparate sources into common dimensions, allowing the coherent interpretation and integration of the data.
- The data set may be large. There are thousands of microarrays each containing thousands of genes. As a result, the signature based integration will also generate a large amount of data. We organize the set of transformed data into a conceptual *cube*. In this paper, we present methods to store and access the data in the cube.
- The proposed framework can also be applicable to other types of experimental data. A typical scientific experiment involves a series of observations made over a set of inter-related variables (e.g., in microarray, genes are variables and

samples are observations.) Moreover, similar experiments typically share some numbers of common variables (e.g., genes, proteins), and the variables are in most cases not independent. The proposed model can exploit such properties.

The remainder of the paper is organized as follows: We first present the signature vector data model in Section 2. Then, in Section 3, we present the signature cube data organization in detail. The related work is discussed in Section 4. Finally, Section 5 concludes the paper and outlines the future research directions.

2   Unified Data Model for Gene Expression Profiles

Figure 1 shows an overview of our signature calculation algorithm, **SigCalc**, and Figure 2 illustrates the signature computation process through an example. **SigCalc** takes as input a microarray table M and a set of k landmark genes. The landmark genes can be selected either manually by the user or automatically by the system. If user did not provide landmarks, the system can automatically select candidate landmark genes. Different techniques can be used. For example, depending on the application, the system may run a feature selection algorithm [1,2] to choose a set of representative genes in the table, or simply choose a random set of genes and use them as landmarks. With random landmarks, the correlation signature model behaves similar to the random projection, a popular dimensionality reduction method [3,4,5,6,7], except that the random projection

Input : Microarray table M ($n \times m$, n genes and m conditions),
              set of k landmark genes $L = \{l_1, ..., l_k\}$
Output: Set of gene signature vectors $S = \{\overrightarrow{sig}(g_1),...,\overrightarrow{sig}(g_n)\}$

for *each gene g_i in M* **do**
    for *each gene l_j in L* **do**
        $d_j \leftarrow dist(\overrightarrow{g_i}, \overrightarrow{l_j})$
    end
    $\overrightarrow{sig}(g_i) \leftarrow [d_1, d_2, ..., d_k]$
end

Fig. 1. SigCalc: signature computation algorithm

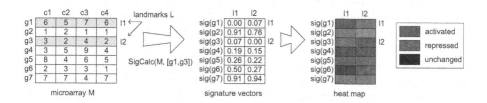

Fig. 2. Example of signature vector computation. Assume $l1$ and $l2$ are regulator genes with similar functions

projects the original high-dimensional space onto a random subspace while the correlation signatures project the original space onto a subspace whose coordinates correspond to the landmark genes. Despite the similarity, the random projection cannot be the solution for the microarray integration problem because the random subspaces projected from different datasets are not generally comparable as each projected subspace consists of random dimensions.

When users provide landmarks to the system, they can either explicitly pass a hand-selected genes to the system, or they can just state what kinds of genes they want the system to use. For the latter case, the system can guide users to make their choices on the group of genes, by providing information about gene annotations, functional groups, known regulator genes, or genes that are involved in a certain pathway, retrieved from some external sources such as GO ontology database (http://www.geneontology.org) and KEGG pathway database (http://www.genome.jp/kegg/).

Once landmark genes are selected, system calculates signature vectors of all genes in the table as shown in Figure 1. **SigCalc** uses a distance function, $dist$, to measure similarities and dissimilarities between gene vectors (rows of M). Any conventional distance metric can be used including standard metrics such as Euclidean or cosine distance, or some variants that are popular in microarray analysis such as correlation distance or mean-expression distance, as defined below.

- *Euclidean Distance:* Given two gene vectors $\overrightarrow{x}$ and $\overrightarrow{y}$, where $\overrightarrow{x} = [a_1, ..., a_n]$ and $\overrightarrow{y} = [b_1, ..., b_n]$, respectively, the Euclidean distance is : $euc(\overrightarrow{x}, \overrightarrow{y}) = \sqrt{(a_1 - b_1)^2 + ... + (a_n - b_n)^2}$.
- *Cosine Correlation:* Given two gene vectors $\overrightarrow{x}$ and $\overrightarrow{y}$, the cosine correlation is: $cos(\overrightarrow{x}, \overrightarrow{y}) = \frac{\sum_{i=1}^{n} a_i b_i}{\sqrt{\sum_{i=1}^{n} a_i^2}\sqrt{\sum_{i=1}^{n} b_i^2}}$. The cosine correlation measures the similarity between gene vectors. For a dissimilarity measure, simply $1 - cos(\overrightarrow{x}, \overrightarrow{y})$.
- *Pearson Correlation:* Given two gene vectors $\overrightarrow{x}$ and $\overrightarrow{y}$, Pearson correlation is: $cor(\overrightarrow{x}, \overrightarrow{y}) = \frac{covariance(\overrightarrow{x}, \overrightarrow{y})}{\sqrt{covariance(\overrightarrow{x}, \overrightarrow{x}) \times covariance(\overrightarrow{y}, \overrightarrow{y})}}$. For a dissimilarity measure, $1 - cor(\overrightarrow{x}, \overrightarrow{y})$.
- *Mean-Expression Distance:* Given two gene vectors, the mean-expression distance is defined as: $dist(\overrightarrow{x}, \overrightarrow{y}) = mean(\overrightarrow{x}) - mean(\overrightarrow{y})$.

Note that the correlation and mean-expression distances are not metrics in a strict sense (e.g., do not satisfy triangular inequality) but introduced here because they are commonly used in practice for microarray analysis. Although Euclidean distance is a common method to represent the similarity or dissimilarity between two vectors, it does not take into account the natural bias of expression level of different types of genes. Some house-keeping genes may naturally express highly while some other genes may always express at a low level. Thus, the distance measure may appear larger for these two types of genes. If we are interested in the fluctuation of the expression levels rather than the absolute gene expression values, then the Euclidean distance measure may not be proper to use. In this case, the correlation metrics could be used.

The mean-expression distance is somewhat simplistic but popular in practice because it gives a natural interpretation of the expression level differences, and can be applicable to the gene vectors with different dimensions. In reality, gene vectors (rows) from different microarray tables almost always have different dimensions (e.g., one table has columns of lymphoblastic leukemia samples and the other has myeloid leukemia samples; number of columns also may differ.) The first three metrics will not work for such comparison. In contrast, all four distance metrics can be used with our model, after transforming the original gene vectors into the corresponding signature vectors.

Now, consider the example in Figure 2. On the left, it shows an input microarray data table M. Suppose the user selected g_1 and g_3 as the two landmarks, l_1 and l_2, respectively. **SigCalc** transforms the original table into a 7×2 table whose rows represent the signature vectors of the corresponding genes in the original table. In this example we used the correlation distance $(0.5 \times (1 - cor(\overrightarrow{x}, \overrightarrow{y}))$ to calculate the signatures. For example, consider $\overrightarrow{sig}(g_7)$ in the signature vector table. It has two entries $[0.91, 0.94]$ representing correlation distances of gene g_7 to the two landmark genes, g_1 and g_3, respectively.

How do we interpret the distance to the landmarks from a gene? What does it exactly mean that the distance is 0.91 or 0.19? The correlation distance ranges from $[0, 1]$, and a distance close to zero implies the two vectors are correlated and a distance close to one implies the two vectors are inversely correlated. If it is 0.5 it means there is no correlation. Now, let us assume that the two landmark genes, l_1 and l_2, are known regulator genes with similar functions. In this example, if a gene's signature vector contains close-to-zero values, it may mean that the gene is *activated* by the two regulator genes. The opposite also holds. The third table from the left of Figure 2 shows the heat map visualizing the activation/repression relations. In our example, g_7 is repressed while g_4 is activated ($\overrightarrow{sig}(g_1)$ and $\overrightarrow{sig}(g_3)$ are also low but they are the landmark genes, and thus ignored.)

A critical precondition that needs to hold to make the proposed approach work is that some genome-wide dependency relations between genes exist and that the relations are conserved across the different experiments, samples, organs, or even across different organisms. In fact, this is a general belief in the biology community. Genes do not act alone: one gene's expression triggers another gene's expression. While most of the dependency relation will remain unchanged, some statistically meaningful changes may be detected from a comparison like *normal cells* vs. *cancerous counterparts*.

One of the main strengths of our approach is the flexibility in landmark selection. The signatures can be further tuned for a specific analysis by choosing landmarks from only the genes that are relevant to the current analysis. For example, suppose one tries to identify how genes behave differently in two sets of cancer samples (e.g., Leukemia and B-cell lymphoma), with respect to only the genes of certain functions (e.g., cell cycle or metabolism). Using our approach, such comparisons become straightforward; we just need to choose landmarks from the genes with cell cycle or metabolism functions.

Our approach also allows flexible cross-validation and analysis. Virtually any expression data sets can be compared provided that the signatures are generated over the common landmarks. One can compare the properties of genes across different tissues (e.g., skin, liver, blood etc.), different clinical stages of cancers (e.g., metastasis vs. primary, recurrent vs. non-recurrent etc.), or can compare across even different organisms (e.g., mouse vs. human; mice and men share 99% of genes [8]).

To demonstrate the efficacy of the proposed model, we conducted the following tests using the leukemia data set published by Golub et al. in [9]. The following experiments were implemented using a statistics package, R [10], and the Bioconductor library [11]. All experiments were performed on a machine with P4 2.4GHz and 1 GB memory running Windows XP Professional.

2.1   Rejecting Null Hypothesis

One of the fundamental questions that we need to address is, *will the gene signature vectors really capture some information that is statistically meaningful?* To answer the question, we first split the leukemia table (1450×47, where 1450 genes tested over 47 Acute Lymphoblastic Leukemia (*ALL*) patient samples; 1450 genes selected out of 7129 genes after filtering out under-expressed genes) into two partitions with randomly selected disjoint sets of 20 samples (1450×20). Then, we computed a separate set of signatures for each partition with 50 common landmark genes (selected by running k-means clustering, using the correlation distance as a dissimilarity metric, with $k = 50$ over the original table, and then choosing the medoids from each resulting cluster), and compared the signatures across the two sets. If it really captures the information, at the very least, the signature vectors of corresponding genes across the two sets should be very similar because they are generated from the same type of samples (patients with same type of cancer).

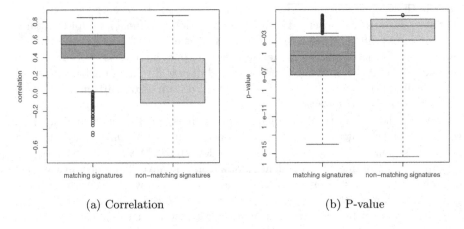

(a) Correlation   (b) P-value

Fig. 3. Correlation and p-value of matching signatures and non-matching signatures

The preliminary test result is shown in Figure 3. Figure 3(a) compares the correlation between matching pairs of signature vectors (for the same gene) and the correlation between random pairs (different genes). The median correlation between the matching pairs was 0.55 while the median correlation between the non-matching pairs was 0.15. Figure 3(b) presents the p-values between the matching and non-matching signature vectors. The median p-value of the matching pairs was 0.000042 while that of the non-matching pairs was just over 0.064. In our context, the p-value states the probability of observing a correlation between two signature vectors *by chance* at the level greater than or equal to the observed correlation. The p-value is calculated by transforming the correlation into t-statistics of N-2 degrees of freedom where N is the number of columns. If a pair's p-value is low we can assume that the correlation value between the pair is statistically significant. On the other hand, a high p-value may suggest that no statistically significant correlation exists between the two signature vectors.

The result shows clear differences in the correlation and the p-values between the pairs of matching and non-matching signature vectors, thereby rejecting *null hypothesis* of the signature vectors carrying no statistically meaningful information.

2.2   Testing Convergence

Another important question to ask is, *does the gene signature converge as more columns (experiments) are added to the signature calculation?* The signatures would converge if it captures some genome-wide properties that are invariant across the experiments. We tested this using the same Leukemia table. First, we calculated the two sets of signature vectors by randomly selecting two disjoint sets of five columns each. We then measured the correlation and p-value of each pair of corresponding signatures across the two sets. We continued this comparison while increasing the number of columns by five in each iteration. The result of this test is shown in Figure 4.

Figure 4(a) shows the correlation between the two sets of signature vectors while Figure 4(b) shows the p-values measured between the two sets. As expected, correlation improves as more numbers of columns were considered in the signature calculation. Similarly, p-values were consistently decreasing as more columns were added. This finding supports our hypothesis that gene signature vector models can be used to combine multiple microarray experiment data and summarize them into one coherent set of signature vectors for further analysis and cross-validation. For example, we can calculate a set of signatures from multiple *ALL* tables simply by juxtaposing the tables and calculate the signatures from the combined table. Similarly, we can calculate a set of signatures for Acute Myeloid Leukemia (*AML*) experiments, and compare the signatures of corresponding genes between *ALL* and *AML* to identify those genes that behave differently in the two cancers and genes that behave similarly across. We can easily extend the analysis to virtually all other cancer types, tissues, metastasis stages, etc.

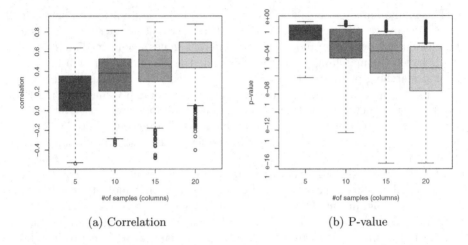

(a) Correlation   (b) P-value

Fig. 4. Changes in correlation and p-value of matching pairs of signatures with the increasing number of samples

2.3   Stability of the Model

Microarray data sets are noisy; individual expression levels are affected by many factors such as different lab protocols (e.g., how long the samples will be hybridized in what temperature), platforms (cDNA or Affymetrix), choice of samples, etc. How well a model generalizes the underlying data is an important factor. In order to test the stability of the model, we examined clustering results from the different sets of signatures computed using different subsets of columns, and measured how consistent the clustering results were across the tests.

Figure 5 shows the results. Figure 5(a) shows the histogram depicting the number of gene pairs falling into the same cluster greater than or equal to 15 times out of 20 total iterations. In each iteration, we randomly selected a 20 column subtable, M_R, from the original table (47 columns), and computed the signature table, M_S, from M_R. We then ran the k-means clustering (w/ k=20) over the two tables, M_R and M_S, and tallied up the pairs that fell into the same cluster. We iterated this 20 times over different sizes of landmarks from 10 to 50. For example, bucket number 20 contains all gene pairs that co-occurred in the same clusters for all 20 times, and similarly bucket number 19 contains the pairs that co-occurred 19 times out of 20. With 10 landmarks, the signature model produced 92 pairs that co-occurred $\geq$ 95% of time (sum of buckets 19 and 20), and 175 pairs for $\geq$ 85% of time. We compared the tallies from M_S and M_R. As shown in Figure 5(a), there were no significant differences between the two results.

Figure 5(b) shows the same result with varying sample sizes (#of columns in M) while fixing the landmark size to 50. Four different sample sizes were tested, including m =10, 20, 30 and 40. Unlike the previous test, the increase of sample size improved the clustering results significantly for both the raw and signature

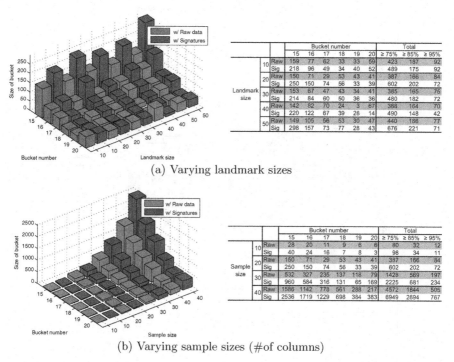

			Bucket number						Total		
			15	16	17	18	19	20	≥ 75%	≥ 85%	≥ 95%
Landmark size	10	Raw	159	77	62	33	33	59	423	187	92
		Sig	218	96	49	34	40	52	489	175	92
	20	Raw	150	71	29	53	43	41	387	166	84
		Sig	250	150	74	56	33	39	602	202	72
	30	Raw	153	67	47	43	34	41	385	165	75
		Sig	214	84	60	50	36	36	480	182	72
	40	Raw	142	82	70	24	3	67	368	164	70
		Sig	220	122	67	39	28	14	490	148	42
	50	Raw	149	105	56	53	30	47	440	186	77
		Sig	298	157	73	77	28	43	676	221	71

(a) Varying landmark sizes

			Bucket number						Total		
			15	16	17	18	19	20	≥ 75%	≥ 85%	≥ 95%
Sample size	10	Raw	28	20	11	9	6	6	80	32	12
		Sig	40	24	16	7	8	3	98	34	11
	20	Raw	150	71	29	53	43	41	387	166	84
		Sig	250	150	74	56	33	39	602	202	72
	30	Raw	532	327	235	137	118	79	1428	569	197
		Sig	960	584	316	131	65	169	2225	681	234
	40	Raw	1586	1142	778	561	288	217	4572	1844	505
		Sig	2536	1719	1229	698	384	383	6949	2694	767

(b) Varying sample sizes (#of columns)

Fig. 5. Testing stability of the model

tables. The rate of improvement, however, was greater with the signature model than with the raw data. For example, although their performance were similar in the 10-sample test, in the 40-sample test, the signature model produced about 50% more numbers of pairs in all three cases of ≥ 75, 85, and 95%.

So far, we tested the stability of the model in terms of the consistency of the clustering results, and showed that the result from the model is at least as stable as that from the raw data. However, the previous tests did not show how much the actual clustering results from the two tables are similar. If the model properly reflects the signals from the original table, the clustering results of the both tables should be similar at least for the genes that were highly agreed upon in the both results. In order to show this, we examined the bucket 20 from the two results. There were 217 gene pairs (105 unique genes) from the raw data and 383 pairs (118 unique genes) from the signatures. Intersecting the two sets of 105 and 118 genes resulted in 49 unique genes. We then selected the corresponding 49 rows from the raw data and from the signature table, and independently ran a hierarchical clustering (w/ "average" agglomeration) over the two sets. As shown in Figure 6, the results were strikingly similar. Each test clustered genes into four main clusters that perfectly overlapped across the two sets.

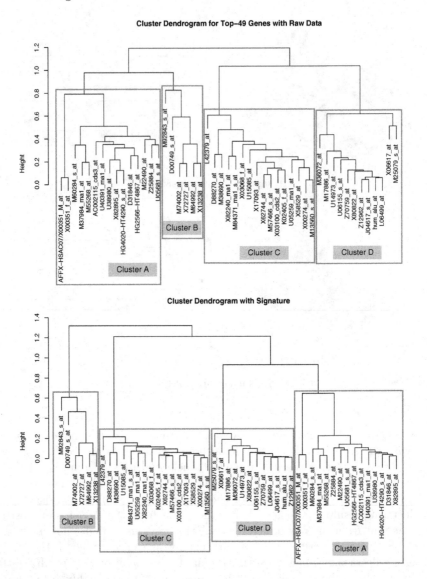

Fig. 6. Comparison of clustering dendrograms of top-49 genes computed using raw data and signatures

3   Gene Signature Cube: A Novel Summary Data Structure for the Global Study

In this section, we present a multi-dimensional data structure, the *gene signature cube*, in order to facilitate efficient storage and retrieval of multiple gene signatures. Figure 7 shows an overview of the signature cube construction and an example cube after the construction.

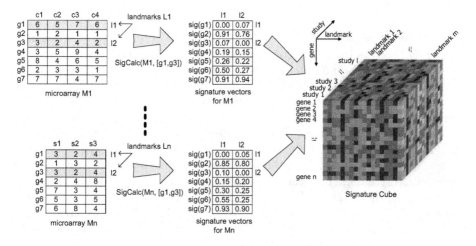

Fig. 7. Overview of signature cube computation and an example gene signature cube

In essence, the gene signature cubes are constructed as follows. Suppose we have k landmark genes in L. For each microarray data set M ($\in \mathcal{M}$) (representing a set of microarrays produced in a study, e.g., a drug response study on colon cancer samples, etc.), we transform each gene $g_i \in M$ into its gene signature vector $\vec{sig}(g_i)$ of k values $[d_{i,1}, d_{i,2}, \ldots, d_{i,k}]$. Let gene l_j be the jth landmark gene. We assume that the gene expression profile of g_i and l_j in M is $\vec{g_i} = [e_{i,1}, e_{i,2}, \ldots, e_{i,n}]$ and $\vec{l_j} = [e_{j,1}, e_{j,2}, \ldots, e_{j,n}]$, respectively. Now $d_{i,j}$ is calculated based on the similarity (or dissimilarity) of $\vec{g_i}$ and $\vec{l_j}$.

After computing gene signature vectors for each gene in every microarray, we can organize them in a gene signature cube. An example of the gene signature cube is shown in Figure 7. The cube consists of three dimensions: genes, landmarks, and studies. Let $\mathcal{G}$ be the set of all distinct genes in all the microarray studies. The gene dimension consists of all genes in $\mathcal{G}$; the landmark dimension consists of all landmark genes in $\mathcal{L}$; and the study dimension consists of all microarray experiments in $\mathcal{M}$. Each entry $C(g_i, l_j, M_r)$ (with gene g_i, landmark l_j, and microarray set M_r) in the cube represents the gene signature value of gene g_i with respect to the landmark l_j at microarray set M_r. It is possible that a microarray does not contain all genes in $\mathcal{G}$. If gene g does not appear in microarray M, then the entries associated with g in M will be set to a special value such as NA. This gene signature cube can be considered as a conceptual representation of the expression profiles of all gene in all heterogeneous microarrays. We can perform further (biological and computational) studies on the gene signature cube.

3.1 Evaluating Different Organizations for Cube Construction

We considered two possible approaches to construct a cube: (1) fully materialized and (2) virtual (on the fly) cube. The fully materialized cube stores precomputed signatures for all genes in all studies in a contiguous file layering values in each signature vectors in a predefined order. Users request signatures for particular genes or set of genes across the studies. In order to correctly retrieve the corresponding vectors from the contiguous cube file, we maintain a separate meta file that contains the information necessary for identifying the begin and end location of each gene in each study.

Another possible way to construct the cube is not to store the precalculated signature vectors but to calculate them on the fly upon request. We refer to this organization as a virtual cube. In order to calculate the signatures on the fly, unlike the materialized approach, we need to store the original microarray tables instead. In our implementation, we used a relational database to store the microarray data. There can be many different ways to organize the expression values in the database. We evaluated two different schemas: (1) multi-table schema and (2) single table schema. With multi-table schema, we created one table for each microarray experiments (e.g., one table for Golub et al.'s leukemia experiments [9] and another for Pomeroy et al.'s brain tumor experiments [12]) while with single-table schema, we just created one big table for storing all experiments. Each study typically consists of 1K-30K genes (rows) and about 5-100 samples (columns). Different studies have different numbers of genes and samples. In order to store them in a single table, we employed a schema with four columns, (study id, gene id, sample id, expression value), and stored each expression value in a separate row. In the multiple table approach, each study (genes × samples) is loaded into a separate table where each row contains all expression values for a corresponding gene in a study.

While the three approaches (the materialized cube and the two virtual cubes) employ different storage schemes, they all export the same API for the upper layer, as follows:

cube[][][] SubCube(genes[], lmarks[], studies[])

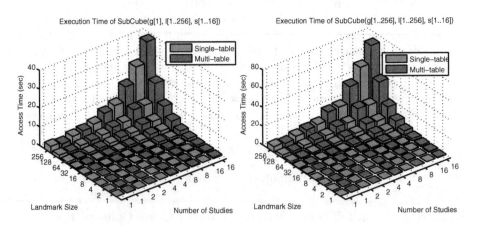

Fig. 8. Execution time for SubCube(g[1], l[1..256], s[1..16])

Fig. 9. Execution time for SubCube(g[1..256], l[1..256], s[1..16])

SubCube accepts three parameters: lists of gene IDs, landmarks, and study IDs. This interface allows users to access any point, vector, matrix, or sub-cube of the signature cube along any dimension.

In our experiment, we evaluated the three organizations with queries of different access patterns (e.g., vector, cube, etc.), on different dimensions (e.g., within or across the studies), and different sizes (e.g., 256 genes from 16 studies). Unlike the virtual cube approaches, the materialized cube can only return values that are precomputed with the preselected landmarks. For fair comparison, we materialized a maximal cube using a complete set of landmarks for each study to ensure that the results returned by SubCube are all same across the three organizations. As a result, the performance of the materialized cube was far worse than the two virtual cubes, both in the time and space complexity. It requires approximately $O(gene^2 \times study)$ space while the virtual cubes requires $O(gene \times sample \times study)$. As for the time complexity (in #of page I/O), the materialized cube in the worst case reads $|g| \times |l| \times |s|$ pages while the virtual cubes read $(|g| + |l|) \times |s|$ pages, where $|g|, |l|$, and $|s|$ are respectively the numbers of genes, landmarks, and studies in the query.

The performance between the two virtual cubes were comparable as shown in Figure 8 and 9. Figure 8 shows the execution time of SubCube(g[1], l[1..256], s[1..16]) where the number of gene is fixed to one while the numbers of landmarks and studies vary 1 to 256 and 1 to 16, respectively. Figure 9 shows the result of SubCube(g[1..256], l[1..256], s[1..16]) where both the numbers of genes and landmarks vary from 1 to 256. Overall, the single-table approach was about 100% faster than the multi-table counterpart. The number of page I/Os of the two models are not significantly different because even with the single-table approach, in most cases, all necessary records for one gene will be found within one page. The performance difference is due to the number of queries issued. In the single-table approach, only one query is issued for each SubCube call while in the multi-table approach, one query is issued for each table (study) being accessed in the call.

4   Background and Related Work

Microarray Analysis: In recent years, microarray gene expression profiles [13, 14] have become a common technique for inferring the relationship or regulation among different genes. There exists a large body of work on microarray data analysis [15, 16, 17, 18, 19, 20, 21]. Typical applications include identification of differentially expressed genes and pathways under changing conditions (e.g. disease related, tissue specific, developmental stage related, etc.) [18, 19], drug development [20], and the functional annotation of genes [21]. Numerous public databases have been created such as ArrayExpress (www.ebi.ac.uk/arrayexpress), Gene Expression Omnibus (www.ncbi.nih.gov/geo), Stanford Microarray Database (genome-www5.stanford.edu), etc.

While most of the previous work on microarray analysis focused on individual microarray data sets, some global studies exploiting large numbers of microarrays have been presented recently. Stuart et al. [22] combined thousands of microarrays to infer conserved genetic modules across the organisms. Segal et al. in [18] exploited large numbers of microarray data to identify condition-specific regulatory modules, and in [19], to discover a module map showing the conditional activity of modules of genes in different types of cancers. Zhou et al. [23] recently introduced a technique, 2nd-order correlation analysis, for integrating heterogeneous microarray data. It works by first computing all pair-wise correlations of genes from each data set (1st-order correlation analysis) and then analyzing the correlation patterns across multiple data sets.

To the best of our knowledge, however, no previous work has ever attempted to build a unified data model that projects large numbers of heterogeneous microarray data into a coherent subspace, offering uniform interpretation and access to the data.

Correlation Signatures and Random Projection: The random projection (RP) is a popular dimensionality reduction method proven to be useful in many application areas including text retrieval [5,6], image processing [5], clustering [24, 25], motif discovery in bioinformatics [26], multimedia indexing [27], just to name a few. Our signature projection method has strong similarity with RP-based approaches. In fact, the correlation signature method is reduced to an RP problem in a cosine-similarity metric space (i.e., if the cosine similarity is used as the distance metric for both the signature and the global distortion computation), with only difference being that RP projects the original high-dimensional space onto a random subspace while the correlation signatures project the original space onto a subspace whose coordinates correspond to the landmark genes. Although RP is known to be generally effective in embedding high-dimensional data into a low-dimensional subspace, it may not solve our problem because the projected random subspaces (from different datasets) are not generally comparable.

5   Conclusion

We studied the problem of integrating and analyzing the heterogeneous microarray data sets and proposed a unified model, *gene signature vector*, and storage organizations, *signature cube*, for microarray data. In our model, a set of genes are chosen as landmarks. The expression of each gene is transformed to a signature vector which represents the correlation between this gene and the set of landmarks. To facilitate the efficient access and retrieval of the signature, we organize the gene signature vectors into a signature cube. Real microarray data sets are used to show the time and space efficiency of the gene signature cube.

Acknowledgments. We gratefully acknowledge many helpful comments from anonymous reviewers.

References

1. Xing, E.P., Jordan, M.I., Karp, R.M.: Feature selection for high-dimensional genomic microarray data. In: ICML '01: Proceedings of the Eighteenth International Conference on Machine Learning, San Francisco, CA, USA, Morgan Kaufmann Publishers Inc. (2001) 601–608

2. Yu, L., Liu, H.: Redundancy based feature selection for microarray data. In: KDD '04: Proceedings of the 2004 ACM SIGKDD international conference on Knowledge discovery and data mining, New York, NY, USA, ACM Press (2004) 737–742

3. Achlioptas, D.: Database-friendly random projections. In: PODS '01: Proceedings of the twentieth ACM SIGMOD-SIGACT-SIGART symposium on Principles of database systems, New York, NY, USA, ACM Press (2001) 274–281

4. Indyk, P., Motwani, R.: Approximate nearest neighbors: towards removing the curse of dimensionality. In: STOC '98: Proceedings of the thirtieth annual ACM symposium on Theory of computing, New York, NY, USA, ACM Press (1998) 604–613

5. Bingham, E., Mannila, H.: Random projection in dimensionality reduction: applications to image and text data. In: KDD '01: Proceedings of the seventh ACM SIGKDD international conference on Knowledge discovery and data mining, New York, NY, USA, ACM Press (2001) 245–250

6. Papadimitriou, C.H., Tamaki, H., Raghavan, P., Vempala, S.: Latent semantic indexing: a probabilistic analysis. In: PODS '98: Proceedings of the seventeenth ACM SIGACT-SIGMOD-SIGART symposium on Principles of database systems, New York, NY, USA, ACM Press (1998) 159–168

7. Johnson, W.B., Lindenstrauss, J.: Extensions of lipschitz mappings into a hilbert space. Amer. Math. Soc. **26** (1984) 189–206

8. Waterston, R.H., et al.: Initial sequencing and comparative analysis of the mouse genome. Nature **420** (2002)

9. Golub, T., Slonim, D., Tamayo, P., Huard, C., Gaasenbeek, M., Mesirov, J., Coller, H., Loh, M., Downing, J., Caligiuri, M., Bloomfield, C., Lander, E.: Molecular classification of cancer: class discovery and class prediction by gene expression monitoring. Science **286** (1999) 531–537

10. Venables, W.N., Smith, D.M.: An Introduction to R. Network Theory Ltd. (2002)

11. Gentleman, Rossini, Dudoit, Hornik: The bioconductor faq. http://www.bioconductor.org/ (2003)

12. et al., S.L.P.: Prediction of central nervous system embryonal tumour outcome based on gene expression. Nature **415** (2002)

13. Schena, M., Shalon, D., Davis, R., Brown.P.O: Quantitative monitoring of gene expression patterns with a complementary dna microarray. Science **270** (1995) 467–470

14. Lockhart, D., Dong, H., Byrne, M., Follettie, M., Gallo, M., Chee, M., Mittmann, M., Wang, C., Kobayashi, M., Horton, H., Brown, E.: Expression monitoring by hybridization to high-density oligonucleotide arrays. Nature Biotechnology **14** (1996) 1675–1680

15. Cheng, Y., Church, G.M.: Biclustering of expression data. In: Proceedings of the International Conference on Intelligent Systems for Molecular Biology (ISMB), San Diego, CA (2000) 93–103 (data sets are available at http://arep.med.harvard.edu/biclustering/).

16. Wang, H., Wang, W., Yang, J., Yu, P.: Clustering by pattern similarity in large data sets. In: sigmod. (2002)

17. Kostka, D., Spang, R.: Finding disease specific alternations in the co-expression of genes. Bioinformatics **20** (2004) 194–199
18. Segal, E., Shapira, M., Regev, A., Pe'er, D., Botstein, D., Koller, D., N., F.: Module networks: identifying regulatory modules and their condition-specific regulators from gene expression data. Nat Genet. **34** (2003) 166–76
19. Segal, E., Friedman, N., Koller, D., Regev, A.: A module map showing conditional activity of expression modules in cancer. Nat Genet. **36** (2004) 1090–8
20. Gerhold, D., Jensen, R., Gullans, S.: Better therapeutics through microarrays. Nature Genetics **32** (2002) 547–551
21. Allocco, D., Kohane, I., Butte, A.: Quantifying the relationship between co-expression, co-regulation and gene function. BMC Bioinformatics **5** (2004)
22. Stuart, J.M., Segal, E., Koller, D., Kim, S.K.: A gene-coexpression network for global discovery of conserved genetic modules. Science **302** (2003) 249–255
23. Zhou, X., Kao, M., Huang, H., Wong, A., Nunez-Iglesias, J., Primig, M., Aparicio, O., Finch, C., Morgan, T., Wong, W.: Functional annotation and network reconstruction through cross-platform integration of microarray data. Nature Biotechnology **23** (2005)
24. Parsons, L., Haque, E., Liu, H.: Subspace clustering for high dimensional data: a review. SIGKDD Explor. Newsl. **6** (2004) 90–105
25. Fern, X.Z., Brodley., C.E.: Random projection for high dimensional data clustering: A cluster ensemble approach. In: In Machine Learning, Proceedings of the International Conference on. (2003)
26. Buhler, J., Tompa, M.: Finding motifs using random projections. In: RECOMB '01: Proceedings of the fifth annual international conference on Computational biology, New York, NY, USA, ACM Press (2001) 69–76
27. Kurimo, M.: Indexing audio documents by using latent semantic analysis and som. In Erkki Oja and Samuel Kaski, editors, Kohonen Maps (1999) 363–374

Knowledge-Based Integrative Framework for Hypothesis Formation in Biochemical Networks

Nam Tran[1], Chitta Baral[1], Vinay J. Nagaraj[2], and Lokesh Joshi[2]

[1] Department of Computer Science and Engineering,
Ira A. Fulton School of Engineering, Arizona State University,
Tempe, AZ 85281, USA
[2] Harrington Department of Bioengineering,
The Biodesign Institute at Arizona State University,
Tempe, AZ 85281, USA

Abstract. The current knowledge about biochemical networks is largely incomplete. Thus biologists constantly need to revise or extend existing knowledge. These revision or extension are first formulated as theoretical hypotheses, then verified experimentally. Recently, biological data have been produced in great volumes and in diverse formats. It is a major challenge for biologists to process these data to reason about hypotheses. Many computer-aided systems have been developed to assist biologists in undertaking this challenge. The majority of the systems help in finding "pattern" in data and leave the reasoning to biologists. Few systems have tried to automate the reasoning process of hypothesis formation. These systems generate hypotheses from a knowledge base and given observations. A main drawback of these knowledge-based systems is the knowledge representation formalism they use. These formalisms are mostly monotonic and are now known to be not quite suitable for knowledge representation, especially in dealing with incomplete knowledge, which is often the case with respect to biochemical networks. We present a knowledge based framework for the general problem of hypothesis formation. The framework has been implemented by extending BioSigNet-RR. BioSigNet-RR is a knowledge based system that supports elaboration tolerant representation and non-monotonic reasoning. The main features of the extended system include: (1) seamless integration of hypothesis formation with knowledge representation and reasoning; (2) use of various resources of biological data as well as human expertise to intelligently generate hypotheses. The extended system can be considered as a prototype of an intelligent research assistant of molecular biologists. The system is available at http://www.biosignet.org.

1 Introduction

Because of the complexity of living systems and the limitation of scientific methods available for the study of those systems, biological knowledge is inherently incomplete. The incompleteness of knowledge constantly manifests itself in unexplainable observations. To account for these novel observations, biologists need to revise or extend the existing knowledge. The revision and extension are first

B. Ludäscher and L. Raschid (Eds.): DILS 2005, LNBI 3615, pp. 121–136, 2005.
© Springer-Verlag Berlin Heidelberg 2005

formulated as hypotheses. After being verified experimentally, a hypothesis is added to existing knowledge and becomes part of the accepted theory.

Recent advances in biological and computational sciences have produced diverse sources of biological data such as: research literature, high-throughput data (e.g. microarray, mass spectrometry), and bioinformatic resources (e.g. interaction databases, biological ontologies). It is a major challenge for biologists to integrate these various data sets to generate hypotheses. Many computer-aided systems have been developed to assist biologists in undertaking this challenge. These systems differ in their goals, namely the automation of generating hypotheses either directly from data or based on knowledge. Although hypothesis generation from data is an important first step, often use of high-level knowledge is necessary to come of with more relevant hypothesis and to narrow down the set of hypothesis. Our work in this paper aims at contributing towards this goal.

Knowledge-based hypothesis generation has been a focus of Artificial Intelligence (AI) research in the past (1; 2). Regarding molecular biology and in particular biochemical networks, the related works in hypothesis generation include HYPGENE (3), HinCyc (4), TRANSGENE (2), GENEPATH (5) and PathoLogic (6). These works are built upon knowledge representation languages that are limited to "monotonic reasoning". In monotonic reasoning, if a proposition p can be concluded from a knowledge base K (denoted by $K \models p$), then p will also be concluded after K is extended with H (i.e, $K \cup H \models p$). However, the contrary is a common phenomena in biology. In that case, p becomes false after the extension of the knowledge base: $K \cup H \not\models p$. Moreover, with the exception of PathoLogic, the related works do not address the integration of multiple data sources (probably because many of the data sources were not been available at that time).

As noted above, making hypotheses from data is important because it creates the foundation to build high-level knowledge. Towards this task, a vast array of computational techniques has been developed (7; 8; 9; 10; 11). The computational systems produce "first-level" knowledge, which should be exploited by large-scale knowledge-based systems for hypothesis formation. It is an important requirement that such large-scale systems should allow for easy updating (referred to as "elaboration tolerance") of the knowledge base when new knowledge becomes available and avoid significant overhauling (or surgery) of the old model or scrapping of the old model and making a new model from scratch. This issue of elaboration tolerance in knowledge representation has been addressed successfully by recent advances in AI research (12).

In this work, we propose a knowledge-based framework for hypothesis formation which is based on non-monotonic reasoning and elaboration tolerant representation. We select the domain of biochemical networks as the test bed, because this domain suffers from largely incomplete knowledge and at the same time, databases and knowledge bases of biochemical networks exist in a great number. We have implemented the framework by extending the BioSigNet-RR knowledge based system (13). We named the new system BioSigNet-RRH, which stands for "Representing, Reasoning and Hypothesizing about Biological Signal

Network". Besides generating hypotheses, the new system also supports ranking of hypotheses and proposes plans for experimental verification.

The rest of the paper is organized as follows. First we discuss representative related works. Then we review basics of knowledge representation and formally define the hypothesis formation problem. We continue with the description of system and methods. Finally, we conclude with a case study of the p53 signal network.

2   Related Works

HYPGENE (3) treated the general problem of hypothesis formation as a *planning problem*. The actions are operators that modify an existing knowledge base and/or assumed initial conditions of an experiment. The goal is to resolve the mismatch between theoretical predictions computed by the knowledge base and experimental observations, with respect to the same initial conditions. The knowledge base was implemented in a frame-based representation language. HYPGENE was proposed to be domain-independent and has been tested on a problem of E.coli gene regulation.

HYPGENE and BioSigNet-RRHtackle the same hypothesis formation problem that arises when an existing theory does not predict an experimental observation. The limitations of HYPGENE lie in methods, which include

- The frame-based representation language is limited to monotonic reasoning. Thus HYPGENE would have difficulty in dealing with incompleteness of biological knowledge.
- Although the biological knowledge is always incomplete, it is currently available in a great volume and in diverse formats. It is unclear how the current knowledge could have been exploited for hypothesis formation in HYPGENE.
- A hypothesis involves the modification of an existing knowledge base and/or assumed initial conditions of an experiment. HYPGENE was restricted to the modification of the initial conditions. This restricted problem amounts to a form of reasoning called *explanation* and studied in (13).

TRANSGENE (2) considered hypothesis formation as diagnosis and redesign of theories. According to this model, when a theory cannot predict an experimental observation, the theory must contain some faulty components that can be found and fixed. TRANSGENE used a "functional representation" language for knowledge representation (14). This representation language was chosen to overcome the limitations of rule based and frame based system. Nevertheless, the language could not allow for non-monotonic reasoning. To sum up, TRANSGENE showed that limitations of knowledge representation language can seriously hinder hypothesis formation. On the other hand, it illustrates that hypothesis formation is intuitive and straightforward in knowledge based framework.

GenePath (5) automated the inference of genetic networks from experimental data. A knowledge base is a genetic network that represents positive and negative influences of a gene on another. Experiments are perturbations to the network, performed by means of gene mutations. A fixed set of inferencing rules was formalized and implemented in GenePath using Prolog. These rules encode heuristic reasoning that are routinely applied by geneticists, namely epistasis analysis. Prior background knowledge are encoded in an initial network. Starting with the initial network, GenePath applies the rules to construct a plausible network as a hypothesis that explains experimental data. GenePath can also propose new experiments for further verification and refinement of hypotheses. Although the knowledge representation and reasoning are simple in GenePath, it has illustrated the important role of expert reasoning in hypothesis formation, and that logic programming provides a straightforward and intuitive representation of human reasoning.

Integrative computational protocols (6; 15; 16) have been proposed for prediction of metabolic and regulatory pathways. They have the general scheme: (1) construct an initial template pathway; (2) fill in missing links in the template, expand the template with new elements, or refine it; (3) verify experimentally the predicted pathway(s). These works integrated various techniques for prediction of missing genes and molecular interactions into functional contexts of pathways. They indicate that more powerful hypotheses can be found by incorporating more background knowledge and reasoning into search.

Cytoscape (17) provided an integration of various resources of molecular interaction data. By means of simulation and visualization, the system is very useful for biologist to identify novel patterns in high-throughput data. Observing novel patterns in data, biologists reason to formulate hypotheses that may explain the patterns; for example as in (18). Cytoscape has alleviated the manual processing of high-throughput information. Nevertheless, in a near future, even the number of such patterns would also become so great that biologists would have difficult to handle such reasoning in their head. Hence, tools such as Cytoscape make the automation of reasoning to formulate hypotheses even more pressing.

HyBrow (19) was designed for computer-aided evaluation of user-defined hypotheses. A hypothesis in the HyBrow system is a set of biological events that are related logically and/or temporally. The knowledge base in HyBrow is a database integration of various data sources (e.g annotated genomic database, microarray expression data). Given a hypothesis, HyBrow checks if the hypothesis conflicts with the knowledge base. It then provides explanation for conflicts as well as suggestions for necessary refinements of the hypothesis. We will discuss later how the output of HyBrow can be useful in the hypothesis formation in BioSigNet-RRH.

Robot Scientist (20) uses machine learning techniques (active learning, decision tree, inductive logic programming) to predict gene function in metabolic networks. The knowledge representation language is a monotonic logical formalism implemented in Prolog. The system is an interesting demonstration of state-of-the-art AI methods, especially machine learning and robotics. However, it is unclear how the system can incorporate elaboration representation and non-monotonic reasoning into hypothesis formation. It is also unclear how this approach can be scaled up to take advantage of multiple sources of biological knowledge.

3   Problem Definition

Before we formally define the hypothesis formation problem, let us review some basic notions of knowledge representation.

3.1   Background of Knowledge Representation

In a computer system, knowledge is represented in a symbolic language with a precise syntax and semantics. For our discussion, we will use the language $\mathcal{A}_T^0$ of BioSigNet-RR (13; 21), but the general principles are applicable to any other knowledge representation formalisms.

The language $\mathcal{A}_T^0$ has an alphabet, and a restricted syntax. The alphabet of $\mathcal{A}_T^0$ consists of a set of Boolean symbols named *fluent* and a set of symbols named *action*. Fluents represent properties of the world, and actions represent mechanisms that cause the state of the world to change. For example, we can have a fluent $high(ligand)$ representing the property that the level of ligand is high. We can have an action $bind(ligand, receptor)$ representing the association of ligand with receptor.

The language $\mathcal{A}_T^0$ consists of three sub-languages: a language for knowledge bases that describe the world, a language for our observations about the world, and a language for queries about the world.

A knowledge base is a set of statements in the following syntax:

$$a \ \textbf{causes} \ f \ \textbf{if} \ f_1, \ \ldots, \ f_k \tag{1}$$

$$g_1, \ \ldots, \ g_m \ \textbf{triggers} \ b \tag{2}$$

$$h_1, \ \ldots, \ h_n \ \textbf{inhibits} \ c \tag{3}$$

where a, b, c are actions, and f_i, g_j, h_k are fluents. Statements of the form (1) are called *causal rule*, which state that if a occurs in the world state s where $f_1, \ldots f_k$ are true, then f will become true in the world state s' resulted from the occurrence of a in s. Statements of the form (2) are called *trigger*, which state that action b has to occur if the preconditions $g_1, \ldots g_m$ hold. Statements of the form (3) are called *inhibition*, which state that action c cannot occur whenever the preconditions $h_1, \ldots h_n$ hold.

Example 1. Let us consider the knowledge base:

$$bind(ligand, receptor) \ \textbf{causes} \ bound(ligand, receptor)$$

$$high(ligand) \ \textbf{triggers} \ bind(ligand, receptor)$$

$$bound(another, receptor) \ \textbf{inhibits} \ bind(ligand, receptor)$$

The knowledge base represents that the association of *ligand* and *receptor* results in *ligand* being bound to *receptor*; that the association occurs when the level of *ligand* is high and that the association is blocked when *receptor* is bound to another molecule.   □

Observations about the world involve properties or action occurrences. To record the observation that a property f is true at time t, we write

$$f \ \textbf{at} \ t.$$

To record the observation that some action a occurs at time t', we write

$$a \textbf{ occurs\_at } t'.$$

The semantics of $\mathcal{A}_T^0$ (21) defines when a set $\mathcal{O}$ of observations is entailed from a knowledge base K and a set I of initial observations. The entailment is usually written as $(K, I) \models O$. For example, let K be the knowledgebase of *ligand* and *receptor*. Let I and O be the following sets of observations

$$I = \{high(ligand) \textbf{ at } 0, \neg bound(another, receptor) \textbf{ at } 0\}$$
$$O = \{bound(ligand, receptor) \textbf{ at } 1\}$$

then $(K, I) \models O$. We also say that the observation O is explained by K, given the initial condition I.

We are now ready to discuss the general problem of hypothesis formation.

3.2   Hypothesis Formation

We take the view that hypothesis formation is a reasoning process to find explanations for "novel" observations. Given a knowledge base K and initial condition I, we call an observation O "novel" with respect to K and I if O is not entailed (i.e. definitely concluded) by (K, I). For example, in the case of K and I as in the previous section, a novel observation is

$$O' = \{\neg bound(ligand, receptor) \textbf{ at } 1\}$$

With the assumption that O' is correct, we need to find explanations for O' by modifying K and I to become K' and I' such that $(K', I') \models O'$. The modification involves expansion and/or revision of the existing knowledge (i.e. K and I).

In this work, we focus on hypothesis formation as the expansion of an existing knowledge base to account for novel observations. This form of reasoning is called abduction, which was introduced by (22; 23) and has been used in various AI applications (24), including abductive logic programming (25; 26; 27; 28), diagnosis (29), planning (30; 31), default reasoning (32; 33; 25), belief revision and update (34). We formally define hypothesis formation as follows.

Definition 1. *Let K be a knowledge base. Let O be some observation that cannot be explained by K, given some initial condition I:*

$$(K, I) \not\models O.$$

A hypothesis space is a pair $(\mathcal{S}_K, \mathcal{S}_I)$, where $\mathcal{S}_K$ is a set of rules and $\mathcal{S}_I$ is a set of observations. A hypothesis is a subset $H \subseteq \mathcal{S}_K$ such that there exists $I' \subseteq \mathcal{S}_I$ satisfying: $(K \cup H, I \cup I') \models O$.   □

A hypothesis formation problem (K, I, O) is to find hypotheses as defined above.

4   System and Methods

The main steps of hypothesis formation in BioSigNet-RRHare: (1) the construction of the hypothesis space $(\mathcal{S}_K, \mathcal{S}_I)$; (2) generation of hypotheses, which includes search for and ranking of hypotheses. The ranking of hypothesis is based on the estimation of the preferences of hypotheses. Hypotheses generated by BioSigNet-RRHare theoretical and thus have to be verified experimentally. Because there are usually many ways to verify a hypothesis and biological experiments are cost sensitive, BioSigNet-RRH provides means to evaluate costs of experiments before they are performed.

We now present these major features of BioSigNet-RRH

4.1   Construction of Hypothesis Space $(\mathcal{S}_K, \mathcal{S}_I)$

In general, the rules and observations of the hypothesis space $\mathcal{S} = (\mathcal{S}_K, \mathcal{S}_I)$ include new fluent and action symbols, which form an additional alphabet. Let us denote the existing alphabet by $\mathbf{A}$ and the new alphabet by $\mathbf{A}^+$. The addition of $\mathbf{A}^+$ and the elements of $\mathcal{S}$ happen together, but we discuss them separately as follows.

Addition of $\mathbf{A}^+$. The elements of the additional alphabet $\mathbf{A}^+$ come from various resources. The representative resources are as follows.

- Biologists define new fluents or actions describing biological properties or processes to be studied. There is also a wide range of techniques to infer the association between biological properties and events, for example Cytoscape (17). If some properties and events are found to be associated with components of the knowledge base, then they would be included as fluents and actions in $\mathbf{A}^+$.
- Automated extraction of biological terms from literature has produced a great resource of biological properties and molecular interactions (35).
- Many protein interaction maps have been constructed by computational and high-throughput biological methods (10; 36). These interaction maps can be used to define new actions.
- Biological ontologies and interaction databases (37; 38; 39) also contain biological properties and reactions as their alphabets.

Construction of $\mathcal{S}_K$. To distinguish the rules of the hypothesis space from the rules of the knowledge base, we call the former *possibilities*.

To include a possibility r in the hypothesis space, we write

$$\mathbf{POSS}[p] : r$$

where p is a non-negative number called the *preference* of r. If we do not want to take into account the preference, or if it is not available, we set $p = 0$. In the next section, we will describe how the preferences are used in ranking hypotheses.

Causal rules can be constructed from interaction databases and biological ontologies (37; 38; 39; 40). There exists no database that contains explicit information regarding triggers and inhibitions. However, there exist datasets from which associations between properties and processes can be found. Presently, we take a simple approach to generate triggers and inhibitions of the hypothesis space: if a set of fluents $f_1, f_2, \ldots f_n$ are found to be associated (or correlated) with an action a, then there are the possibilities that

$$\textbf{POSS}[p]: \quad f_1, f_2, \ldots f_n \textbf{ triggers } a$$
$$f_1, f_2, \ldots f_n \textbf{ inhibits } a$$

where the number p is either estimated from the data, or defined by biologists.

We can also to take advantage of data integration efforts such as HyBrow (19). Recall that HyBrow aides in manual construction of sets of biological events that are consistent with respect to an integrated database. Such as set of events can be used as suggestions for possibilities.

Example 2. Consider a simple set of events output by HyBrow: "Gal2p transports galactose into the cell. In the cytoplasm, galactose activates Gal3p. Gal3p binds to the promoter of the Gal1 gene" (19). Based on this set of events, there can be the following possibilities:

$$high(Gal2p) \textbf{ triggers } trans(Gal2p, galact)$$
$$trans(Gals2p, galact) \textbf{ causes } in(galact, cyto)$$
$$in(galact, cyto) \textbf{ triggers } activates(Gal3p)$$
$$activates(Gal3p) \textbf{ causes } active(Gal3p)$$
$$active(Gal3p) \textbf{ triggers } binds(Gal3p, Gal1\_promoter)$$

Such rules are possible elements of $\mathcal{S}_K$.   □

Construction of $\mathcal{S}_I$. We declare possible unknown factors in the initial conditions as follows

- f may be true or false initially: **POSS initial** f.
- a may occur initially: **POSS initial** a.

4.2   Generation of Theoretical Hypotheses

The reasoning in BioSigNet-RR is implemented using AnsProlog, a non-monotonic logic programming language (12). The semantics of AnsProlog is *stable model semantics*. For example, the AnsProlog program

$$a \leftarrow not\ b$$
$$b \leftarrow not\ a$$

has 3 models $\{a\}$, $\{b\}$ and $\{a, b\}$. The models $\{a\}$ are $\{b\}$ stable, while $\{a, b\}$ is not. Stable models are minimal with respect to the $\subseteq$ ordering on sets.

The hypothesis generation in BioSigNet-RRHis also implemented using Ans-Prolog. A hypothesis - a set of rules - is extracted from a stable model of the AnsProlog implementation. Intuitively, we want to find hypotheses as simple as possible. The minimality of stable models has an important role towards this goal.

The ranking of hypotheses is based on the following partial ordering.

Definition 2. *Let γ be some scoring function for hypotheses. A hypothesis H is more* preferred *than a hypothesis H', written as $H \prec H'$, if $H \subset H'$ and $\gamma(H) \geq \gamma(H')$.*

A hypothesis H is *maximally* preferred, if there exists no hypothesis H' such that $H' \prec H$. We now explain how BioSigNet-RRHgenerates hypotheses that are maximally preferred. To ensure the minimality of hypotheses with respect to the $\subseteq$ relation search heuristics are added in the form of AnsProlog rules. Some examples of heuristics are:

- A trigger is added only if it is the only cause of some action occurrence that is needed to explain the novel observations.
- An inhibition is added only if it is the only blocker of some triggered action at some time.

The implementation of these heuristics is straightforward, and they can function as a plug-in component of BioSigNet-RRH.

The γ scoring function is currently defined such that it can be maximized using a built-in feature of the AnsProlog engine.

Let r be an element in the hypothesis space given by

$$\mathbf{POSS}[p] : r$$

Let $pref(r) = p$. The function $\gamma(H)$ is defined as the sum of the preferences of the rules in H; that is,

$$\gamma(H) = \sum_{r \in H} pref(r)$$

4.3   Guidance for Experimental Verification

Because of the incompleteness of biological knowledge, hypotheses can only be verified using some plausibility measure. In general, a hypothesis is accepted as a theory when there are enough experimental evidences supporting it. Thus, biologists would like to carry out as many experiments as possibile for the verification of a hypothesis. In reality, the set of possible experiments are seriously constrained by resources such as time and available techniques. Hence, it is desirable to perform only experiments that require a minimal available resource but produce a maximal information.

In this section, we propose a model of guidance for experimental verification.

Let us represent a wet-lab experiment in the abstract form (I, O), where I is the set of initial conditions of the experiment, and O is the set of observed outcomes.

Definition 3. *Let K be a knowledge base and H be a hypothesis. Let (I, O) be a experiment. We say that (I, O) is an* evidence *for the hypothesis H, if O can be explained by $K \cup H$ given I: $(K \cup H, I) \models O$.*

Example 3. Let $K = \{a \textbf{ causes } g\}$ and $H = \{f \textbf{ triggers } a\}$. Let $I_1 = \{f \textbf{ at } 0, \neg g \textbf{ at } 0\}$, $O_1 = \{g \textbf{ at } 1\}$. Let $I_2 = \{\neg f \textbf{ at } 0, \neg g \textbf{ at } 0\}$, $O_2 = \{\neg g \textbf{ at } 1\}$. Then (I_1, O_1) and (I_2, O_2) are evidences for the hypothesis H, but only (I_2, O_2) is an evidence for the hypothesis $\emptyset$. $\square$

There are two important measures of an experiment, namely its cost and its information content. Let us denote these measure as $cost(I, O)$ and $info(I, O)$. Given a hypothesis H, the objective is to find a set E of evidences for H that has minimal cost and maximal information content. Let us simply define:

$$cost(E) = \sum_{(I,O) \in E} cost(I, O)$$

$$info(E) = \sum_{(I,O) \in E} info(I, O)$$

An initial condition such as f **at** 0 can be achieved by some wet-lab operation and can be associated with some cost. We then define

$$cost(I) = \sum_{x \in I} cost(x)$$

Biological observations are achieved by means of measurements, which also have associated costs. Hence, we define

$$cost(O) = \sum_{y \in O} cost(y)$$

Finally, $cost(I, O) = cost(I) + cost(O)$.

Let $\Omega(K, I)$ be the maximal observations that can be entailed from K, given I. That is, $(K, I) \models \Omega(K, I)$ and for all ω, if $(K, I) \models \omega$ then $\omega \subseteq \Omega(K, I)$. We define the information content of (I, O) as the deviation (or distance) of O from $\Omega(K, I)$. The distance between two sets of observations in turn is defined based on the distance between their elements.

We now present the p53 signal network as a case study to illustrate our theoretical methods to automate the process of hypothesis formation.

5   Case Study

First, we describe the biology the p53 network in parallel with its knowledge-based representation.

5.1   p53 Signal Network

The p53 protein plays a central role as a tumor suppressor and is subjected to tight control through a complex mechanism involving several proteins. The key aspects of the p53 network are as follows.

Tumor suppression by p53: The p53 protein has three main functional domains; the N terminal transactivator domain, the central DNA-binding domain and a C terminal domain that recognizes DNA damage. The binding of the transactivator domain to the the promoters of target genes activates pathways to lead to a reversible arrest of the cell cycle, prevention of genomic instability or apoptosis and thus protects the cell from cancer (41). The ability to suppress tumors is retained when the interacting partners of p53 do not inhibit the functionality of the transactivator domain.

fluent $bound(dom(p53, N))$

action $grow(tumor)$

$high(p53)$ **inhibits** $grow(tumor)$

$high([p53 : P]), not\ bound(dom(p53, N))$ **inhibits** $grow(tumor)$

(The keywords **fluent** and **action** are used to declare fluent and action symbols in BioSigNet).

Interaction between Mdm2 and p53: Mdm2 binds to the transactivator domain of p53, thus inhibiting the p53 induced tumor suppression. The binding of Mdm2 to p53 also causes changes in the protein concentration levels.

fluent $high(p53), high(mdm2), high([p53 : mdm2])$

action $bind(p53, mdm2)$

$bind([p53 : mdm2])$ **causes** $bound(dom(p53, N))$

$high(p53), high(mdm2)$ **triggers** $bind(p53, mdm2)$

$bind(p53, mdm2)$ **causes** $high([p53 : mdm2]),$

$bind(p53, mdm2)$ **causes** $\neg high(p53), \neg high(mdm2)$

Mdm2 induced degradation of p53: Under normal physiological conditions, p53 levels remain low due to rapid and constant turnover. The short half life of p53 is due to the formation of a complex with Mdm2 that gets targeted for ubiquitin dependent proteosomal degradation.

action $degrade(p53, mdm2)$

$high([p53 : mdm2])$ **triggers** $degrade(p53, mdm2)$

$degrade(p53, mdm2)$ **causes** $\neg high([p53 : mdm2])$

Upregulation of p53: The elevated levels of p53 may be a result of upregulation of p53 gene expression, increased transcript stability, enhanced translation of p53 mRNA (42), or post-translational modifications of the p53 protein which favor a prolonged half life and increased activity (43).

For the case study, we consider the upregulation of p53 expression, which is represented as follows.

$upregulate(mRNA(p53))$ **causes** $high(mRNA(p53))$

$high(mRNA(p53))$ **triggers** $translate(p53)$

$translate(p53)$ **causes** $high(p53)$

Stress: UV, ionizing radiation, and chemical carcinogens cause stress. Stress can induce the upregulation of p53.

$high(UV)$ **triggers** $upregulate(mRNA(p53))$

Stress can induce changes in expression of tumor related genes, (e.g. cmyc), which result in uncontrolled cell division (tumor).

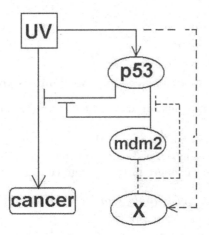

Fig. 1. A hypothesis in p53 interaction network. The → represents trigger. The ⊣ represents inhibition. The solid and dash lines represent known and hypothetical interactions, respectively

$high(UV)$ **triggers** $alter(expr(cmyc))$

$alter(expr(tumorgenes))$ **causes** $altered(expr(cmyc))$

$altered(expr(cmyc))$ **triggers** $grow(tumor)$

$grow(tumor)$ **causes** $tumorous$

Given the theory of the p53 network, a hypothesis formation problem arises as follows.

5.2   The Problem

X is a tumor-suppressor gene. Mutants of X are highly susceptible to cancer. We would like to hypothesize on the various possible influences of X on the p53 pathway.

Thus, we have the hypothesis problem (K, I, O), where K is the knowledge base of p53 biology, and I is the initial condition

$$I = \{null(X) \textbf{ at } 0\}$$

and O is the observation

$$O = \{\textbf{eventually } tumorous\}$$

(Here, **eventually** F is a logical proposition denoting that some property F will be true at some future time).

We need to extend K with H such that there exists I' satisfying: $(K \cup H, I \cup I') \models O$.

5.3   Hypothesis Formation

Construction of the Hypothesis Space. First, we show how various possibilities can be found and included in the hypothesis space. In the following, the literature means (41; 42; 43).

There may be functional similarities between X and p53: X is a tumor suppressor, so we have a prior knowledge that X may play the same effects as p53 in stressed cells, which is captured by the following possibilities:

POSS : $high(UV)$ **triggers** $upregulate(mRNA(X))$

$upregulate(mRN(X))$ **causes** $high(mRNA(X))$

$high(mRNA(X))$ **triggers** $translate(X)$

$translate(X)$ **causes** $high(X)$

Stress may induce high level of X: Data from the literature show that the levels of protein X is found to be higher in cells subjected to stress. Consequently, it is possible that stress induces the upregulation of X expression. That is,

POSS : $high(UV)$ **triggers** $upregulate(mRNA(X))$

X or p53 may induce upregulation of the other: There are observations from the literature that high levels of X are concomitant with elevated levels of p53. Thus, it is possible that a high level of X induces the upregulation of p53, or vice versus.

POSS : $high(X)$ **triggers** $upregulate(mRNA(p53))$

$high(p53)$ **triggers** $upregulate(mRNA(X))$

X may interact with the known proteins in the network: The possible interactions are $bind(p53, X)$ and $bind(mdm2, X)$. The possibile properties are the protein levels and the domains of p53. By associating a possible action with possible effects, we form possibilities such as

POSS : $bind(p53, X)$ **causes** $bound(dom(p53, N))$

$bind(p53, X)$ **causes** $\neg bound(dom(p53, N))$

That is, binding of X to p53 may or may not affecting the transactivator domain. *X may influence (trigger/inhibit) other interactions:* We consider all the possibilities of X's influences on the interactions in the network, which results in

POSS : $high(X)$ **influences** $upreg(mRNA(p53))$

$high(X)$ **influences** $translate(p53)$

$high(X)$ **influences** $bind(p53, mdm2)$

(where **influences** stands for either **triggers** or **inhibits**).

Hypotheses Generation. We present representative examples of the hypotheses generated by BioSigNet-RRH.

– *X is a negative regulator of Mdm2:* Stress induces high expression of X. X binds to Mdm2 and this complex is rapidly degraded by proteolysis. Scavenging of Mdm2 arrests the proteolyis p53 (Fig. 1). The important elements of the hypothesis are:

$high(UV)$ **triggers** $upregulate(mRNA(X))$

$high(X), high(mdm2)$ **triggers** $bind(X, mdm2)$

- *X directly influences p53 protein stability:* X binds to p53 protein at a domain different from the transactivator domain, so p53 is stabilized (formation of Mdm2-p53 complex is prevented) and still functional as tumor suppressor. The important elements of the hypothesis are:

$$high(X), high(p53) \textbf{ triggers } bind(p53, X)$$

$$bind(p53, X) \textbf{ causes } \neg bound(dom(p53, N))$$

The non-monotonicity of the framework manifests itself in the results. The knowledge base in Section 5.1 predicts that cancer will finally occur due to high level of UV (stress). After being extended with the hypothesis described in Fig. 1., the new knowledge base predicts that cancer will not occur, given the presence of UV.

The presented study is incomplete in the sense that changes in the regulation of p53 also occurs as a result of stress induced damage to DNA. Due to the elaboration tolerance feature, we could start by first constructing a small initial knowledge base, then incrementally adding more knowledge. We have also represented simple rules with only one or two preconditions. More elaborated representation and the results on experiments with ranking can be found at the system's Website.

6   Conclusion

We have presented a general framework for the automation of hypothesis formation in systems biology. We considered the hypothesis formation problem in the context of knowledge representation and reasoning. We implemented an initial system by extending BioSigNet-RR. The advantages of our approach includes: (1) hypothesis formation is defined as a form of reasoning and is implemented using AnsProlog, which is an elaboration tolerant and non-monotonic representation and reasoning language; (2) it provides a mean to integrate various resources of biological knowledge; (3) it is a high-level approach to hypothesis formation that is necessary for building an intelligent system to aid biologists.

Our work is a proof-of-concept and substantial works remain for the scaling-up the system for real-world applications. An immediate task is to automate the construction of the hypothesis space. Besides, we identify many important future works. First, it is important to allow for declaration and instantiation of "similarity" background knowledge; such as gene homology, or the similarity of relationships between proteins or biological processes. Next, we want to explore different models for ranking hypotheses. We will explore how AnsProlog with preferences can be applied for ranking hypotheses. Finally, we have restricted to the hypothesis formation as knowledge extension. Hypothesis formation based on knowledge revision is an important next development.

References

[1] Shrager, J., Langley, P.: Computational Models of Scientific Discovery and Theory Formation. Morgan Kaufmann (1990)

[2] Darden, L.: Recent work in computational scientific discovery. In: Proceedings of the Nineteenth Annual Conference of the Cognitive Science Society. (1997) 161–166

[3] Karp, P.D.: Design methods for scientific hypothesis formation and their application to molecular biology. Machine Learning **12** (1993) 89–116

[4] Karp, P.D., Ouzounis, C., Paley, S.: HinCyc: A Knowledge Base of the Complete Genome and Metabolic Pathways of H. influenzae. In: Proc. of ISMB. (1996)

[5] Zupan, B., et al: Genepath: a system for inference of genetic networks and proposal of genetic experiments. Artif. Intell. Med. **29** (2003) 107–30

[6] Karp, P.D., Paley, S., Romero, P.: The pathway tools software. Bioinformatics **18** (2002) S225–S232

[7] Yaffe, M.B., Leparc, G.G., Lai, J., Obata, T., Volinia, S., Cantley, L.C.: A motif-based profile scanning approach for genome-wide prediction of signaling pathways. Nat Biotechnol. **19** (2001) 348–53

[8] Gomez, S., Rzhetsky, A.: Towards the prediction of complete protein - protein interaction networks. In: Pacific Symposium on Biocomputing 2000 (PSB 2000). (2002) 413–24

[9] Obenauer, J.C., Cantley, L.C., Yaffe, M.B.: Scansite 2.0: proteome-wide prediction of cell signaling interactions using short sequence motifs. Nucleic Acids Research **31** (2003) 3635–41

[10] Valencia, A., Pazos, F.: Computational methods for the prediction of protein interactions. Curr Opin Struct Biol. **12** (2003) 368–73

[11] Salwinski, L., Eisenberg, D.: Computational methods of analysis of proteinprotein interactions. Current Opinion in Structural Biology **13** (2003) 377–382

[12] Baral, C.: Knowledge representation, reasoning and declarative problem solving. Cambridge University Press (2003)

[13] Baral, C., Chancellor, K., Tran, N., Tran, N., Berens, M.: A knowledge based approach for representing and reasoning about signaling networks. Bioinformatics 20 (Suppl 1) (2004) i15–i22

[14] Sembugamoorthy, V., Chandrasekaran, B.: Functional Representation of Devices and Compilation of Diagnostic Problem-Solving Systems. Experience, Memory and Reasoning (1986) 47–73

[15] Osterman, A., Overbeek, R.: Missing genes in metabolic pathways: a comparative genomics approach. Current Opinion in Chemical Biology **7** (2003) 238–251

[16] Su, Z., Dam, P., Chen, X., Olman, V., Jiang, T., Palenik, B., Xu, Y.: Computational inference of regulatory pathways in microbes: an application to phosphorus assimilation pathways in synechococcus sp. wh8102. In Gribskov, M., Kanehisa, M., Miyano, S., Takagi, T., eds.: Genome Informatics. Volume 14. (2003) 3–13

[17] Shannon, P., Markiel, A., Ozier, O., Baliga, N.S., Wang, J.T., Ramage, D., Amin, N., Schwikowski, B., Ideker, T.: Cytoscape: A Software Environment for Integrated Models of Biomolecular Interaction Networks. Genome Research **13** (2003) 2498–2504

[18] Baliga, N.S., Bjork, S.J., Bonneau, R., Pan, M., Iloanusi, C., Kottemann, M.C., Hood, L., DiRuggiero, J.: Systems Level Insights Into the Stress Response to UV Radiation in the Halophilic Archaeon Halobacterium NRC-1. Genome Res. **14** (2004) 1025–1035

[19] Racunas, S.A., Shah, N.H., Albert, I., Fedoroff, N.V.: HyBrow: a prototype system for computer-aided hypothesis evaluation. Bioinformatics **20** (2004) i257–264

[20] King, R., et. al.: Functional genomic hypothesis generation and experimentation by a robot scientist. Nature **427** (2004) 247–52

[21] Tran, N., Baral, C.: Reasoning about triggered actions in AnsProlog and its application to molecular interactions in cells. In: Proc. of KR 2004. (2004) 554–563

[22] Peirce, C.: Collected papers of Charles Sanders Peirce, Vol. 1-8. Havard University Press, Cambridge, MA (1931-1958)

[23] Peirce, C.: Reasoning and the Logic of Things. Havard University Press, Cambridge, MA (1992)

[24] Poole, D., Mackworth, A., Goebel, R.: Computational Intelligence. Oxford University Press, Oxford (1998)

[25] Kakas, A., Kowalski, R., Toni, F.: The role of abduction in logic programming. Handbook of logic in Artificial Intelligence and Logic Programming (1998) 235–324

[26] Kakas, Antonis, C., Van Nuffelen, B., Denecker, M.: A-system : Problem solving through abduction. In: Proc. of the IJCAI. Volume 1. (2001) 591–596

[27] Denecker, M., Kakas, A.C.: Abduction in Logic Programming. In: Computational Logic: Logic Programming and Beyond. (2002) 402–436

[28] Doherty, P., Kertes, S., Magnusson, M., Szalas, A.: Towards a Logical Analysis of Biochemical Pathways. In: Proc. of JELIA. (2004)

[29] Reiter, R.: A theory of diagnosis from first principles. Artificial Intelligence 13 (1980) 81–132

[30] Allen, J., Kautz, H., Pelavin, R., Tenenberg, J.: Reasoning about plans. Morgan Kaufmann, San Mateo,CA (1991)

[31] Missiaen, M., Bruynooghe, L., Denecker, M.: CHICA: A planning system based on event calculus. J. Logic Comput. 5 (1995) 579–602

[32] Poole, D.: A logical framework for default reasoning. Artificial Intelligence 36 (1988) 27–48

[33] Eshghi, K., Kowalski, R.: Abduction compated with negation as failure. In: Proc. 6th Inter. Conf. in Logic Programming. (1989) 234–255

[34] Boutilier, C.: Abduction to plausible causes: An even based model of belief update. Artificial Intelligence 83 (1996) 143–166

[35] Rzhetsky, A., Iossifov, I., Koike, T., Krauthammer, M., Kra, P., Morris, M., Yu, H., Duboue, P.A., Weng, W., Wilbur, W.J., Hatzivassiloglou, V., Friedman, C.: Geneways: a system for extracting, analyzing, visualizing, and integrating molecular pathway data. J. of Biomedical Informatics 37 (2004) 43–53

[36] Bouwmeester, T., et. al.: A physical and functional map of the human TNF-alpha/NF-kappaB signal transduction pathway. Nat. Cell. Biol. 6 (2004) 97–105

[37] Xenarios, I., Rice, D.E., Salwinski, L., Baron, M.K., Marcotte, E.M., Eisenberg, D.: Dip: The database of interacting proteins. Nucleic Acids Research 28 (2000) 289–91

[38] Bader, G.D., Donaldson, I., Wolting, C., Ouellette, B.F.F., Pawson, T., Hogue, C.W.V.: Bind the biomolecular interaction network database. Nucleic Acids Research 29 (2001) 242–245

[39] Demir, E., Babur, O., Dogrusoz, U., Gursoy, A., Ayaz, A., Gulesir, G., Nisanci, G., Cetin-Atalay, R.: An ontology for collaborative construction and analysis of cellular pathways. Bioinformatics 20 (2004) 349–356

[40] Joshi-Tope, G., Gillespie, M., Vastrik, I., D'Eustachio, P., Schmidt, E., de Bono, B., Jassal, B., Gopinath, G., Wu, G., Matthews, L., Lewis, S., Birney, E., Stein, L.: Reactome: a knowledgebase of biological pathways. Nucl. Acids Res. 33 (2005) D428–432

[41] Michael, D., Oren, M.: The p53 and Mdm2 families in cancer. Curr. Opin. Genet. Dev. 12 (2002) 53–59

[42] Hamid, T., Kakar, S.: PTTG/securin activates expression of p53 and modulates its function. Mol. Cancer. 3 (2004) 18

[43] Bode, A.M., Dong, Z.: Post-translational modification of p53 in tumorigenesis. Nat. Rev. Cancer. 4 (2004) 793–805

Semantic Correspondence in Federated Life Science Data Integration Systems

Malika Mahoui[1], Harshad Kulkarni[2], Nianhua Li[2],
Zina Ben-Miled[2], and Katy Börner[3]

[1] School of Informatics, IUPUI, Walker Plaza, 719 Indiana Avenue, suite WK307,
Indianapolis, IN 46202, USA
1-317-278-9205
mmahoui@iupui.edu
[2] Department of Electrical and Computer Engineering, IUPUI
{hkulkarn, niali, zmiled}@iupui.edu
[3] School of Library and Information Science, Indiana University
katy@indiana.edu

Abstract. For execution of complex biological queries, data integration systems often use several intermediate data sources because the domain coverage of individual sources is limited. Quality of intermediate sources differs greatly based on the method used for curation, frequency of updates and breadth of domain coverage, which affects the quality of the results. Therefore, integration systems should provide data provenance; i.e. information about the path used to obtain every record in the result. Furthermore, since query capabilities of web-accessible sources are limited, integration systems need to support refinement queries of finer granularity issued over the integrated data. However, unlike the individual sources, integration systems have to handle the absence of data and conflicts in the integrated data caused by inconsistencies among the sources. This paper describes the solution proposed by BACIIS, the Biological and Chemical Information Integration System, for providing data provenance and for supporting refinement queries over integrated data. Semantic correspondence between records from different sources is defined based on the links connecting these data sources including cross-references. Two characteristics of semantic correspondence, namely degree and cardinality, are identified based on the closeness of the links that exist between data records and based on the mappings between domains of data records respectively. An algorithm based on semantic correspondence is presented to handle absence of data and conflicts in the integrated data.

1 Introduction

The rapid development of experimental biology has led to the emergence of a large number of web-accessible biological data sources [1]. Together, these data sources cover a wide range of subjects and data types. But each individual data source often focuses on a specific subject area; and thus represents only a fraction of all the

B. Ludäscher and L. Raschid (Eds.): DILS 2005, LNBI 3615, pp. 137–144, 2005.

available data. Cross-references are provided between different data sources to connect data into a network. In addition of cross-references, integration systems [2, 3] use field values in records produced by one data source to connect to another data source to address complex queries. So a query plan can contain chains of links connecting several sources. One important issue is data quality control. Due to factors such as the method of annotation, update frequency and overall coverage of the subject area, not all of these sources are trusted equally. The quality of a data source will affect both the quality of data fields and the quality of cross-references. It is therefore important to specify the source of every piece of result data, and to aggregate records with cross-references to each other.

Query capabilities of web-accessible data sources are limited and it is often not possible to use every characteristic of biological entities in the query predicate. Therefore, most initial queries are not specific enough and their results contain several unwanted records. In such cases, once the integrated result of an initial query is available, the integration system can allow scientists to issue refining queries. However, unlike the initial query, while processing a refining query, complex inter-relationships among records from different sources must be considered. Most integration systems assume that different sources cover different characteristics of the biological entities and hence, do not deal with absence of data or contradictory data [4]. However, this does not represent the true nature of relationships among records and consequently, the results of such systems are not complete and reliable. In reality different sources have significant overlap of information and data inconsistencies are present in the overlapping portions due to different methods of curating the data. Therefore, to process refining queries in a comprehensive and correct manner, we must assume an overlapping coverage of the global schema by different sources and deal with the data absence and inconsistency.

The objective of this paper is to describe the solution proposed by BACIIS, the Biological and Chemical Information Integration System [5-7], for providing data provenance for result records and for supporting queries over integrated results. Section 2 briefly introduces BACIIS system and its main data integration features. Section 3 defines the concept of semantic correspondence and its characteristics. In section 4, processing of refinement queries over the integrated data is discussed and an algorithm is presented to handle the conflicts in the integrated data.

2   BACIIS: An Ontology Augmented Database Integration System

BACIIS (Biological and Chemical Information Integration System) is a highly coupled federation of life science web-databases. It uses a mediator-wrapper approach, augmented with a knowledge base. The wrapper extracts information from a given remote data source. The mediator transforms data from its format in the source database to the internal format used by the integration system. The BACIIS knowledge base has two components: the ontology and data source schema. The ontology provides a method for mapping differences in terminology to a common term that is recognized throughout the domain. In addition to syntactic reconciliation, the ontology is used for semantic reconciliation as well as a global schema in BACIIS. Global queries are built by using concepts from the ontology. These global queries are decomposed within BACIIS into database specific sub-queries. The query planner in BACIIS [7] identifies the data sources that can answer the sub-queries based on the description of the data sources that is included in database specific data source schema. The results of the graph planner is a graph where nodes represent data sources and an edge is present between two nodes if a link can be established between the corresponding data sources (see section 3.1). Finally, each database is associated with a specific wrapper and these wrappers are responsible for executing the sub-queries on the web databases and retrieving the result.

3 Semantic Correspondences Among Heterogeneous Data

In section 3.1, the concept of semantic correspondence is explored in the context of integration of web-accessible life science data sources. Two characteristics of semantic correspondence, degree and cardinality, are then introduced in section 3.2 and 3.3. Degree is a measure of how closely two data records from different databases correspond with each other. Cardinality is a measure of domain mapping between two real world objects with some semantic correspondence.

3.1 Concept of Semantic Correspondence (SC)

The issue of semantic correspondence between two objects that have significant representational differences was examined in [8]. It also provides a way to distinguish between different degrees of semantic correspondence using factors like the context, abstraction, domains and the state of objects. However, in the context of integrating domain specific data from autonomous, heterogeneous and semi-structured sources, we maintain that the SC is established between two records when field values of one record can be used to identify the other record. Sometimes, the link between records is explicitly given by the data sources. For example, SwissProt records provide hyperlinks to related records in PDB. This is similar to the concept of hyperlink authority explored in [9]. However hyperlinks are not the only way to establish SC. Consider the case of BIND [10], which does not have explicit hyperlinks to SwissProt records. However, BIND records contain attributes 'protein-name' and 'organism-name', which can be combined to identify a protein sequence record from SwissProt. The roles of 'protein-name' and 'organism-name' here are similar to the role of foreign keys in relational databases.

The idea of SC can be illustrated by an example query: "Which protein family does chaperonin hsp60 precursor in Arabidopsis thaliana belong to? What is its coding gene sequence? What are the 3D structures of proteins that belong to the same family?" The predicate of this query has two constraints (i.e., Protein Name = chaperonin HSP60 precursor and Organism Name = Arabidopsis thaliana), and the output requires four characteristics (protein family, coding gene sequence, and 3D structure). To the best of our knowledge, no individual life science data source can answer the above query directly due to limited query capabilities and domain coverage [7]. Information from multiple data sources has to be combined together for a complete answer. Figure 1 shows one possible query plan and some results for illustration purpose. The predicate criteria 'protein-name' and 'organism-name' are combined and submitted to SwissProt, which provides the sequence-info part of the output. The SwissProt data record matching the protein name, also provide hyperlinks to related data records in GenBank and PROSITE. These sources provide the gene sequence information and the pattern description part of the output, respectively. Finally, PROSITE data records provide hyperlinks to the related PDB data records, which contain the 3-D structure part of the output. Thus, the result of this query consists of data records that are obtained from four different sources.

Consider the SC between SwissProt record P29197 and the PROSITE record PS00296 in figure 1. This semantic correspondence is established because the SwissProt record has a hyperlink to the PROSITE record. In terms of domain knowledge, this SC denotes the fact that the protein represented by the sequence in SwissProt record belongs to the family represented by the PROSITE record. Similarly, the SC between SwissProt and GenBank records denotes the domain knowledge that the protein represented by the SwissProt record is a product of the gene represented by the GenBank record.

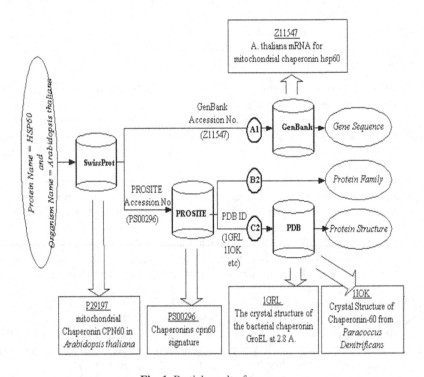

Fig. 1. Partial result of a query

3.2 Degree of Semantic Correspondence

Now consider the PROSITE and GENBANK records in figure 1. Do they have SC among them? In terms of domain knowledge, the protein family represented by PROSITE record and the gene represented by the GENBANK record, both are definitely related to the sequence represented by the SwissProt record, i.e., these records represent different characteristics of the same protein. Therefore, GENBANK and PROSITE records do have certain SC. However, this SC is not as strong as the one that links GENBANK and SwissProt records; because there are neither direct hyperlinks nor matching field values between the GENBANK and PROSITE records. Formelly, we define two degrees of SC: strong SC and weak SC.

Strong SC (SSC): Two data records are said to have strong SC, if they are linked directly either by matching field values or by hyperlinks. These data records are immediate neighbors in the query plan. For example, the SwissProt and PROSITE records mentioned in the example above, have strong SC, as do SwissProt and GenBank records.

Weak SC (WSC): Two data records are said to have weak SC, if they are connected using a chain of SSC that travels through at least one other data source. These data records are connected but not immediate neighbors in the query plan. Records connected by WSC may represent different characteristics of the same biological entity. However, WSC is just a possibility and its validity must be confirmed using some other means as explained in the next section.

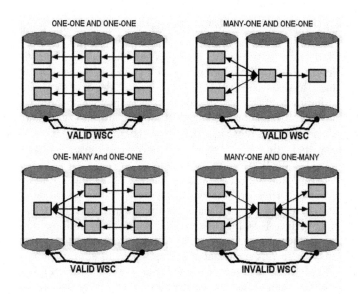

Fig. 2. Domain mapping and validity of WSC

3.3 Cardinality of Semantic Correspondence

According to the above definitions, SwissProt record P29197 and PDB record 1GRL in figure 1 are connected by WSC because both of them have SSC with PROSITE record PS00296. However it is misleading to connect them (P29197 and 1GRL) together because they represent two different proteins. On the other hand, the WSC between GenBank record Z11547 and PROSITE record PS00296 makes more sense because the gene record and the protein family record are both characteristics of the protein represented by the connecting SwissProt record. In other words, the WSC between SwissProt and PDB records is invalid while the WSC between PROSITE and GENBANK is valid. The validity of WSC between two records thus depends on whether or not we can biologically pair them with each other.

This can be determined from the mapping of domains of the biological entities involved. For example, each protein record will have a corresponding Gene record that it can be biologically paired with; however, each protein family record can be paired with several corresponding protein records. Figure 2 shows the possible cases of domain mappings and corresponding validity or invalidity of the WSC. The mapping between the domain of intermediate source and its neighbors is the most important factor in deciding validity of WSC. If the intermediate record maps to multiple records with both its neighbors, its SC has a plural cardinality; otherwise it has singular cardinality. When an intermediate record has plural cardinality, we cannot reliably pair its neighbors and the WSC between them is labeled invalid.

4   Refinement Query Processing over Integrated Data

Query processing capabilities of web-accessible data sources are limited and not every field in the record can be used in the predicate. For example, it is not possible to use the field 'induction' as predicate in the initial query. Therefore, there will be many records in the result of the initial query with values of 'induction' different than the desired value. However, since BACIIS now has all the data locally, it can apply the additional criterion to that data regardless of the sources' capabilities. In general, BACIIS can provide refinement query capabilities of arbitrary granularity over the global schema and process those queries over the integrated data. For example, refinement query 'induction=heat shock' will only keep those records that contain 'heat shock' in field 'induction', and their related records.

Given the rich population of biological databases available online, it is not surprising that some portions of the domain be covered by multiple sources. For example, protein sequence information is available from several sources such as SwissProt, PIR, etc. Since BACIIS collects information from multiple data sources, it may get multiple values about the same data field from records of different sources. Those values may be inconsistent, but it is impossible to eliminate the wrong ones automatically. So, BACIIS will present all the data to users by default. If a user wants to further refine the result based on the value of one data field, inconsistent field values may cause a problem.

Consider the following query issued to BACIIS "What is the GENE ontology classification of protein featured for the protein phytochrome B in Arabidopsis thaliana?" Along with many others, the result for this query contains the following three records: At2g18790 from TIGR, NF00659007 from iProClass and 1005515 from TAIR. Figure 5 shows the cross-references among these records and using those along with domain mapping information; we can state that there is a valid WSC between the iProClass record and the TAIR record.

Now, consider the following refinement query issued on this result: "cellular component = membrane". From the data source schema, BACIIS finds out that only

TIGR and iProClass records can be directly evaluated for this predicate. Therefore, TAIR record's selection is completely dependent on its having valid semantic correspondence with a record that can be evaluated. However, the value of this field is different in both of such records, where iProClass record satisfies the predicate and TIGR does not. This inconsistency of data can be attributed to the different methods of annotation employed by the two sources. Nevertheless, the TAIR record now corresponds to one record that satisfies the predicate and another that doesn't. However, BACIIS has to take into account the relative degree of SC between these records and since the TAIR record has a SSC with the mismatching TIGR record, its WSC with iProClass record should be considered invalid and it should not be included in the result.

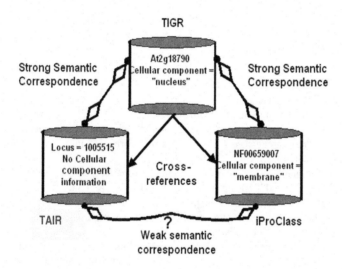

Fig. 3. Invalid WSC due to Data Inconsistency

To solve this problem, we propose an algorithm that first finds out all the records that can be directly evaluated for the predicate, and marks them as either valid or invalid. And then, the algorithm uses semantic correspondence to evaluate other records. For each record that has not been marked yet, if its predecessor is not marked invalid and if any of its neighbors are marked valid, then that record itself becomes valid. The rest of the records are invalid. Thus, the validity of a record for the refinement query is based on its being part of an unbroken chain of records in a path expression. Therefore, records with SSC to invalid records are eliminated.

5   Conclusion

In this paper two challenges were addressed; providing provenance for records in integrated data and processing queries over integrated data in a semantically meaningful way. The concept of semantic correspondence was introduced for heterogeneous data obtained using several query paths. Two characteristics of semantic correspondence were also defined. First, the degree of semantic correspondence which represents the closeness of entities represented by different records and second, the cardinality which represents the mapping between domains of entities.

Data quality in biological data sources varies greatly based on several factors. Therefore, integrating data from overlapping data sources may generate results with missing data items or results that contain inconsistencies. The algorithm provided in this paper deals with these conflicts based on the characteristics of semantic correspondence among the records. It makes no assumption about the correctness of any data source involved. Furthermore, by removing semantically distant records from the integrated data, it achieves a better consistency for the integrated results.

References

[1] A. D. Baxevanis, The Molecular Biology Database Collection: 2003 update, Nucleic Acids Res., 31(1): 1-12, 2003.

[2] E. M. Zdobnov, R. Lopez, R. Apweiler, and T. Etzold, "The EBI SRS server-recent developments", Bioinformatics, 18(2): 368-73, 2002.

[3] C. A. Goble, R. Stevens, G. Ng, S. Bechhofer, N. W. Paton, P. G. Baker, M. Peim, and A. Brass, "Transparent access to multiple bioinformatics information sources", IBM Systems Journal, 40(2): 532-552, 2001.

[4] T. Hernandez, and S. Kambhampati, "Integration of Biological Sources: Current Systems and Challenges Ahead", To appear in SIGMOD Record, Vol 33, No3, September 2004.

[5] Z. Ben Miled, O. Bukhres, Y. Wang, N. Li, M. Baumgartner, and B. Sipes, "Biological and Chemical Information Integration System", Network Tools and Applications in Biology, Genoa, Italy, May 2001

[6] Z. Ben Miled, Y. Webster, N. Li, and Y. Liu, "An Ontology for the Semantic Integration of Life Science Web Databases," International Journal of Cooperative Information Systems, Vol. 12, N0.2, 2003

[7] Z. Ben-Miled, N. Li, G. Kellett, B. Sipes, and O. Bukhres, "Complex Life Science Multidatabase Queries", Proceedings of the IEEE, 90(11), 2002.

[8] A. Sheth, and V. Kashyap, "So Far (Schematically) yet So Close (Semantically)", Proceedings of the MT DS-5 Conference on Semantics of Interoperable Database Systems, Lorne, Australia, Elsvier Publishers, November 1992;

[9] J. Kleinberg., "Authoritative sources in a hyperlinked environment", Proc. 9th ACM-SIAM Symposium on Discrete Algorithms, 1998. Extended version in Journal of the ACM 46(1999). Also appears as IBM Research Report RJ 10076, May 1997

[10] [http://www.blueprint.org/bind/bind.php]

Assigning Unique Keys to Chemical Compounds for Data Integration: Some Interesting Counter Examples

Greeshma Neglur[1], Robert L. Grossman[2], and Bing Liu[3]

[1] Laboratory for Advanced Computing,
University of Illinois at Chicago, Chicago, IL 60607
`neglur@lac.uic.edu`
[2] Laboratory for Advanced Computing,
University of Illinois at Chicago, Chicago, IL 60607
`grossman@uic.edu`
[3] Department of Computer Science,
University of Illinois at Chicago, Chicago, IL 60607
`liub@cs.uic.edu`

Abstract. Integrating data involving chemical structures is simplified when unique identifiers (UIDs) can be associated with chemical structures. For example, these identifiers can be used as database keys. One common approach is to use the Unique SMILES notation introduced in [2]. The Unique SMILES views a chemical structure as a graph with atoms as nodes and bonds as edges and uses a depth first traversal of the graph to generate the SMILES strings. The algorithm establishes a node ordering by using certain symmetry properties of the graphs. In this paper, we present certain molecular graphs for which the algorithm fails to generate UIDs. Indeed, we show that different graphs in the same symmetry class employed by the Unique SMILES algorithm have different Unique SMILES IDs. We tested the algorithm on the National Cancer Institute (NCI) database [7] and found several molecular structures for which the algorithm also failed. We have also written a python script that generates molecular graphs for which the algorithm fails.

1   Introduction

The volume of biological data, especially chemical structures, is increasing at an unprecedented rate. There are numerous databases today that store chemical substances and thousands of chemical structures are being added to these databases each year. For example, the Chemical Abstracts Service (CAS) alone has more than 71,285,000 records, while the NCI database has close to 250,251 chemical structures. In general, each database uses a different method of assigning keys to the chemical compounds. For example, many databases assign keys based upon the order the compound was added to the database. For this reason, determining whether a compound has been entered into a database more than once or comparing chemical structures across databases is difficult.

B. Ludäscher and L. Raschid (Eds.): DILS 2005, LNBI 3615, pp. 145–157, 2005.
© Springer-Verlag Berlin Heidelberg 2005

This paper is concerned with data integration techniques that use the structural features of chemical compounds to assign unique IDs (UIDs). Using UIDs it is relatively simple to compare chemical structures across different databases, something which facilitates the discovery of new drugs and therapeutic treatments.

In this paper, we consider two schemes for assigning UIDs: Unique SMILES [2] and Universal Chemical Keys (UCKs) [20]. Although Unique SMILES are widely deployed and very useful in practice, we show that the algorithm described in [2] does not lead to unique IDs. We emphasize that the Unique SMILES as deployed by the Daylight Chemical Information System is an enhanced version of the algorithm described in [2], but, as far as we know, there is not a published version of this algorithm.

We believe that our paper makes the following research contributions:

1. We show that the Unique SMILES IDs although extremely useful are not unique.
2. We describe some common circumstances leading to the non-uniqueness of Unique SMILES IDs.

This paper is organized as follows: Section 2 describes related work. Sections 3-4 described one popular technique of assigning IDs to chemical compounds called Unique SMILES [2]. Sections 5-6 explain why Unique SMILES doesn't always generated UIDs. Sections 7 provides some counter examples. The final section summarizes the reason behind the failure of the unique SMILES algorithm and suggests alternate techniques for data integration of chemical compound databases using UIDs.

2   Related Work

The International Union of Pure and Applied Chemistry (IUPAC) rules [13] have been use for several decades. However, these names are growing more complicated and causing inconsistencies and mistakes as compounds become more and more complex [14]. To deal with this problem, the IUPAC has initiated a project [15] to assign unique keys known as IUPAC Chemical identifiers (INChI) to chemical compounds. This approach is based in part on graph theory. The chemical identifiers are alphanumeric text strings obtained from the molecular graph of the compound and are designed so that the chemical structure can be recovered from the UID. However, the details are not yet published.

The most common approach for integrating information about chemical compounds across databases is to use a unique key assigned by one of the databases, such as an acquisition-based or Chemical Abstracts Service (CAS) based registry numbers, as the foreign key for the other databases. For example, the NCI database stores the corresponding CAS registry number for its chemical compounds. Integrating databases in this way is labor intensive and does not easily scale.

Another approach is to view the molecular structures as a graph and to compare them directly using a graph isomorphism algorithm. There are several algorithms [11,12,17] which test for graph isomorphism. The problem with this approach is the amount of computing required to compare two structures. A more important problem is that just identifying two graphs as isomorphic does not directly provide a UID.

Several graph based techniques to solve the problem of assigning unique keys to chemical structures are known. For example, Randic and coworkers [16] developed a technique that canonically orders the adjacency matrix to produce an ID. Another popular method to discriminate molecular graphs is by means of graph invariants and vertex-in-graph invariants. One such method is the Morgan algorithm [18] which uses extended sum connectivities to distinguish atoms in a molecule. Another molecular graph canonizer is MOLGEN-CID [19].

In contrast, the Universal Chemical Key (UCK) algorithm [20] enumerates all paths up to a specified depth d in the molecular graph, lexicographically orders them, and concatenates them to produce an ID. These strings are long and cannot be used to recover the graphs. On the other hand, it is easy to use them to integrate distributed bioinformatics databases [20]. For databases of chemical compounds examined to date, a depth of $d = 3$ or 4 produces UIDs.

3   The Unique SMILES Algorithm

SMILES [1] (Simplified Molecular Input Line Entry System) is a popular chemical notation system used for computerized processing of chemical information. SMILES is a string obtained by enumerating the atomic symbols and bond types via a depth-first tree-traversal of a molecular graph, where, as usual, the nodes represent atoms and the edges represent bonds.

The problem with all such approaches is that there is no natural order to nodes in a molecular graph, and different depth-first traversals will result from different starting points. This means that there may be more than one correct SMILES string obtained from the same molecular graph. For this reason, SMILES strings, which in general are not unique, cannot be used as database keys.

To overcome this disadvantage, the creators of SMILES came up with a 2-stage algorithm called CANGEN [2] to generate a unique SMILES string for a given molecular structure. The first stage, CANON, involves CANonicalization of the structure represented as a molecular graph. The second stage, GENES, GENerates the unique SMILES notation as a depth-first traversal of the canon-icalized molecular graph.

For most chemical structures the CANGEN algorithm as described in [2] generates UIDs. However, as we show below by counter examples, there are exceptions. These exceptions need not be complicated. See Section 7. The reason is simple: if the graph is symmetric enough, it is possible for the CANON stage of the Unique SMILES algorithm to generate different canonical labels for the nodes of the molecular graph. This results in several different Unique SMILES strings.

The Unique SMILES algorithm consists of the following two stages [2]:

1. The CANON stage labels a molecular graph with canonical labels. Each atom/node is given a numerical label on the basis of its topology.

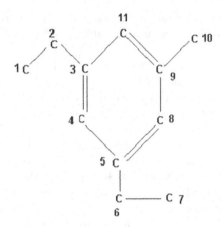

Fig. 1. Molecular graph of 3,5 di-ethyl toluene. NSC number 62141

2. The GENES stage generates unique SMILES notation as a tree representation of the graph. GENES selects the starting atom and makes branching decisions by referring to the canonical labels as needed.

The algorithm and its non-uniqueness will be explained with the example of chemical compound 3,5 di-ethyl toluene, It is stored in the NCI database with NSC number 62141. The molecular graph of this compound is described in Figure 1, where the number beside each atom is just assigned for brevity to refer to the atom in describing the following steps of the algorithm.

4   The CANON Stage of Unique SMILES

Node ordering for the generation of unique SMILES is obtained by developing topological symmetry classes, using the product of corresponding primes as illustrated below.

Graph Invariants. The algorithm claims that a set of six atomic invariants is sufficient for the purpose of obtaining a unique notation for simple SMILES. (More invariants are added for cases like Absolute SMILES to differentiate between structural and stereo-isomers).

The set is described below in descending order of priority :

1. number of connections
2. number of non-hydrogen bonds
3. atomic number
4. sign of charge
5. absolute charge
6. number of attached hydrogen

For the molecular graph in Figure 1, the initial atomic invariants for the atoms is described in Table 1(a) row labeled 'A'.

Rank Equivalence. The algorithm replaces the initial node invariant values by smaller numbers based on their sorted order to avoid numerical overflow since there is nothing intrinsically meaningful in their specific values. The row labeled 'B' in Table 1(a) describes the initial ranks.

Products of Primes. To obtain a canonical ordering of the nodes, and to obtain and identify all the symmetry classes of the nodes, an extended connectivity method using the product of the corresponding primes is used. This method is essentially used only to break ties between the initial node ordering to obtain a canonical order of the nodes.

The corresponding primes for the atoms of the molecular graph are described in the row labeled ' B* ' in Table 1(b). The product of the corresponding primes, which is the product of the primes associated with the atoms adjacent to a given atom, is displayed in the row labeled 'C' of Table 1(b).

Notice that node '10' was initially ranked '1' and appeared to belong to the same symmetry class as the other two nodes (1,7) with rank '1' when actually it did not, but by using the product of the corresponding primes we have been able to break the tie. (row 'D' of Table 1(b)).

By further following the steps of the algorithm as described in the unique SMILES algorithm [9], we obtain the final node partitioning as in Table 2: (the details of the steps are described in Table 1).

Table 1. Perception of Topological Symmetry classes for 3,5 di-ethyl toluene

Node id	1	2	3	4	5	6	7	8	9	10	11
(a) Initial atomic invariants											
A	1,01, 06,0, 0,3	2,02, 06,0, 0,2	4,04, 06,0, 0,0	3,03, 06,0, 0,1	4,04, 06,0, 0,0	2,02, 06,0, 0,2	1,01, 06,0, 0,3	3,03, 06,0, 0,1	4,04, 06,0, 0,0	1,01, 06,0, 0,3	3,03, 06,0, 0,1
B	1	2	4	3	4	2	1	3	4	1	3
(b) Classification by product of primes											
B*	2	3	7	5	7	3	2	5	7	2	5
C	3	14	75	49	75	14	3	49	50	7	49
D	1	3	6	4	6	3	1	4	5	2	4
D*	2	5	13	7	13	5	2	7	11	3	7
E	5	26	245	169	245	26	5	143	147	11	143
F	1	3	7	5	7	3	1	4	6	2	4
F*	2	5	17	11	17	5	2	7	13	3	7
G	5	34	385	289	385	34	5	221	147	13	221
H	1	3	7	5	7	3	1	4	6	2	4

Table 2. Invariant partitioning and symmetry classes of nodes

Canonical label	1	2	3	4	5	6	7
Node ids	1,7	10	2,6	8,11	4	9	3,5

5   Explanation of Non-uniqueness

Breaking Ties. We have observed that the extended connectivity method using the product of corresponding primes was able to generate 7 different symmetry classes. Since the highest rank/label (7) is smaller than the number of nodes (11), there is more than one atom in certain symmetry classes. To avoid an arbitrary decision among these atoms in a given symmetry class, the algorithm proceeds to define a next step called 'breaking ties'. In this step, all the ranks of the atoms are doubled and the value of the first (lowest valued) atom that is tied is reduced by one. This set is then treated as a new invariant set and the previous algorithm for generating an invariant partitioning is repeated until the highest rank is equal to the number of nodes.

This concept of *double-and-tie-break* works for certain highly symmetric structures like cubane (consisting of eight carbon atoms at the vertices of a cube) irrespective of the initial ordering of the nodes. However, for our example in Figure 1, this ends up generating different canonical orderings of the graph resulting in different unique SMILES strings.

In our example following the *double-and-tie-break* step, we detect the first tie among the nodes with id's 1,7. We need to reduce the first lowest valued atom (out of nodes with id's 1,7) that is tied by one. In our example since we can have two starting nodes, and the notion of 'first' in this case is ambiguous, we can either choose node '1' or '7'. The algorithm fails to establish a mechanism of preference within the nodes belonging to the same symmetry class. It assumes that choosing any of the nodes within a symmetry class will result in the same unique SMILES string. This assumption works for certain regular graphs, however for graphs similar to our example, it does not work as desired.

For our example, by merely changing the input order of the nodes we can choose either node with id '1' or '7' as the first lowest valued atom and reduce its rank by '1', totally changing the start node for the depth-first traversal (DFT). If the graph was entered as shown in Figure 2, we would have ended up choosing the node with id '7' of Figure 1 as the first node and reduced its rank making it the start node for DFT.

By choosing the node with id '1' of Figure 1 as the first lowest valued atom to break the tie and continuing the algorithm, we obtain a canonical ordering as in Table 3. (This is just one of the many canonical orderings we can obtain, and is explained later).

However, if we had chosen the node with id '7' of Figure 1 as the first node, and continuing the algorithm one of the many canonical orderings we would obtain is shown in Table 4. This will be the case if we had input the graph as in Figure 2.

The problem of establishing an order within a given symmetry class can be solved for a limited enough collection of molecular graphs by considering more chemical/topological characteristics to distinguish between these atoms and establish a precedence order.

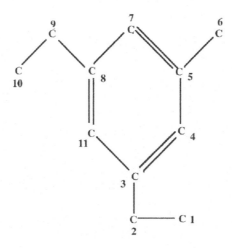

Fig. 2. Alternate input graph of 3,5 di-ethyl toluene

Table 3. One of the final canonical orderings choosing node with id 1 of Figure 1

Node id	1	2	3	4	5	6	7	8	9	10	11
Canonical label	1	4	10	8	11	5	2	7	9	3	6

Table 4. One of the final canonical orderings choosing node with id 7 of Figure 1

Node id	1	2	3	4	5	6	7	8	9	10	11
Canonical label	2	4	10	8	11	5	1	7	9	3	6

Obtaining the Unique SMILES String via GENES. By following the CANON process we have obtained a canonicalization of the graph. According to the CANON process, the nodes with the same rank are supposed to belong to the same symmetry class. The GENES process treats this structure as a tree and generates a SMILES string by Depth-First Traversal.

1. *Initial node selection*: The lowest canonical numbered atom is chosen as the starting point and it becomes the root of the Depth-First Traversal tree.

2. *Branching decision*: The following two rules apply :
   (a) Branch to double or triple bond in the ring if one exists or
   (b) Branch to the lower canonically numbered atom.

   In this particular case, we observe that we can have two initial node selections, resulting in two different depth-first traversal trees from the two different canonical orderings described in Table 3 and Table 4.

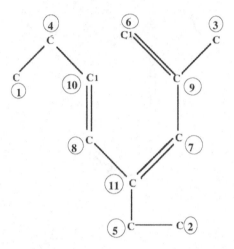

Fig. 3. Depth-first traversal associated with the initial node '1' and the canonical labeling described in Table 3. This gives the Unique SMILES **CCC1=CC(=CC(=C1)C)CC**

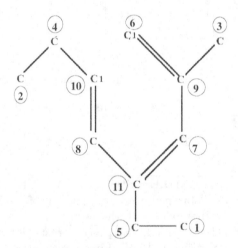

Fig. 4. Depth-first traversal associated with the initial node '7' and the canonical labeling described in Table 4. This gives the Unique SMILES **CCC1=CC(=CC(=C1)CC)C**

- The USMILES with start node id '1' is CCC1=CC(=CC(=C1)C)CC. The node IDs are described in Figure 1. The canonical labeling is described in Table 3. The depth-first traversal is described in Figure 3.
- The USMILES with start node id '7' is CCC1=CC(=CC(=C1)CC)C. The node IDs are described in Figure 1. The canonical labeling is described in Table 4. The depth-first traversal is described in Figure 4.

– The UCK algorithm generated using the web-service at [10], generates one unique key for this molecular graph, which is 85C7DC186897FD83D8ECB6B 167D988BE.

6 Experimental Studies

We have written a Python program implementing the CANGEN algorithm. The program takes as input an adjacency list of the molecular graph (described in [8]) and generates all possible unique SMILES strings for the graph. Once an invariant partitioning is obtained and it is determined that there is more than one node in any symmetry class, the script permutes the individual nodes within a symmetry class and generates all possible node selections.

For the example above, since there are 4 symmetry classes, we will get 16 different final invariant partitionings. Once we obtain these different partitionings we proceed to break ties in each of them and continue the remaining steps of the algorithm. Not all of the 16 final canonical orderings obtained from these different invariant partitionings generate different SMILES strings — only a subset of these generate different unique SMILES strings. In our example only two of the different final canonical orderings (Table 3 and Table 4) generate two different unique SMILES strings.

A web interface to this program can be accessed at [6].

7 Examples from the NCI Database

Here are some counter examples found in the NCI Database [7]. A web interface to these counter examples can be accessed at [6]. For each of the examples in this section:

1. We verified that the two different unique SMILES strings obtained map onto the same molecular graph via the on-line implementation [4] of the depict algorithm [3] provided by Daylight software [4].
2. We also verified this using another on-line implementation [5] of the CANGEN algorithm provided by the cactus service of the NCI chemical structure database. Using this service, one can input a SMILES string and get the unique SMILES for it.

NSC ID 4420. Here are two different Unique SMILES strings for N, N-Diallylmelamine with NSC id 4420:

– NC1=NC(=NC(=N1)N(CC=C)CC=C)N
– NC1=NC(=NC(=N1)N)N(CC=C)CC=C

See Figure 5 for the molecular graph. The unique key generated by UCK for this compound is: 020A134950962096577666701295295E.

NSC ID 10392. Here are two different Unique SMILES strings for 2, 4-Mesitylenediamine with NSC ID 10392.

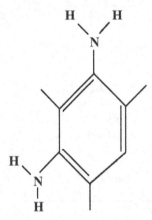

Fig. 5. Structural formula of N, N-Diallylmelamine with NSC id 4420

Fig. 6. Structural formula of 2,4-Mesitylenediamine with NSC id 10392

– CC1=C(N)C(=C(N)C(=C1)C)C
– CC1=CC(=C(N)C(=C1(N))C)C

See Figure 6 for the molecular graph. The unique key generated by UCK is F61473AE54FEC1737F7D15590650BBA2.

NSC ID 1889. Here are two different Unique SMILES strings for Pentamethyl-benzene with NSC ID 1889.

– CC1=C(C)C(=C(C)C(=C1)C)C
– CC1=CC(=C(C)C(=C1(C))C)C

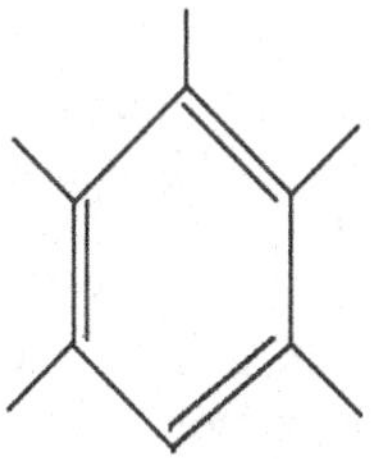

Fig. 7. Structural formula of Pentamethylbenzene with NSC ID 1889

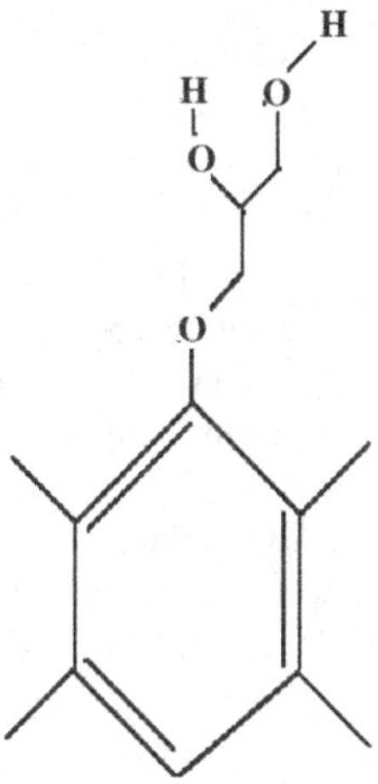

Fig. 8. Structural formula of 3-(3,5-dimethylphenoxy)-1,2-propanediol with NSC id 25239

See Figure 7 for the molecular graph. The unique UCK key generated by UCK: 1C5659F3ED5E10F02310455B56649849.

NSC ID 25239. Here are two different Unique SMILES strings for 3-(3,5-dimethylphenoxy)-1,2-propanediol with NSC id 25239.

- CC1=CC(=CC(=C1)C)OCC(O)CO
- CC1=CC(=CC(=C1)OCC(O)CO)C

See Figure 8 for the molecular graph. The unique key generated by UCK: AFD17D1BB28847F4FFAAD8C744A268AE.

8   Summary and Conclusion

Data integration involving chemical compounds is greatly aided by attaching unique IDs to chemical compounds. This is especially important when working with distributed bioinformatics data.

It has been recognized for some time that common names for chemicals, IUPAC names, CAS numbers, and general SMILES strings do not provide a good infrastructure for data integration. In this paper, we show that Unique SMILES strings [2] are also not a good foundation for data integration.

As the examples in the section above show, there are relatively simple chemical compounds that do not have Unique SMILES IDs. We have computed additional counter examples using our python script and these can be accessed at [6].

The CANGEN component of the Unique SMILES algorithm starts with a set of graph invariants and uses these to generate a canonical ordering of the nodes. This is then used as a basis for a depth-first traversal of the graph to generate the Unique SMILES string. Unfortunately, there is no set of invariants known that can distinguish all possible graph asymmetries that arise with the molecular graphs in common databases, such as the NCI database.

Although the Universal Chemical Key (UCK) algorithm [20] does not generate easy to interpret strings, it does generate unique keys for common databases such as the NCI database.

This suggests a strategy of using UCK like strings as keys to integrate distributed bioinformatics data, supplemented by SMILES-like strings that are easier to interpret.

References and Notes

1. David Weininger, SMILES, a Chemical Language and Information System 1: Introduction to Methodology and Encoding Rules, Medicinal Chemistry Project, Pomona College, 1988.
2. David Weininger, Arthur Weininger and Joseph L. Weininger, SMILES 2: Algorithm for Generation of Unique SMILES Notation, Daylight Chemical Information Systems, Irvine, California 92714, 1989. *Note* that although the Unique SMILES implementation has been changed by the Daylight Chemical Information System, this appears to be the most recent publication describing the algorithm.
3. David Weininger, SMILES 3: Depicting Graphical Depiction of Chemical Structures, Daylight Chemical Information Systems, New Orleans, Louisiana.
4. A SMILES to graph translation can be found at http://www.daylight.com/daycgi/depict.
5. A SMILES to UNIQUE SMILES translation can be found at http://cactus.nci.nih.gov/services/translate/.
6. More counter examples can be found at the web site http://ncdm171.lac.uic.edu/neglur/USMILES/USMILES.html.
7. NCI database, retrieved from http://129.43.27.140/ncidb2/ on March 2, 2005.

8. Sample adjacency list used -
   ```
   {1:[['C',1,6,'O',3],[[1,2]]],
    2:[['C',2,6,'O',2],[[1,1],[1,3]]],
    3:[['C',4,6,'O',0],[[1,2],[2,4],[1,11]]],
    4:[['C',3,6,'O',1],[[2,3],[1,5]]],
    5:[['C',4,6,'O',0],[[1,4],[1,6],[2,8]]],
    6:[['C',2,6,'O',2],[[1,5],[1,7]]],
    7:[['C',1,6,'O',3],[[1,6]]],
    8:[['C',3,6,'O',1],[[2,5],[1,9]]],
    9:[['C',4,6,'O',0],[[1,8],[1,10],[2,11]]],
    10:[['C',1,6,'O',3],[[1,9]]],
    11:[['C',3,6,'O',1],[[1,3],[2,9]]]}
   ```

9. CANON Algorithm (Extract from Reference [2])-
   (1) Set the atomic vector to initial invariants.
       Go to step 3.
   (2) Set vector to product of primes corresponding to
       neighbors' ranks.
   (3) Sort vector, maintaining stability over
       previous ranks.
   (4) Rank atomic vector.
   (5) If not invariant partitioning, go to step 2.
   (6) On first pass, save partitioning as symmetry classes.
   (7) If highest rank is smaller than number of nodes,
       break ties, go to step 2.
   (8)... else done

10. See http://bioweb.dataspaceweb.org/chemicalKeys, retrieved on March 2, 2005.
11. T. Beyer and A. Proskurowski, Symmetries in graph coding, in Proceedings of Northwest 1976 ACM–CIPS Pacific Regional Symposium, 198-203 (1976)
12. C. B HM and A. Santolini, A quasi-decision algorithm for the p-equivalence of two matrices. ICC Bull. 8, 1 (1964), 57-69.
13. IUPAC, Nomenclature of Organic Chemistry, Pergamon Press, Oxford, 1979
14. Klin M. H., Lebedev O. V., Pivina T. S., Zefirov N. S., Nonisomorphic cycles of maximum length in a series of chemical graphs and the problem of application of IUPAC nomenclature rules. MATCH, 1992, 27, 133-151.
15. See http://www.iupac.org/projects/2000/2000-025-1-800.html, retrieved on March 2, 2005.
16. Milan Randic, Gregory M. Brissey, Charles L. Wilkins: Computer perception of topological symmetry via canonical numbering of atoms. Journal of Chemical Information and Computer Sciences 21(1): 52-59 (1981)
17. B. McKay, Practical Graph Isomorphism, Congr. Numer. 30, 45-87, 1981.
18. H. L. Morgan, The Generation of a Unique Machine Description for Chemical Structures – A Technique Developed at Chemical Abstracts Service. , J. Chem. Doc., 1965 , 5 , 107-113.
19. J. Braun, R. Gugisch, A. Kerber, R. Laue, M. Meringer, C. Rcker: MOLGEN-CID, A Canonizer for Molecules and Graphs Accessible through the Internet, Journal of Chemical Information and Computer Sciences 44, 542-548, 2004.
20. Robert Grossman, Donald Hamelberg, Pavan Kasturi, and Bing Liu, Experimental Studies of the Universal Chemical Key (UCK) Algorithm on the NCI Database of Chemical Compounds, Proceedings of the 2003 IEEE Computer Society Bioinformatics Conference (CSB 2003), IEEE Computer Society, Los Alamitos, California, pages 244-250

Integrating and Warehousing Liver Gene Expression Data and Related Biomedical Resources in GEDAW

E. Guérin[1], G. Marquet[2], A. Burgun[2], O.Loréal[1], L. Berti-Equille[3],
U. Leser[4], and F. Moussouni[1]

[1] INSERM U522 CHU Pontchaillou, 35033 Rennes, France
[2] EA 3888 LIM, Faculté de Médecine 35043 Rennes, France
[3] IRISA, Campus Universitaire de Beaulieu, 35042 Rennes, France
[4] Dep. for Computer Science, Humboldt-Universität, 10099 Berlin, Germany

Abstract. Researchers at the medical research institute Inserm U522[1], specialized in the liver, use high throughput technologies to diagnose liver disease states. They seek to identify the set of dysregulated genes in different physiopathological situations, along with the molecular regulation mechanisms involved in the occurrence of these diseases, leading at mid-term to new diagnostic and therapeutic tools. To be able to resolve such a complex question, one has to consider both data generated on the genes by in-house transcriptome experiments and annotations extracted from the many publicly available heterogeneous resources in Biomedicine. This paper presents GEDAW, a gene expression data warehouse that has been developed to assist such discovery processes. The distinctive feature of GEDAW is that it systematically integrates gene information from a multitude of structured data sources. Data sources include: i) XML records of GENBANK to annotate gene sequence features, integrated using a schema mapping approach, ii) an inhouse relational database that stores detailed experimental data on the liver genes and is a permanent source for providing expression levels to the warehouse without unnecessary details on the experiments, and iii) a semi-structured data source called BioMeKE-XML that provides for each gene its nomenclature, its functional annotation according to Gene Ontology, and its medical annotation according to the UMLS. Because GEDAW is a liver gene expression data warehouse, we have paid more attention to the medical knowledge to be able to correlate biology mechanisms and medical knowledge with experimental data. The paper discusses the data sources and the transformation process that is applied to resolve syntactic and semantic conflicts between the source format and the GEDAW schema.

1 Introduction

In human health and life science, the rapid emergence of new biotechnological platforms for high throughput investigations in genome, transcriptome and proteome, prompts further advances in information management techniques to take in charge the

---

[1] Regulation of functional balances of normal and pathological liver.

B. Ludäscher and L. Raschid (Eds.): DILS 2005, LNBI 3615, pp. 158–174, 2005.
© Springer-Verlag Berlin Heidelberg 2005

data and knowledge generated by these technologies. A tremendous amount of bio-medical data is continuously deposited by scientists in public Web resources, and is in return searched by other scientists to interpret results and generate and test hypothesis.

The management of these data is challenging, mainly because : (i) data items are rich and heterogeneous: experiment details, raw data, scientific interpretations, images, literature, etc. ii) data items are distributed over many heterogeneous data sources rendering a complex integration, iii) data are speculative and subject to errors and omissions within these data sources, and bio-data quality is difficult to evaluate, and iv) bio-medical knowledge is constantly morphing and in progress..

This paper reports on our experience in building GEDAW: an object-oriented Gene Expression Data Warehouse to store and manage relevant information for analyzing gene expression measurements [12]. GEDAW (Gene Expression DAta Warehouse) aims on studying in silico liver pathologies by using expression levels of genes in different physiopathological situations enriched with annotations extracted from the variety of the scientific sources and standards in life science and medicine.

A comprehensive interpretation of a single gene expression measurement requires the consideration of the available knowledge about this gene, including its sequence and promoters, tissue-specific expression, chromosomal location, molecular function(s) and classification, biological processes, mechanisms of its regulation, expression in other pathological situations or other species, clinical follow-ups and, increasingly important, bibliographic information. Beyond the process of data clustering, this knowledge provides representations that can help the scientist to address more complex questions and suggest new hypothesis, leading in our context to a clearer identification of the molecular regulation mechanisms involved in the occurrence of liver diseases and at mid-term to new diagnostic and therapeutic tools.

The required knowledge is spread world-wide and hosted on multiple heterogeneous resources. Manually navigating them to extract relevant information on a gene is highly time-comsuming and error-prone. Therefore, we have physically integrated into GEDAW a number of important sources in life science and medicine that are structured or semi-structured. Our final objective is to propose a more systematic approach to integrate data on liver genes and to organize and analyze them within a target question - which is in our case specific to an organ and a pathological state. This is a complex task, with the most challenging questions being: i) bio-knowledge representation and modeling, ii) semantic integration issues and iii) integrated bio-data analysis.

Building a scientific data warehouse to store microarray expression data is a well studied problem. Conceptual models for gene expression are for instance discussed in [18].The Genomic Unified Schema (GUS) integrates diverse life science data types including microarray data, and a support of data cleansing, data mining and complex queries analyses, thus making it quite generic [2]. The warehouse of [11] focuses on storing as possible details on the experiments and the technologies used. In GEDAW we only focus on the result of an experiment, i.e., expression measurements. No further experimental details are stored within the warehouse. The Genome Information Management System (GIMS) in which one of the authors has been participating, allows the storage and management of microarray data on the scale of a genome, making GIMS, in contrast to GEDAW, a genome-centric rather than gene-centric data warehouse [9]. Finally, [10] describe the GeneMapper Warehouse for expression data integrating a number of genomic data sources. In contrast, GEDAW has a focus on medical and "knowledge-rich" data sources.

1.1 Architecture for BioData Integration

GEDAW is a gene-centric data warehouse devoted to the study of liver pathologies using a transcriptome approach. New results from medical science on the gene being studied are extremely important to correlate gene expression patterns to liver phenotypes. To connect to this information, we take advantage of the recent standards developed in the medical informatics domain, i.e., the UMLS knowledge base [3].

GEDAW schema includes three major divisions: (i) gene and gene features along with transcripts and gene products division, (ii) expression measurements of liver genes division generated by in-house experiments and (iii), universal vocabularies and ontologies division. As illustrated in Figure 1, to store the gene expression division a local relational database has been built, as a repository of array data storing as many details as possible on the methods used, the protocols and the results obtained. It is a MIAME (Minimum Information About Microarray Experiment) compliant source [6].

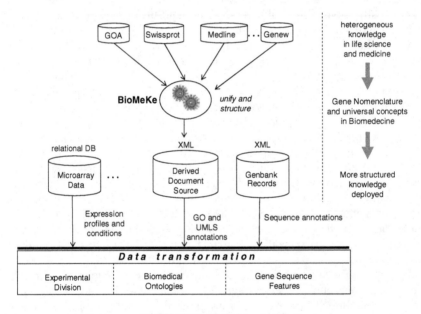

Fig. 1. GEDAW System Architecture

The sources currently integrated are spread world wide and hosted on different representation systems, each having its own schema. XML records from the GENBANK [7] have been used to populate the gene sequence features division into GEDAW.

Explicit relationships associating genes and their expression profiles with diseases are also extremely needed to understand the pathogenesis of the liver. For this purpose, we use the system BioMeKE [8,17] to curate the ontology division of each expressed gene with relative concepts in life science and medicine. The BioMEdical

Knowledge Extraction module (BioMeKE) includes the Unified Medical Language System® (UMLS) covering the whole biomedical domain, and the Gene Ontology™ (GO) that focuses on genomics. It includes additional terminologies, as that provided by the HUman Genome Organisation (HUGO) Gene Nomenclature Committee (HGNC) to resolve synonymy conflicts [19]. An XML document that annotates each gene by exploring these biomedical terminologies is derived from BioMeKE. It is then parsed and integrated into the warehouse.

1.2 Contribution

The aim of this paper is to share our experience on designing and implementing an integration process for biomedical data in the presence of syntactic and semantic conflicts. Other aspects such as biological data quality controlling, mining and refreshing will be described elsewhere.

1.3 Outline

An overview on the biological background and the questions that motivate the design of GEDAW are given in the next section. In section 3, the provenance, content and the format of the structured resources used for integration in GEDAW are described. In section 4, the integration process along with a brief schema design is presented. The data mapping rules that have been defined for instances conciliation and cleansing during the integration process are also presented. The generic interface used for queries composition and execution is tackled in section 5. Section 6 concludes and presents the perspectives of our future works.

2   Biological Background and Motivations

Transcriptome is the study of the transcriptional response of the cell to different environment conditions such as, growth factors, chemicals, foods treatments, genetic disturbance, etc. The cell may response by an excessive expression or repression of certain genes in two different situations, for example normal vs. pathologic.

2.1 Transcriptome Experiments

In the liver framework, the objective of transcriptome experiments is to emphasize both co-expressed genes and gene networks in a specific pathology within the hepatocyte.

To determine whether a single gene is expressed is a routine task for a biologist, but this process becomes more complicated because the data generated are massive. DNA-chips are indeed used and thousands of genes are deposited on a two dimensional grid. The experiment generating thousands of data points requires an efficient processing of the storage and the management of data. The key question is: which of (and why?) the deposited genes are abnormally expressed in the injured tissues? Each gene is represented by a spot, and its expression level is measured by means of the spot intensity. This same gene does have other multiple features, recorded in World Wide Web resources, and that must be considered to answer such questions.

2.2 Biomedical Issues Underlying Data Integration

To study experimental data, the scientist expects an integrated environment that captures his own experimental data enriched with information and expertise on the expressed genes. Beyond the process of clustering expression measurements in gene clusters, such an integrated environment should allow him to better focus on the scientific interpretation derived from such a clustering that reveals such clusters.

Together with the collected gene data, the integrated environment should be able to answer questions that need an integration of knowledge from the biological level to the pathological level. Below we give three types of questions that scientists frequently ask and that cannot be answered by simple SQL queries, but require the application of data mining techniques.

1 The set of genes that have seen their expression modified in a given condition?
2 Within this set, is there a subset of genes that are co-regulated?
3 What are the elements that may explain a parallel (or opposite) modulation of certain genes: membership to a functional class, homologies occurring in their peptides sequences, or in their nucleic sequences particularly in the promoting region?

Scientists may need to go thoroughly into sequences (question 3.) of the co-expressed genes for discovering common motifs, because genes sharing similar expression profiles must share transcription regulation mechanisms that include common transcription factors. They also need to go thoroughly into disease information and clinical follows-up in order to find out correlations between particular mutants' phenotypes and expression patterns. The integrated environment should also be able to answer questions such as:

1 Is there any correlation between gene expression levels and a certain pathological phenotype?
2 What is the set of genes for which a dysregulation characterizes a pathological sample by indicating a gravity level, a prognostic factor, a sensitivity level or on a contrary a resistance to a certain treatment ?

Respective genes annotations that comes from the UMLS knowledge-base and the Gene Ontology, along with gene expression profiles, are used to proceed such questions. Relative conceptual terms in both ontologies are extracted from the unified document-source, derived by BioMeKE.

2.3 GEDAW: An Object-Oriented Environment for Integrating Liver Genes Data

Considering the different integration issues previously described, an object oriented data warehouse called GEDAW (Gene Expression DAta Warehouse) has been designed for integrating and managing : i) data being produced on the expressed genes in public databanks and literature, ii) normalized experimental data produced by Microarray experiments and iii) complementary biological, genomic, and medical data.

3 Data Resources

Searching across heterogeneous distributed biological resources is increasingly difficult and time-consuming for biomedical researchers. Bioinformatics is coming to the forefront to address the problem of drawing effectively and efficiently information from a growing collection of multiple and distributed databanks. Several resources can be used to instantiate the liver warehouse GEDAW. We describe here the ones that have been selected for having the most appropriate properties, enabling a systematic extraction of gene attributes: 1) experiment resources, 2) genomic databanks and 3) ontological resources. We demonstrate for each selected resource, its provenance, content, structure and which gene attributes are being extracted.

3.1 Experimental Resources

To not burden the warehouse, a MIAME compliant relational database has been built independently (Figure2), in order to store and manage experimental microarray data [12]. This database stores as much as possible details on the microarray experiments, including the techniques used, protocols, samples and results obtained (ratios and images).

We will not go in further details concerning this database, except saying that it acts as a permanent source of expression levels delivered by in-house transcriptome experiments on injured liver tissues, and provides facilities to select and export data. Part of those data is exported to the data warehouse.

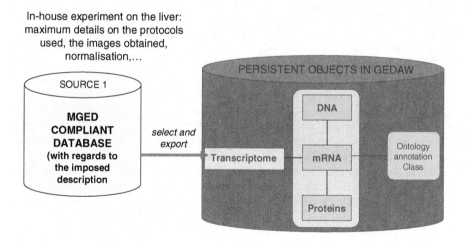

Fig. 2. An external source to manage liver transcriptome experiments

3.2 Genomic Databanks Resources

In order to perform consistent analyses on the expressed genes, the integration of the precise pre-existing annotations of their sequences is necessary. Sequence data to consider include: 1) the DNA sequence and sequence components : known promoters, known transcription binding sites, introns, exons, known regulators, 2) the mRNA sequence, sequence components and alternative transcripts and 3) functional proteins. Being conscious that an exhaustive gene annotation is available for a limited number of genes, it is however helpful to infer new knowledge on yet unknown co-expressed genes.

Data describing genomic sequences are available in several public databanks via Internet: banks for nucleic acids (DNA, RNA), banks for protein (polypeptides, proteins) such as SWISS-PROT , generalist or specialized databanks such as GENBANK, EMBL (European Molecular Biology Laboratory), and DDBJ (DNA DataBank of Japan). Each databank record describes a sequence with its several annotations.

As an example, the description of the Homosapiens Hemochromatosis gene HFE, which mutation causes a genetic liver disease having the same name is given in GENBANK. The description of this gene is available in both HTML[2] and XML[3] formats. An XML format that focused on the sequence of HFE gene is also available[4].

Each record is also identified by a unique accession number and may be retrieved by key-words. Annotations include the description of the sequence: its function, its size, the species for which it has been determined, the related scientific publications (authors and references) and the description of the regions constituting the sequence (start codon, stop codon, introns, exons, ORF, etc.). GENBANK (with more than 20 million records of different sequences) [7] is one of the first banks that propose XML format for its records with a well-defined DTD specifying the structure and the domain terminology for the records of genes and submitted sequences.

3.3 Ontological Resources

Relating genotype data on genes with their phenotype during the integration process is essential to be able to associate gene expression levels to a pathological phenotype.

Tremendous web resources provide such information for a given gene. But their heterogeneity is a major obstacle for a consistent semantic integration. They are numerous and continually evolving, the number of biomolecular entities is very large, the names of biological entities are associated with synonymy: a gene can have multiple aliases (synonyms) in addition to its official symbol, and genes that are functionally different across species may have the same name (ambiguity) [14,20], different databases organize data according to different schemas and use different vocabularies. Shared ontologies are used to conciliate and to attain as much as possible data conflicts. Various standards in life science have been developed to provide domain knowledge to be used for semantically driven integration of information from different sources.

3.3.1 Gene Ontology

Gene Ontology™ (GO) is an ontology for molecular biology and genomics [13]. The three hierarchies of GO are molecular function (F), biological process (P) and cellular component (C). GO terms are used as attributes of gene products to provide information about the molecular functions, the biological processes, and the cellular compo-

---

[2] www.ncbi.nlm.nih.gov/entrez/viewer.fcgi?db=nucleotide&val=1890179

[3] www.ncbi.nlm.nih.gov/entrez/viewer.fcgi?db=nucleotide&list_uids=1890179&dopt=xml

[4] www.ncbi.nlm.nih.gov/entrez/viewer.fcgi?db=nucleotide&list_uids=1890179&dopt=gbx

nents related to the gene product. In our context of high throughput transcriptome experiments, we use GO to annotate the genes expressed in different situations in the liver. Furthermore, GO is broadly used by public databanks to annotate genes. Therefore, it has become a standard and plays an important role in biomedical research, by making possible to draw together information from multiple resources. To illustrate with an example, to the ceruloplasmin concept (a gene involved in iron transport, having a central role in iron metabolism and is secreted in plasma by hepatocytes) is associated the set of concepts in each hierarchy of GO ontology (Table 1).

Table 1. Ceruloplasmin annotations in Gene Ontology

Molecular function	Biological process	Cellular Component
Multicopper Feoxidase iron Transport mediator	Iron homeostasis	Extracellular space

3.3.2 UMLS Knowledge Base

The UMLS is developed by the US National Library of Medicine. It comprises two major inter-related components: the Metathesaurus®, a large repository of concepts (around 900,000 concepts), and the Semantic Network, a limited network of 135 Semantic Types [3]. The Metathesaurus is built by merging existing vocabularies, including Medical Subject Headings (MeSH), which is used to index biomedical literature in MEDLINE, and GO. In the Metathesaurus, synonymous terms are clustered under a same concept, each having a Concept Unique Identifier (CUI). To the ceruloplasmin concept is associated the CUI:C0007841 and a set of synonymous terms (Table 2a) (2003AC release of the UMLS).

Although the UMLS was not specifically developed for bioinformaticists, it includes also terminologies such as the NCBI taxonomy, OMIM terminology and GO that are of great interest for biologists. It also includes the MeSH, which is used to index MEDLINE abstracts. Therefore, the UMLS is a means to integrate resources since it integrates (repetition) terminologies that are used to represent data in various resources. The second motivation is that the UMLS contains 12 million relations among the Metathesaurus concepts. The source vocabularies provide hierarchical relations. RO (Other Relation) relations associate concepts from different kinds, such as diseases and tissues, or diseases and kinds of cells. In addition, co-occurrences in MEDLINE are also represented in the UMLS [3]. The last motivation is that the UMLS includes an upper level ontology of the biomedical domain (the UMLS Semantic Network) made of 135 Semantic Types. Each Metathesaurus concept is assigned to one or more Semantic Types. Three major relations are then concerned and extracted for each concept from UMLS:

- Parent concept (Table 2b): the parents of ceruloplasmin concept illustrate hierarchical relations in UMLS.
- Related concepts in diseases (Table 2c), tissues or kind of cells.
- Co-occurrences in Medline concepts (Table 2d), each with an additional numeric frequency.

Table 2. Ceruloplasmin annotations extracted from UMLS

Synonymous	Parents concepts	Related concepts	Co-occurred Concepts in MEDLINE
Ceruloplasmin alpha(2)-Ceruloplasmin Ceruloplasmin Ferroxidase Ceruloplasmin Oxidase CP - Ceruloplasmin Fe(II):oxygen oxidoreductase ferroxidase <1>	Alpha-Globulins Acute-Phase Proteins Carrier Proteins Alpha-Globulins Metalloproteins Oxidoreductases Enzyme	Copper Menkes Kinky Hair Syndrome copper oxidase Serum Ceruloplasmin Test Ceruloplasmin Serum Decreased Ceruloplasmin measurement	Copper Iron Antioxidants Hepatolenticular Degeneration Ferritin Brain Liver Superoxide Dismutase
(a)	(b)	(c)	(d)

3.3.3 Other Resources: Terminologies

At present, an additional terminology is mainly used to manage heterogeneity in naming genes, gene products or diseases, as well as in identifying items in different databanks. Given a term or a gene symbol, lexical knowledge is needed to deal with synonyms and find the corresponding concept. Available resources in the biomedical domain include the Genew database developed by the Human Gene Nomenclature Committee to provide approved names and symbols for genes, as well as previous gene names and symbols [19].

3.3.4 Mapping Ontologies into GEDAW

The use of ontologies and terminologies terms as attributes values for genes has been made possible by the joint application project BioMeKE [17]. A local consistent support into BioMeKE system of the terminologies described above enables the extraction of respective nomenclature and conceptual terms in biology and medicine, given a gene name, a symbol, or any gene relative identifier in biomedical databanks. To navigate through these resources, a set of JAVA functions have been developed to:

- Find all the synonyms of a term and all the identifiers of a gene or gene product in Genew and the UMLS Metathesaurus,
- Provide the cross-references between a gene and a protein (e.g. SWISS-PROT ID) from Genew.
- Represent the different paths to reach the information about a gene or a gene product via all the available cross-references.
- Search for information about a gene or a gene product, i.e. the set of concepts related to this gene in GO (molecular function, biological process and cellular component) and the set of concepts related to the gene in UMLS including chemicals and drugs, anatomy, and disorders.

These annotations are then considered by the expert, filtered and stored within the warehouse for further classifications using gene expression profiles. Because the aim of this paper is not to describe BioMeKE but rather to introduce its general scope and outputs, we will not go in further details. We suggest the reader to get further details in another paper devoted to this application [8,17].

```
<biomeke_annotation>
  <biomeke_annotation_nomenclature>
  ^<seq-id_locuslink>1356</seq-id_locuslink>                          Gene
  <seq-id_hgnc>2295</seq-id_hgnc>                             nomenclature
<seq-name_hgnc>ceruloplasmin (ferroxidase)</seq-name_hgnc>
<seq-symbol_hgnc>CP</seq-symbol_hgnc> <seq-aliases_hgnc></seq-aliases_hgnc>
<seq-id_omim>117700</seq-id_omim>
<seq-id_refseq>NM_000096</seq-id_refseq>
<seq-id_swissprot>P00450</seq-id_swissprot>
  <seq-id_pubmed></seq-id_pubmed>
</biomeke_annotation_nomenclature>                                   GO
<biomeke_GO_annotation_list>                                   annotations
  <biomeke_GO_annotation-type value="molecular function">
<biomeke_GO_annotation>
<GO-accession>GO:0004322</GO-accession>
<GO-name>ferroxidase activity</GO-name>
<GO-evidence>TAS</GO-evidence> . . . etc
</biomeke_GO_annotation>
<biomeke_UMLS_annotation_list>                                     UMLS
<biomeke_UMLS_annotation-name>                                  annotations
          <UMLS_name_search> Ceruloplasmin </UMLS_name_search>
          <UMLS_CUI_search>C0007841 </UMLS_CUI_search>
</biomeke_UMLS_annotation-name>
<biomeke_UMLS_annotation-semantic-type value = " Amino Acid, Peptide, or Protein">
          <biomeke_UMLS_annotation-relation value = "Parent">
                  <biomeke_UMLS_annotation>
                  <UMLS-name>acute phase protein 2</UMLS-name>
                  </biomeke_UMLS_annotation>          . . . etc
          <biomeke_UMLS_annotation-relation value = "other relations">
                  <biomeke_UMLS_annotation>
                  <UMLS-name>Metalloproteins</UMLS-name>
                  </biomeke_UMLS_annotation>          . . . etc
          <biomeke_UMLS_annotation-relation value = "Co-occurences">
                  <biomeke_UMLS_annotation>
                  <UMLS-name>ATP phosphohydrolase</UMLS-name>
                  <UMLS-freq>4</UMLS-freq>
                  . . . etc
```

Fig. 3. BioMeKE-xml document to annotate the ceruloplasmin Gene

To annotate each expressed gene, BioMeKE delivers an XML document (Figure 3) to be parsed, transformed and stored into GEDAW within the Ontology_annotation Class. This document-source standing as a structured data source derived by BioMeKE.

4   Bio-data Integration

Designing a single schema that integrates syntactically and semantically the whole heterogeneous life science data sources is still a challenging question. Integrating the source schemas is presently the most commonly used approach in literature [15]. By restricting ourselves to structured or semi-structured data sources, we have been able to use a schema mapping approach with the GAV paradigm [16]. In our context, schema mapping is the process of transforming data conforming to a source schema to the corresponding warehouse schema by the definition of a set of mapping rules. The data sources include : i) GENBANK for the genomic features of the genes recorded in XML format, ii) conceptual annotations derived from the biomedical ontologies and terminologies using BioMeKE outputs as XML documents, iii) and gene expression measurements selected from the in-house relational database.

By using a mapping approach from one source at a time, we have minimized as much as possible the problem of identification of equivalent attributes between sources, whereas the problem of duplicate detection is still important. Identifying identical objects in the biomedical domain is a complex problem, since in general the meaning of "identity" cannot be defined properly. In most applications, even the identical sequences of two genes in different organisms are not treated as a single object. In GENBANK, each sequence is treated as an entity in its own, since it was derived using a particular technique, has particular annotation, and could have individual errors. For example, there are more than 10 records for the same DNA segment of the HFE gene. Thus, classical duplicate detection methods [22] do not suffice. Duplicate detection and removal is usually performed either using a simple similarity threshold approach, as in the case of GEDAW, or based on manual intervention for each single object, such as in RefSeq. Data submission to public biological databanks is often a rather unformalized process that usually does not include name standardization or data quality controls. Erroneous data may be easily entered and cross-referenced. Even if a tool like LocusLink[5] proposes a cluster of records, across different biological databanks, as being semantically related, biologists still must validate the correctness of the clustering and resolve value differences among the records.

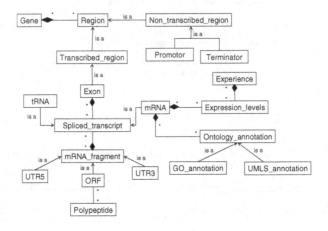

Fig. 4. GEDAW UML Conceptual schema

In GEDAW, a unique schema (Figure 4) has been defined to describe different aspects of a gene, to which has been added an ontological annotation class associated to each gene transcript. The stored ontological annotations represent the more specialized concepts associated to the genes. The ontology annotation class used for storing the terms from both medical and biological terminologies includes attributes like: ontology and annotation type along with category, value and description attributes of a term. These attributes are extracted by parsing the XML files delivered by BioMeKE. At the schema-level, the problem of format heterogeneity makes necessary to trans

---

[5] www.ncbi.nlm.nih.gov/LocusLink

form data, so that they conform to the data model used by our warehousing system. Information sources consist of sets of XML files, while the GEDAW target schema is object-oriented. This translation problem is inherent in almost all data integration approaches, but becomes much more complex in the biological domain because the potentially different (and not formalized yet) biological interpretations of schema elements and the fact that, together with the current state of knowledge, schemas and interpretations tend to evolve quickly and independently in the different sources.

In order to define an appropriate data aggregation of all the available information items, data conflicts have to be resolved using rules for mapping the source records and conciliating different values recorded for a same concept. Mapping rules are defined to allow the data exchange from the public databanks into GEDAW (Figure 5). Apart from experimental data, public information items are automatically extracted by scripts using the DTD (Document Type Definition) of the data source translated into the GEDAW conceptual data model.

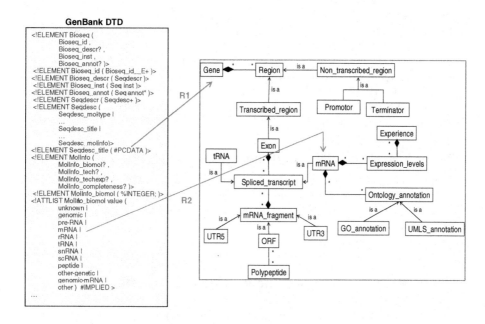

Fig. 5. Example of mapping rules between GENBANK DTD and GEDAW schema

Three categories of mapping rules are proposed: 1) structural mapping rules, 2) semantic mapping rules and 3) cognitive mapping rules according to the different knowledge levels and perspectives for biological interpretation.

The structural mapping rules are defined at the schema level according to the GEDAW model by identifying the existing correspondences with relevant DTD elements (e.g., the *Seqdesc_title* element in GENBANK DTD is used to extract the name "name" of the gene and the *MolInfo_biomol* value its type of molecule with respectively structural mapping rules R1 and R2 in Figure 5). Then, the records of interest are selectively structured and data are extracted.

Semantic and cognitive mapping rules are used for data unification at the instance level: several rules may use available tools for determining analogies between homologous data (such as sequence alignment, for example): the result of the BLAST algorithm (implemented in a set of similarity search programs for Basic Local Alignment Search Tool) allows considering that two sequences match. The nomenclature section provided by BioMeKE (Figure 3) is also considerably used to conciliate duplicate records. More semantic mapping rules have been built using this information during the process of integration. For example, the Locus-ID is used to cluster submitted sequences associated to a same gene (cross-referenced in LocusLink) and the official gene name along with its aliases to relate different gene appearance with different names, in literature for example.

Let us consider three distinct selectively structured records we may obtain from GENBANK databank by querying the DNA sequence for gene HFE. A first record identified by the accession number AF204869 describes a partial sequence (size = 3043) of the HFE gene with no annotation but one relevant information item about the position of the promoter region. A second record identified by the accession number AF184234 describes a partial sequence (size = 772) of the protein precursor of HFE gene with a detailed but incomplete annotation. The third record identified by the accession number Z92910 describes the complete sequence (size = 12146) of the HFE gene with a complete annotation. In this example, BLAST(sequence(Z92910), sequence(AF184234))=100% indicates the sequence in both records are perfectly homologous and can be merged. Cognitive mapping rules may be used in this example for conciliating data such as:

R3 : Descriptive Inclusion: record(Z92910) contains record(AF184234)
R4 : Position Offset: position(Z92910.exon)=6364+position(AF184234.exon)

In our context a liver cDNA microarray corresponding to 2479 cDNA clones spotted onto glass slides has been designed. The data unification process described above has lead to identify 612 distinct genes on the 2479 deposited clones. A complete integration of 10 hybridization experiments took around one day runtime, with around 11 Mbytes charged database size.

5   Integration Results Construction and User Interface

Now to recapitulate, the integration process of transcriptomic data into GEDAW is operated in four steps. During the first step, to the probes (or clones) used by in-house experiments, is associated a set of gene names, in terms of accession numbers of similar sequences in GENBANK along with textual descriptions. The second step is in charge of selecting the set of experiments for which the researcher wishes to integrate and analyse the experiments results, and then of loading expression levels measured for these genes. For each gene having its expression levels in different physiopathological situations already stored in GEDAW, the full annotation of the sequence associated to this gene is loaded from GENBANK by XML transformation to Objects. BioMeKE is launched in Step 4 to bring for each integrated gene its nomenclature and its ontological annotations in life science from Gene Ontology and in medicine from UMLS. In step 5, the results are delivered to the expert, for a filtering phase using either predefined mapping rules, output nomenclature, or simply his expertise, to eliminate duplicate records of genes.

When the user poses a query, the whole integration results for each gene are brought in. Further refinements on these data can be operated, by selecting for example genes having expression levels between a minimum value and a maximum value, those belonging to a given biological process or co-occurring in Medline with a given concept, or having a known motif in their mRNA sequences and co-located on a same chromosome. It could be also a conjunction of these criteria. In Figure 6, we show an example of a query composed in the generic java-based interface we have developed for GEDAW. Resulting sets are presently browsed using either FastObjects interface, or delivered as Textfiles to the expert for further analyses.

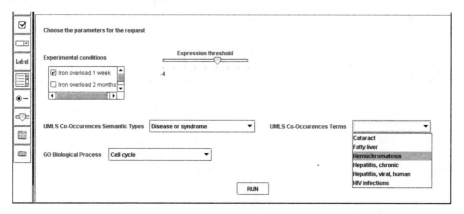

Fig. 6. Example of Query Composition

6   Conclusion

The GEDAW system presented in this paper allows massive importation of biological and medical data into an object-oriented data warehouse that supports transcriptome analyses specific to the human liver. This paper focused on the relevant genomic, biological and medical resources that have been used to build GEDAW. The integration process of the full sequence annotations of the genes expressed is described. It is performed by parsing and cleaning the corresponding XML description in GENBANK, transforming the recorded genomic items to persistent objects and storing them in the warehouse. This process is almost systematic because another aspect related to the conciliation of duplicate records has been added. Elements of formalization of expertise rules for mapping such data were given. This ongoing work is still a difficult problem in information integration in life science and has not yet satisfied answers by classical solutions proposed in existing mediation systems.

In order to lead strong analysis on expressed genes and correlate expression profiles to liver biology and pathological phenotype, a second way of annotation has been added to the integration process. We chose to integrate Gene Ontology, due to its available biological annotations in the most used bio-computer resources, mainly Swissprot, GENBANK, Ensembl, TrEMBL and LocusLink databanks. It is also referenced in other relevant ontologies, like MGED [21]. More important is our consideration during integration of the medical annotations of the genes from UMLS, a well considered knowledge base in Medical Informatics [3,4,5]. These ontological annotations have been delivered by BioMeKE within the semi-structured document source BioMeKE-xml. Also, because a gene may have different appearances with different names in several bio-data banks and literature the approved nomenclature of the gene and its synonyms have been collected in BioMeKE-xml. This information is also a pre-requisite to resolve the problem of duplicate records.

An exhaustive integrated tool that facilitates access to diverse data on the expressed genes is then provided to the researcher. Intensive querying of the integrated database using OQL queries has been conducted with multiple criteria on genes attributes. Current investigations are focusing on the application of advanced data mining techniques for a combined analysis of expression levels on genes with enriched annotations, and functional similarities are likely to reveal authentic clusters of genes.

With regards to the limits of our warehousing approach, it is relevant as long as data integration from the heterogeneous sources in Biomedicine and their refreshment in the warehouse stay feasible automatically and with a reasonable performance. One argument in favor of actually storing data in GEDAW instead of dynamically linking to the corresponding sources concerns reproducibility purposes, i.e., being able to analyze several gene expression data in reference to the same domain knowledge at different times. BioMeKE system provides domain knowledge useful for acquiring information from diverse resources. It is intended to be an ontology-based mediation system that continuously supplies the gene expression warehouse with a homogeneous access to multiple data sources in Biomedicine. A filtering task is nevertheless performed by the expert on the delivered annotations before their storage in the warehouse by using multiple criteria, like the frequency information of a concept co-occurrences in Medline.

The standard ontologies such as GO and UMLS continue to evolve. They are physically supported by BioMeKE system rather than accessed via the web, making possible their refinement to expert knowledge in specific sub-domains like the liver or the iron metabolism. An interesting point to quote is the acquisition of news concepts and relationships from the analyses operated on the transcriptome data. Expressive and formal representation of this new biomedical knowledge will then be gradually added to the domain, allowing the expansion of queries on transcriptomic data.

Acknowledgements. This work was supported by grants from Region Bretagne (20046805) and inter-EPST. Emilie Guérin was supported by a MRT fellowship and grants from Region Bretagne.

References

[1]   Achard, F., Vaysseix, G. and Barillot, E. (2001) XML, bioinformatics and data integration, Bioinformatics, 17(2), 115-125.

[2]   Babenko V, Brunk B, Crabtree J, Diskin S, Fischer S, Grant G, Kondrahkin Y, Li L, Liu J, Mazzarelli J, Pinney D, Pizarro A, Manduchi E, McWeeney S, Schug J, Stoeckert C.(2003) GUS The Genomics Unified Schema A Platform for Genomics Databases. http://www.gusdb.org/

[3]   Bodenreider O. The Unified Medical Language System (UMLS): integrating biomedical terminology. Nucleic Acids Res. 2004 Jan 1;32(Database issue):D267-70.

[4]   Bodenreider O, Burgun A. Aligning Knowledge Sources in the UMLS: Methods, Quantitative Results, and Applications. Medinfo. 2004;2004:327-31.

[5]   Bodenreider O, Mitchell JA, McCray AT. (2002) Evaluation of the UMLS as a terminology and knowledge resource for biomedical informatics. Proc AMIA Symp. 2002; : 61-5.

[6]   Brazma A, Hingamp P, Quackenbush J, Sherlock G, Spellman P, Stoeckert C, Aach J, Ansorge W, Ball CA, Causton HC, Gaasterland T, Glenisson P, Holstege FC, Kim IF, Markowitz V, Matese JC, Parkinson H, Robinson A, Sarkans U, Schulze-Kremer S, Stewart J, Taylor R, Vilo J, Vingron M. Minimum information about a microarray experiment (MIAME)-toward standards for microarray data. Nat Genet. 2001 Dec;29 (4):365-71.

[7]   Benson D.A, Karsch-Mizrachi I, Lipman D.J, Ostell J, and Wheeler D.L. GENBANK: update, Nucleic Acids Res., Jan 2004; 32: 23 - 26.

[8]   Burgun A, Bodenreider O, Le Duff F, Moussouni F, Loréal O. Representation of roles in biomedical ontologies : a case study in functional genomics. JAMIA (supl), Proc. AMIA 2002 Symp, 86-90

[9]   Cornell M, Paton NW, Wu S, Goble CA, Miller CJ, Kirby P, Eilbeck K, Brass A, Hayes A, Oliver SG (2001) GIMS - a data warehouse for storage and analysis of genome sequence and functional data. Proc. 2nd IEEE International Symposium on Bioinformatics and Bioengineering (BIBE) 15-22.

[10] Do, H.-H. and Rahm, E. (2004). "Flexible Integration of Molecular-biological Annotation Data: The GenMapper Approach". EDBT'04, Heraklion, Greece, Springer LNCS.

[11] Fellenberg K, Hauser N.C, Brors B, Hoheisel J.D, and Vingron M. Microarray data warehouse allowing for inclusion of experiment annotations in statistical analysis, Bioinformatics, Mar 2002; 18: 423 - 433.

[12] Guerin E., Marquet G., Moussouni F., Burgun A., Mougin F., Loréal O. Deployment of heterogeneous ressources of genomic, biological and medical knowledge on the liver to build a datawarehouse. Proc. ECCB 2003, pp. 59-60

[13] Harris MA et. al. Gene Ontology Consortium. The Gene Ontology (GO) database and informatics resource. Nucleic Acids Res. 2004 Jan 1;32(Database issue):D258-61.

[14] Kashyap V, Sheth A. (1996) Schematic and semantic similarities between database objects: a context –based approach. Int. J. Very Large Data Bases, 5(4): 276-304

[15] Lakshmanan L, Sadri F, Subramanian I, : On the logical Foundation of Schema Integration and Evolution in Heterogeneous Database Systems. DOOD International Conference (1993) 81-100

[16] Maurizio Lenzerini. Data integration: a theoretical perspective. In Proc. of PODS 2002.

[17] Marquet G, Burgun A, Moussouni F, Guerin E, Le Duff F, Loreal O. BioMeKE: an ontology-based biomedical knowledge extraction system devoted to transcriptome analysis. Stud Health Technol Inform. 2003;95:80-5.

[18] Paton N.W, Khan S.A, Hayes A, Moussouni F, Brass A, Eilbeck K, Goble C.A, Hubbard S.J, and Oliver S.G. Conceptual modelling of genomic information, Bioinformatics, Jun 2000; 16: 548 - 557.

[19] Povey S, Lovering R, Bruford E, Wright M, Lush M, Wain H. (2001) The HUGO Gene Nomenclature Committee (HGNC).Hum Genet.;109(6):678-80

[20] Tuason O, Chen L, Liu H, Blake JA, Friedman C.(2004) Biological nomenclatures: a source of lexical knowledge and ambiguity. Pac Symp Biocomput. 2004;:238-49.

[21]   MGED Microarray Gene Expression Data (MGED). A guide to microarray experiments – an open letter to the scientific journals. Lancet. 2002 Sep 28;360(9338):1019

[22]   Galhardas, H., Florescu, D., Sasha, D., Simon, E. and Saita, C.-A. (2001). "Declarative Data Cleaning: Model, Language, and Algorithms". 27th Conference on Very Large Database Systems, Rome, Italy.

Information Integration and Knowledge Acquisition from Semantically Heterogeneous Biological Data Sources

Doina Caragea[1,4], Jyotishman Pathak[1,4], Jie Bao[1,4], Adrian Silvescu[1,4], Carson Andorf[1,3,4], Drena Dobbs[2,3,4], and Vasant Honavar[1,2,3,4]

[1] AI Research Laboratory, Department of Computer Science, 226 Atanasoff Hall
[2] Department of Genetics, Development and Cell Biology, 1210 Molecular Biology
[3] Bioinformatics and Computational Biology Program, 2014 Molecular Biology
[4] Computational Intelligence, Learning and Discovery Program,
214 Atanasoff Hall Iowa State University, Ames, IA 50011
honavar@cs.iastate.edu

Abstract. We present INDUS (Intelligent Data Understanding System), a federated, query-centric system for knowledge acquisition from autonomous, distributed, semantically heterogeneous data sources that can be viewed (conceptually) as tables. INDUS employs ontologies and inter-ontology mappings, to enable a user or an application to view a collection of such data sources (regardless of location, internal structure and query interfaces) as though they were a collection of tables structured according to an ontology supplied by the user. This allows INDUS to answer user queries against distributed, semantically heterogeneous data sources without the need for a centralized data warehouse or a common global ontology. We used INDUS framework to design algorithms for learning probabilistic models (e.g., Naive Bayes models) for predicting GO functional classification of a protein based on training sequences that are distributed among SWISSPROT and MIPS data sources. Mappings such as EC2GO and MIPS2GO were used to resolve the semantic differences between these data sources when answering queries posed by the learning algorithms. Our results show that INDUS can be successfully used for integrative analysis of data from multiple sources needed for collaborative discovery in computational biology.

1 Introduction

Ongoing transformation of biology from a data-poor science into an increasingly data-rich science has resulted in a large number of autonomous data sources (e.g., protein sequences, structures, expression patterns, interactions). This has led to unprecedented, and as yet, largely unrealized opportunities for large-scale collaborative discovery in a number of areas: characterization of macromolecular sequence-structure-function relationships, discovery of complex genetic regulatory networks, among others.

B. Ludäscher and L. Raschid (Eds.): DILS 2005, LNBI 3615, pp. 175–190, 2005.

Biological data sources developed by autonomous individuals or groups differ with respect to their ontological commitments. These include assumptions concerning the *objects* that exist in the *world*, the *properties* or *attributes* of the objects, *relationships* between objects, the possible *values* of attributes, and their *intended meaning*, as well as the *granularity* or *level of abstraction* at which objects and their properties are described [17]. Therefore, *semantic differences* among autonomous data sources are simply unavoidable. Effective use of multiple sources of data in a given context requires reconciliation of such semantic differences. This involves solving a data integration problem. Development of sound approaches to solving the information integration problem is a prerequisite for realizing the goals of the Semantic Web as articulated by Berners-Lee et al. [5]: seamless and flexible access, integration and manipulation of semantically heterogeneous, networked data, knowledge and services.

Driven by the semantic Web vision, there have been significant community-wide efforts aimed at the construction of ontologies in life sciences. Examples include the Gene Ontology (www.geneontology.org) [2] in biology and Unified Medical Language System (www.nlm.nih.gov/research/umls) in heath informatics. Data sources that are created for use in one context often find use in other contexts or applications (e.g., in collaborative scientific discovery applications involving data-driven construction of classifiers from semantically disparate data sources [9]). Furthermore, users often need to analyze data in different contexts from different perspectives. Therefore, there is no single privileged ontology that

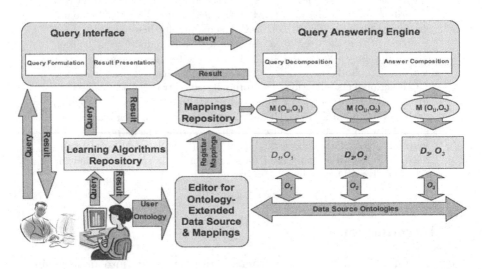

Fig. 1. INDUS: a system for information integration and knowledge acquisition from semantically heterogeneous distributed data. Queries posed by the user are answered by a query answering engine, which uses mappings between the user ontology and the data source ontologies to resolve semantic differences. A user-friendly editor is used to specify ontologies and mappings between ontologies

can serve all users, or for that matter, even a single user, in every context. Effective use of multiple sources of data in a given context requires flexible approaches to reconciling such semantic differences from the user's point of view.

Against this background, we have investigated a federated, query-centric approach to information integration and knowledge acquisition from distributed, semantically heterogeneous data sources, from a user's perspective. The choice of the federated, query-centric approach was influenced by the large number and diversity of data repositories involved, together with the user-specific nature of the integration tasks that need to be performed. Our work has led to INDUS, a system for information integration and knowledge acquisition (see Figure 1). INDUS relies on the observation that both the information integration and knowledge acquisition tasks can be reduced to the task of answering queries from distributed, semantically heterogeneous data sources. We associate ontologies with data sources and users and show how to define mappings between them. We exploit the ontologies and the mappings to develop sound methods for flexibly querying (from a user perspective) multiple semantically heterogeneous distributed data sources in a setting where each data source can be viewed (conceptually) as a single table [10, 9].

The rest of the paper is organized as follows: Section 2 introduces the problem we are addressing more precisely through an example. Section 3 describes the design and the architecture of INDUS. Section 4 demonstrates how INDUS can be used for knowledge acquisition tasks using as an example a simple machine learning algorithm (Naive Bayes). We end with conclusions, discussion of related work and directions for future work in Section 5.

2   Illustrative Example

The problem that we are wish to address is best illustrated by an example. Consider several biological laboratories that independently collect information about *Protein Sequences* in connection to their *Structure* and *Function*. Suppose that the data D_1 collected by a first laboratory contains human proteins and it is described by the attributes *Protein ID, Protein Name, Protein Sequence, Prosite Motifs* and *EC Number* (stored as in Table 1). The data D_2 collected by a second laboratory contains yeast proteins and it is described by the attributes *Accession Number AN, Gene, AA Sequence, Length, Pfam Domains,* and *MIPS Funcat* (stored as in Table 2). A data set D_3 collected by a third laboratory contains both human and yeast proteins and it is described by the attributes *Entry ID, Entry Name, Organism, CATH Domains* and *CATH Classes* corresponding to the domains (stored as in Table 3).

Consider a biologist (user) U who wants to assemble a data set based on the data sources of interest D_1, D_2, D_3, from his or her own perspective, where the representative attributes are *ID, Source, AA composition* (a.k.a. amino acid distribution, i.e. number of occurrences of each amino acid in the amino acid sequence corresponding to the protein), *Structural Classes* and *GO Function*. This requires the ability to manipulate the data sources of interest from the

Table 1. Data D_1 containing human proteins collected by a laboratory Lab_1

Protein ID	Protein Name	Protein Sequence	Prosite Motifs	EC Number
P35626	Beta-adrenergic receptor kinase 2	MADLEAVLAD VSYLMAMEKS ...	RGS PROT_KIN_DOM PH_DOMAIN	2.7.1.126 Beta-adrenergic receptor kinase
Q12797	Aspartyl/asparaginyl beta-hydroxylase	MAQRKNAKSS GNSSSGSGS ...	TPR TPR_REGION TRP	1.14.11.16 Peptide-aspartate beta-dioxygenase
Q13219	Pappalysin-1	MRLWSWVLHL GLLSAALGCG ...	SUSHI	3.4.24.79 Pappalysin-1
...	...	...	...	...

Table 2. Data D_2 containing yeast proteins collected by a laboratory Lab_2

AN	Gene	AA Sequence	Length	Pfam Domains	MIPS Funcat
P32589	SSE1	STPFGLDLGN NNSVLAVARN ...	692	HSP70	16.01 protein binding
P07278	BCY1	VSSLPKESQA ELQLFQNEIN ...	415	cNMP_binding RIIa	16.19.01 cyclic nucleotide binding (cAMP, cGMP, etc.)
...	...	...	...	...	...

Table 3. Data D_3 containing human and yeast proteins collected by a laboratory Lab_3

Entry ID	Entry Name	Organism	CATH Domains	CATH Classes
P35626	ARK2_HUMAN	Human	1omwB0 1omwG0	Mainly beta Few Sec. Struct.
Q12797	ASPH_HUMAN	Human	not known	not known
Q13219	PAPPA_HUMAN	Human	1jmaB1 1jmaB2	Mainly beta Mainly beta
P32589	HS78_YEAST	Yeast	1dkgA1 1dkgA2 1dkgB1	Alpha beta Mainly alpha Alpha beta
P07278	KAPR_YEAST	Yeast	1cx4A1 1dkgA2	Alpha beta Alpha beta
...	...	...	...	...

user's perspective. However, the three data sources differ in terms of semantics from the user's perspective. In order to cope with this heterogeneity of semantics, the user must observe that the attributes *Protein ID*, *Accession Number* and *Entry ID*, in the three data sources of interest, are similar to the user attribute

ID; the attribute *Protein Sequence* in the first data source and the attribute *AA Sequence* in the second data source are also similar and they can be used to infer the user attribute *AA Composition* (by counting the number of occurrences of each amino acid in the corresponding AA sequence); similarly, the attributes *EC Number* and *MIPS Funcat* are similar to the user attribute *GO Function*; finally, the attributes *Organism* and *CATH Classes* in the third data source are similar to the attributes *Source* and *Structural Classes* in the user view.

Therefore, to assemble the user data, one would need to project the data in D_1 (with respect to the attributes *Protein ID, Protein Sequence* and *EC Number*) and the data in D_2 (with respect to *AN, AA Sequence,* and *MIPS Funcat*) and take the union D_{12} of the resulting sets; then the third data set D_3 needs to be projected with respect to the attributes *Entry Name, Organism,* and *CATH Classes*. The cross-product with respect to the common attribute *ID*, between D_{12} and D_3 represents the data that the user is interested in. Notice that all these operations can be written as a query whose result is $D_U = (project(D_1) \cup project(D_2)) \times project(D_3)$. However, before the query can be executed, the semantic differences between values of similar attributes must be resolved.

To establish the correspondence between values that two similar attributes can take, we need to associate types with attributes and map the domain of the type of an attribute to the domain of the type of the corresponding attribute (e.g., *AA Sequence* to *AA Composition* or *EC Number* to *GO Function*). We assume that the type of an attribute can be a standard type such as a collection of values (e.g., amino acids, Prosite motifs, etc.), or it can be given by a simple hierarchical ontology (e.g., species taxonomy). Figure 2 shows examples of (simplified) attribute value hierarchies for the attributes *EC Numbers, MIPS Funcat,* and *GO Function* in the data sources D_1, D_2 and the user perspective.

Examples of semantic correspondences in this case could be: *EC 2.7.1.126* in D_1 is equivalent to *GO 0047696* in D_U, *MIPS 16.01* in D_2 is equivalent to *GO 0005515* in D_U and *MIPS 16.19.01* is equivalent to GO *0016208* in D_U. On the other hand, *EC 2.7.1.126* in D_1 is lower than (i.e., hierarchically below) *GO 0004672* in D_U, or for that matter *EC 2.7.1.126* is higher than *GO0004672*. Similarly, *MIPS 16.19.01* in D_2 is lower than *GO 0017076* in D_U, and so on. Therefore the integrated user data D_U could look like in Table 4, where the semantic correspondences have been applied.

In general, the user may want to answer queries such as *the number of human proteins that are involved in kinase activity* from the integrated data or even to infer models based on the data available in order to use them to predict useful information about new unlabeled data (e.g., protein function for unlabeled proteins). INDUS, the system that we have developed in our lab, can be used to answer such queries against distributed, semantically heterogeneous data sources without the need for a centralized data warehouse or a common global ontology. We will describe INDUS in more detail in the next section.

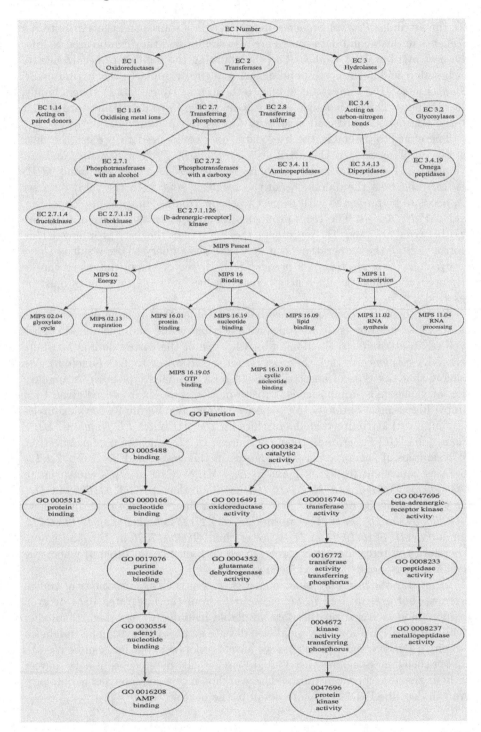

Fig. 2. Ontologies associated with the attributes *EC Number*, *MIPS Funcat* and *GO Function* that appear in the data sources of interest D_1, D_2 and D_U

Table 4. Integrated user data D_U

ID	Source	AA composition	Struct. Classes	GO Funct. Class
P35626	Human	7 3 9 14 $\cdots$	Mainly beta Few Sec. Struct.	0047696:beta-adrenergic-receptor kinase activity
Q12797	Human	5 1 7 12 $\cdots$	not known	0004597: peptide-aspartate beta-dioxygenase activity
Q13219	Human	10 8 6 15 $\cdots$	Mainly beta Mainly beta	0008237: metallopeptidase activity
P39708	Yeast	13 17 18 11 $\cdots$	Alpha beta Alpha beta Mainly alpha	0005515: protein binding
Q01574	Yeast	23 16 8 1 $\cdots$	Mainly alpha Mainly alpha	0016208: AMP binding
$\cdots$	$\cdots$	$\cdots$	$\cdots$	$\cdots$

3   INDUS Design and Architecture

A simplified version of INDUS architecture is shown in Figure 1. As can be seen, several related distributed and semantically heterogeneous *data sources* (servers) can be available to *users* (clients) who may want to query the data sources through a *query interface*. Each user has his or her own view of the domain of interest reflected by a *user ontology*. The system provides default user ontologies (e.g., *GO Function*) and mappings from the data source ontologies to the user ontology (e.g., from *AA Sequence* to *AA Composition* or from *EC Number* to *GO Function*) in a *mapping repository*. However, a user-friendly *ontology and mapping editor* is also available for users if they need to design or modify their own ontologies or mappings (for example, if they need to explore different mappings such as *AA Sequence* to *AA composition* or *AA sequence* to *hydrophobic* versus *hydrophilic AA Composition*).

Once a query is posed by the user, it is sent to a *query answering engine* which acts as a middleware between clients and servers. The query answering engine has access to the data sources in the system and also to the set of mappings available. Thus, when the query answering engine receives a user query, it decomposes this query according to the distributed data sources, maps the individual queries to the data source ontologies, then it composes the results to sub-queries into a final result that is sent back to the user.

The main features of INDUS include:

(1) A clear distinction between data and the semantics of the data: this makes it easy to define mappings from data source ontologies to user ontologies.

(2) User-specified ontologies: each user can specify his or her ontology and mappings from data source ontologies to the user ontology; there is no single global ontology.

(3) A user-friendly ontology and mappings editor: this can be easily used to specify ontologies and mappings; however, a predefined set of ontologies and mappings are also available in a repository.

(4) Knowledge acquisition capabilities: if the information requirements of an algorithm for knowledge acquisition from data (e.g., learning algorithm) can be formulated as statistical queries [10], then such an algorithm can be easily linked to INDUS, making it an appropriate tool for information integration as well as knowledge acquisition tasks.

Some of these features are shared by other systems developed independently, e.g., BioMediator [25]. In the remaining of this section we describe the first three features into more detail, while in the next section we show how INDUS can be used to infer Naive Bayes models.

3.1   Ontology Extended Data Sources

Suppose that the data of interest are distributed over the data sources $D_1, \cdots, D_p$, where each data source D_i contains only a fragment of the whole data D.

Let D_i be a distributed data set described by the set of attributes $\{A_1^i, \cdots, A_n^i\}$ and $O_i = \{\Lambda_1^i, \cdots, \Lambda_n^i\}$ an ontology associated with this data set. The element $\Lambda_j^i \in O_i$ corresponds to the attribute A_j^i and defines the type of that particular attribute. The type of an attribute can be a standard type (e.g., types such as Integer or String; the enumeration of a set of values such as Prosite motifs; etc.) or a hierarchical type, which is defined as an ordering of a set of terms (e.g., the values of the attribute EC number) [6]. Of special interest to us are *isa* hierarchies over the values of the attributes that describe a data source, also called *attribute value taxonomies* (see Figure 2).

The schema S_i of a data source D_i is given by the set of attributes $\{A_1^i, \cdots, A_n^i\}$ used to describe the data together with their respective types $\{\Lambda_1^i, \cdots, \Lambda_n^i\}$ defined by the ontology O_i, i.e., $S = \{A_1 : \Lambda_1, \cdots, A_n : \Lambda_n\}$. We define an *ontology-extended data source* as a tuple $\mathcal{D}_i = <D_i, S_i, O_i>$, where D_i is the actual data in the data source, S_i is the schema of the data source and O_i is the ontology associated with the data source. In addition, the following condition needs also to be satisfied: $D_i \subseteq \Lambda_1^i \times \cdots \times \Lambda_n^i$, which means that each attribute A_j^i can take values in the set Λ_j^i defined by the ontology O_i.

3.2   User Perspective

Let $<D_1, S_1, O_1>, \cdots, <D_p, S_p, O_p>$ be an ordered set of p ontology-extended data sources and U a user that poses queries against the heterogeneous data sources $D_1, \cdots, D_p$. A user perspective is given by a user ontology O_U and a set of semantic correspondences SC between terms in $O_1, \cdots, O_p$, respectively, and terms in O_U. The semantic correspondences can be at attribute level (or schema level), e.g., $A_j^i : O_i \equiv A_l^U : O_U$, or at attribute value level (or attribute type level), e.g., $x:O_i \leq y:O_U$ (x is semantically subsumed by y), $x:O_i \geq y:O_U$ (x semantically subsumes y), $x:O_i \equiv y:O_U$ (x is semantically equivalent to y), $x:O_i \neq y:O_U$ (x is semantically incompatible with y), $x:O_i \approx y:O_U$ (x is semantically compatible with y) [7,21].

We say that a set of ontologies $O_1, \cdots, O_p$ are integrable according to a user ontology O_U in the presence of the semantic correspondences SC if there exist p partial injective mappings $\psi_1, \cdots, \psi_p$ from $O_1, \cdots, O_p$, respectively, to O_U with the following two properties [9, 6]:

(a) For all $x, y \in O_i$, if $x \preceq y$ in O_i then $\psi_i(x) \preceq \psi_i(y)$ in O_U (order preservation property);
(b) For all $x \in O_i$ and $y \in O_U$, if $(x : O_i \ op \ y : O_U) \in SC$, then $\psi_i(x) \ op \ y$ in the ontology O_U (semantic correspondence preservation property).

In general, the set of mappings can be (semi-automatically) inferred from the set of semantic correspondences specified by the user [9].

3.3   Ontology-Extended Data Sources and Mappings Editor

In many practical data integration scenarios, the ontologies associated with data sources are not explicitly specified in a form that can be manipulated by programs. In such cases, it is necessary to make *explicit*, the implicit ontologies associated with the data sources before data integration can be performed. In addition, users need to be able to specify the user ontology and the semantic correspondences between user ontology and data source ontologies (used later to generate a set of semantics preserving mappings). To address this need, we have developed a user-friendly editor for editing data source descriptions (associated with ontology extended data sources) and for specifying the relevant semantic correspondences (a.k.a., interoperation constraints).

The current implementation of our data source editor provides interfaces for:

(a) Defining attribute types or *isa* hierarchies (attribute value taxonomies) or modifying a predefined set of attribute types.
(b) Defining the schema of a data source by specifying the names of the attributes and their corresponding types.
(c) Defining semantic correspondences between ontologies associated with the data sources and the user ontology.
(d) Querying distributed, semantically heterogeneous data sources and retrieving and manipulating the results according to the user-imposed semantic relationships between different sources of data.

Figure 3 shows the interface that allows specification of semantic correspondences between two data sources. The leftmost panel shows an ontology extended schema associated with a data source, which includes the hierarchical type ontologies associated with attributes. The second panel shows the available semantic correspondences. The third panel shows the ontology extended schema associated with the user data. The user can select a term in the first schema, the desired semantic correspondence, and a term in the second schema. The user-specified semantic correspondences that are used to infer consistent mappings-specified are shown on the rightmost panel. The ontologies and mappings defined using the user-friendly editor in INDUS are stored in a repository

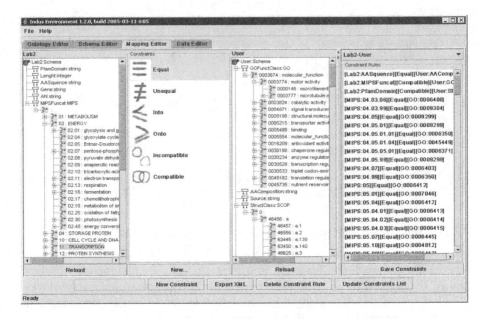

Fig. 3. Editor for defining ontology-extended data sources and semantic correspondences between two ontology-extended data sources

that is available to the query answering engine. INDUS contains a list of predefined mappings (e.g., mappings from *EC Number* to *GO Function* or from *AA Sequence* to *AA Composition*). Some of these functions are procedural (e.g., procedure that maps an *AA Sequence* to *AA Composition*), others represent the enumeration of a list of mappings between values (e.g., *EC Number* to *GO Function*). Furthermore, the user is given the freedom to define new mappings or modify the existing ones according to his or her own needs. For example, if the user wants to map *AA Sequence* to *AA Composition* and this mapping does not exist in the repository, then the user can easily upload the corresponding procedure through the editor interface. Also if a user decides to use a modified version of a pre-defined mapping function, that particular function can be loaded into the editor from the repository and edited according to the user needs.

4   Learning Classifiers for Assigning Protein Sequences to Gene Ontology Functional Families

Caragea et al. [10] have shown that the problem of learning classifiers from distributed data can be reduced to the problem of answering queries from distributed data by decomposing the learning task into an information integration component in which the information needed for learning (i.e., *sufficient statistics*) is identified and gathered from the distributed data and a hypothesis generation component, which uses this information to generate a model.

Assigning putative functions to novel proteins and the discovery of sequence correlates of protein function are important challenges in bioinformatics. In what follows, we will show how a biologist interested in learning models for predicting the *GO Function* of unlabeled proteins based on data coming from SWISSPROT and MIPS databases, can use the tools provided by INDUS to achieve this task.

4.1   Data and Problem Specification

We consider again the data sources described in our illustrative example. Because the user is interested in learning to predict the *GO Function* of a protein based on the information contained in the amino acid sequence, the data of interest to the user can be seen as coming from two horizontal fragments as in Table 5 (where the data set D_1 is assembled from SWISSPROT and the data set D_2 is assembled from MIPS).

Table 5. Horizontal data fragments that are of interest to a biologist

	Protein ID	Protein Sequence	EC Number
	P35626	MADLEAVLAD VSYLMAMEKS $\cdots$	2.7.1.126 Beta-adrenergic...
D_1	Q12797	MAQRKNAKSS GNSSSSGSGS $\cdots$	1.14.11.16 Peptide-aspartate...
	$\cdots$	$\cdots$	$\cdots$

	AC	AA Sequence	MIPS Funcat
	P32589	STPFGLDLGN NNSVLAVARN $\cdots$	16.01 protein binding
D_2	P07278	VSSLPKESQA ELQLFQNEIN $\cdots$	16.19.01 cyclic nucleotide bind.
	$\cdots$	$\cdots$	$\cdots$

Typically a user (e.g., a biologist) might want to infer probabilistic models (e.g., Naive Bayes) from the available data. Using INDUS the user defines the semantic correspondences between the data source attributes *Protein ID* in D_1, *AC* in D_2 and the user attribute *ID*; *Protein Sequence* in D_1, *AA Sequence* in D_2 and *Sequence* in O_U; and *EC number* in D_1, *MIPS catfun* in D_2 and *GO Function* in the user perspective. Furthermore, the user can use predefined mappings between the values of semantically similar attributes (e.g., mappings from *EC Number* and *MIPS Funcat* to *GO function*) or modify existing mappings according to the user's view of the domain.

We will briefly review the Naive Bayes model, identify sufficient statistics for learning Naive Bayes models from data and show how these sufficient statistics can be computed from distributed, heterogeneous data using INDUS query answering engine.

4.2   Classification Using a Probabilistic Model

Suppose we have a probabilistic model α for sequences defined over some alphabet Σ (which in our case is the 20-letter amino acid alphabet). The model α specifies for any sequence $\overline{S} = s_1, \cdots, s_n$ the probability $P_\alpha(\overline{S} = s_1, \cdots, s_n)$ according to the probabilistic model α. We can construct such a probabilistic model and explore it as a classifier using the following (standard) procedure:

- For each class c_j train a probabilistic model $\alpha(c_j)$ using the sequences belonging to class c_j.
- Predict the classification $c(\overline{S})$ of a novel sequence $\overline{S} = s_1, \cdots, s_n$ as given by: $c(\overline{S}) = \arg\max_{c_j \in C} P_\alpha(\overline{S} = s_1, \cdots, s_n | c_j) P(c_j)$

The Naive Bayes classifier assumes that each element of the sequence is independent of the other elements given the class label. Consequently, $c(\overline{S}) = \arg\max_{c_j \in C} P_\alpha \prod_{i=1}^{n} P_\alpha(s_1 | c_j) \cdots P_\alpha(s_n | c_j) P(c_j)$. Note that the Naive Bayes classifier for sequences treats each sequence as though it were simply a bag of letters and it calculates the number of occurences $\sigma(s_i | c_j)$ of each letter in a sequence given the class of the sequence as well as the number of sequences $\sigma(c_j)$ belonging to a particular class c_j. These frequency counts completely summarize the information needed for constructing a Naive Bayes classifier, and thus, they constitute *sufficient statistics* for Naive Bayes classifiers [10]. An algorithm for learning probabilistic models from data can be described as follows:

(1) Compute the frequency counts $\sigma(s_i | c_j)$ and $\sigma(c_j)$.
(2) Generate the probabilistic model α given by these frequency counts.

The query answering engine receives queries such as $q(\sigma(s_i | c_j))$ and $q(\sigma(c_j))$ asking for frequency counts, it decomposes them into subqueries $q_k(\sigma(s_i | c_j))$ and $q_k(\sigma(c_j))$ according to the distributed data sources D_k $(k = 1, p)$ and maps them to the data source ontologies. Once the individual results are received back, the query answering engine composes them into a final result by adding up the counts returned by each data source. Thus, there is no need to bring all the data to a central place. Instead queries are answered from distributed data sources viewed from a user's perspective.

Experimental results on learning probabilistic models for assigning protein sequences to gene ontology functional families are reported by our group in [1]. They show that INDUS can be successfully used for integrative analysis of data from multiple sources needed for collaborative discovery in computational biology.

5   Summary, Discussion and Further Work

5.1   Summary

We have presented INDUS, a federated, query-centric approach to answering queries from distributed, semantically heterogeneous data sources. INDUS assumes a clear separation between data and the semantics of the data (ontologies) and allows users to specify ontologies and mappings between data source ontologies and user ontology. These mappings are stored in a mappings repository to ensure their re-usability and are made available to a query answering engine. The task of the query answering engine is to decompose a query posed by a user into subqueries according to the distributed data sources and compose the results into a final result to the intial user query.

In previous work [10] we have shown that learning algorithms can be decomposed into an information extraction component and a hypothesis generation component. This decomposition makes it possible to see learning algorithms as pseudo-users that pose queries to the query answering engine in order to gather the information that they need for generating the models that they output. Modular implementations of several learning algorithms have been linked to INDUS, thus obtaining algorithms for learning classifiers from distributed, semantically heterogeneous data sources. We have demonstrated how we can use INDUS to obtain algorithms for learning Naive Bayes models for predicting the functional classification of a protein based on training sequences that are distributed among several distributed, semantically heterogeneous data sources.

An initial version of INDUS software and documentation are available at: http://www.cild.iastate.edu/software/indus.html.

5.2   Discussion

There is a large body of literature on information integration and systems for information integration. Davidson et al. [12] and Eckman [13] survey alternative approaches to data integration. Hull [19] summarizes theoretical work on data integration. Several systems have been designed specifically for the integration of biological data sources. It is worth mentioning SRS [15], K2 [29], Kleisli [11], IBM's DiscoveryLink [18], TAMBIS [28], OPM [22], BioMediator [25], among others.

Systems such as SRS and Kleisli do not assume any data model (or schema). It is the user's responsability to specify the integration details and the data source locations, when posing queries. Discovery Link and OMP rely on schema mappings and the definition of views to perform the integration task. TAMBIS and BioMediator make a clear distinction between data and the semantics of the data (i.e., ontologies) and take into account semantic correspondences between ontologies (both at schema level and attribute level) in the process of data integration.

Most of the above mentioned systems assume a predefined global schema (e.g., Discovery Link, OMP) or ontology (e.g., TAMBIS), with the notable exception of BioMediator, where users can easily tailor the integrating ontology to their own needs. This is highly desirable in a scientific discovery setting where users need the flexibility to specify their own ontologies.

While some of these systems can answer very complex queries (e.g., BioMediator), others have limited query capabilities (e.g, SRS which is mainly an information retrieval system). Furthermore, for some systems it is very easy to add new data sources to the system (e.g., SRS or Kleisli, where new data source wrappers can be easily developed), while this is not easy for other biological integration systems (e.g., Discovery Link or OMP, where the global schema needs to be reconstructed).

Finally, while some systems (e.g., SRS, BioMediator) provide support for biological information retrieval tools (such as BLAST or FASTA), to the best of our knowledge none of them are linked to machine learning algorithms that can be used for data analysis, classification or prediction.

On a different note, there has been a great deal of work on ontology development environments. Before developing INDUS editor, off-the-shelf alternatives such as IBM's Clio [14] or Protege [24] were considered, but they proved insufficient for our needs. Clio provides support only for schema mapping, but not for hierarchical ontology mapping. Protege is a purely knowledge base constructing tool (including ontology mappings). It does not provide support for the association of ontologies with data, data management or queries over the data. Furthermore, neither of these systems allow procedural mappings (a.k.a., conversion functions), which are essential for data integration.

Of particular interest to ontology-based information integration is work on modular ontolgies. Ontolingua [17, 16] and ONION [23] support manipulation of modular ontologies. Calvanese et al. [8] proposed a view-based mechanism for ontology integration. However, a global ontology is typically unavailable in information integration from loosely linked, distributed, semantically heterogeneous data. We have explored a description logic based approach to modular design and reuse of ontologies, specification of inter-ontology semantic correspondences, and mappings [4]. However, support for asserting and reasoning with partially specified semantic correspondences between local ontologies and localized reasoning in distributed description logic is lacking.

In terms of learning from distributed, semantically heterogeneous data, while there is a lot of work on distributed learning (see [20] for a survey), there has been little work on learning classifiers from semantically heterogeneous, distributed data. Ontology extended relational algebra [6] provides a framework within which users can specify semantic correspondences between names and values of attributes and obtain answers to relational queries. This approach has been extended in our work on INDUS to handle more general statistical queries across semantically heterogeneous data sources [9].

5.3   Further Work

Our approach has been applied successfully to scenarios where the ontologies associated with some attributes are given by tree structured *isa* hierarchies. It is desirable to extend our work to the more general case where the hierarchies are directed acyclic graphs, as this case is more often encountered in practice.

As Protege [24] is the most popular tool for creating knowoledge bases, in the future INDUS will allow users to import ontologies that are edited using Protege.

In our current framework, we assume that each data source can be seen as a single table. It is of interest to extend INDUS to scenarios where each data sources can be conceptually viewed as a set of inter-related (possibly hierarchical) tables. This requires a framework for asserting semantic correspondences between tables and relations across multiple ontologies (see [14]). In this context, recent work on description logics for representing and reasoning with ontologies [3, 27], distributed description logics [7] as well as ontology languages, e.g., web ontology language (OWL) [26] are of interest. These developments, together with our work on INDUS, set the stage for making progress on the problem of integration of a collection of semantically heterogeneous data sources where each data source can be conceptually viewed as a set of inter-related tables in its full generality.

Acknowledgements. This work was funded in part by grants from the National Science Foundation (IIS 0219699) and the National Institutes of Health (GM 066387).

References

1. C. Andorf, A. Silvescu, D. Dobbs, and V. Honavar. Learning classifiers for assigning protein sequences to gene ontology functional families. In *Fifth International Conference on Knowledge Based Computer Systems (KBCS 2004)*, India, 2004.
2. M. Ashburner, C. Ball, J. Blake, D. Botstein, H. Butler, J. Cherry, A. Davis, K. Dolinski, S. Dwight, J. Eppig, M. Harris, D. Hill, L. Issel-Tarver, A. Kasarskis, S. Lewis, J. Matese, J. Richardson, M. Ringwald, G. Rubin, and G. Sherlock. Gene ontology: tool for unification of biology. *Nature Genetics*, 25(1):25–29, 2000.
3. F. Baader and W. Nutt. Basic description logics. In F. Baader, D. Calvanese, D. McGuinness, D. Nardi, and P. F. Patel-Schneider, editors, *The Description Logic Handbook: Theory, Implementation, and Applications*, pages 43–95. Cambridge University Press, 2003.
4. J. Bao and V. Honavar. Collaborative ontology building with wiki@nt - a multi-agent based ontology building environment. In *Proceedings of the Third International Workshop on Evaluation of Ontology based Tools, at the Third International Semantic Web Conference ISWC*, Hiroshima, Japan, 2004.
5. T. Berners-Lee, J. Hendler, and O. Lassila. The Semantic Web. *Scientific American*, May 2001.
6. P. Bonatti, Y. Deng, and V. Subrahmanian. An ontology-extended relational algebra. In *Proceedings of the IEEE Conference on Information Integration and Reuse*, pages 192–199. IEEE Press, 2003.
7. A. Borgida and L. Serafini. Distributed description logics: Directed domain correspondences in federated information sources. In *Proceedings of the Intenational Conference on Cooperative Information Systems*, 2002.
8. D. Calvanese, G. D. Giacomo, and M. Lenzerini. A framework for ontology integration. In *Proceedings of the international semantic web working symposium*, pages 303–316, Stanford, USA, 2001.
9. D. Caragea, J. Pathak, and V. Honavar. Learning classifiers from semantically heterogeneous data. In *Proceedings of the International Conference on Ontologies, Databases, and Applications of Semantics for Large Scale Information Systems*, 2004.
10. D. Caragea, A. Silvescu, and V. Honavar. A framework for learning from distributed data using sufficient statistics and its application to learning decision trees. *International Journal of Hybrid Intelligent Systems*, 1(2), 2004.
11. J. Chen, S. Chung, and L. Wong. The Kleisli query system as a backbone for bioinformatics data integration and analisis. *Bioinformatics*, pages 147–188, 2003.
12. S. Davidson, J. Crabtree, B. Brunk, J. Schug, V. Tannen, G. Overton, and C. Stoeckert. K2/Kleisli and GUS: experiments in integrated access to genomic data sources. *IBM Journal*, 40(2), 2001.
13. B. Eckman. A practitioner's guide to data management and data integration in bioinformatics. *Bioinformatics*, pages 3–74, 2003.

14. B. Eckman, M. Hernndez, H. Ho, F. Naumann, and L. Popa. Schema mapping and data integration with clio (demo and poster). In *Proceedings of the International Conference on Intelligent Systems for Molecular Biology (ISMB 2002)*, Edmonton, Canada, 2002.

15. T. Etzold, H. Harris, and S. Beulah. SRS: An integration platform for databanks and analysis tools in bioinformatics. *Bioinformatics Managing Scientific Data*, pages 35–74, 2003.

16. R. Fikes, A. Farquhar, and J. Rice. Tools for assembling modular ontologies. In *The Fourteenth National Conference on Artificial Intelligence*, 1997.

17. T. Gruber. Ontolingua: A mechanism to support portable ontologies.

18. L. Haas, P. Schwarz, P. Kodali, E. Kotlar, J. Rice, and W. Swope. DiscoveryLink: a system for integrated access to life sciences data sources. *IBM System Journal*, 40(2), 2001.

19. R. Hull. Managing semantic heterogeneity in databases: A theoretical perspective. In *PODS*, pages 51–61, Tucson, Arizona, 1997.

20. H. Kargupta and P. Chan. *Advances in Distributed and Parallel Knowledge Discovery*. AAAI/MIT, 2000.

21. A. Kementsietsidis, M. Arenas, and R. J. Miller. Mapping data in peer-to-peer systems: Semantics and algorithmic issues. In *Proceedings of the ACM SIGMOD International Conference on Management of Data*, pages 325–336, 2003.

22. A. Kosky, I. Chen, V. Markowitz, , and E. Szeto. Exploring heterogeneous biological databases: Tools and applications. In *Proceedings of the 6th International Conference on Extending Database Technology (EDBT98), Lecture Notes in Computer Science Vol. 1377, Springer-Verlag*, 1998.

23. P. Mitra, G. Wiederhold, and M. Kersten. A graph-oriented model for articulation of ontology interdependencies. In *Conference on Extending Database Technology*, Konstanz, Germany, 2000.

24. N. F. Noy, R. W. Fergerson, and M. A. Musen. The knowledge model of protege-2000: Combining interoperability and flexibility. In *Second International Conference on Knowledge Engineering and Knowledge Management (EKAW'2000)*, Juan-les-Pins, France, 2000.

25. R. Shaker, P. Mork, J. S. Brockenbrough, L. Donelson, and P. Tarczy-Hornoch. The biomediator system as a tool for integrating biologic databases on the web. In *Proceedings of the Workshop on Information Integration on the Web (held in conjunction with VLDB 2004)*, Toronto, ON, 2004.

26. M. Smith, C. Welty, and D. McGuinness. *OWL Web Ontology Language Guide*. W3C Recommendation, 2004.

27. S. Staab and R. Studer. *Handbook on Ontologies*. International Handbooks on Information Systems Springer, 2004.

28. R. Stevens, C. Goble, N. Paton, S. Becchofer, G. Ng, P. Baker, and A. Bass. Complex query formulation over diverse sources in tambis. *Bioinformatics*, pages 189–220, 2003.

29. V. Tannen, S. Davidson, and S. Harker. The information integration in K2. *Bioinformatics*, pages 225–248, 2003.

Cluster Based Integration of Heterogeneous Biological Databases Using the AutoMed Toolkit

Michael Maibaum[1], Lucas Zamboulis[2], Galia Rimon[2], Christine Orengo[1], Nigel Martin[2], and Alexandra Poulovassilis[2]

[1] Department of Biochemistry and Molecular Biology,
University College London, Gower Street, London WC1E 6BT
[2] School of Computer Science and Information Systems, Birkbeck College,
University of London, London WC1E 7HX

Abstract. This paper presents an extensible architecture that can be used to support the integration of heterogeneous biological data sets. In our architecture, a clustering approach has been developed to support distributed biological data sources with inconsistent identification of biological objects. The architecture uses the AutoMed data integration toolkit to store the schemas of the data sources and the semi-automatically generated transformations from the source data into the data of an integrated warehouse. AutoMed supports bi-directional, extensible transformations which can be used to update the warehouse data as entities change, are added, or are deleted in the data sources. The transformations can also be used to support the addition or removal of entire data sources, or evolutions in the schemas of the data sources or of the warehouse itself. The results of using the architecture for the integration of existing genomic data sets are discussed.

1 Introduction

This paper presents work on an architecture for integrating biological data sources, and reports our experience in applying it to an existing application aimed at providing an integrated sequence/structure/function resource that supports analysis, mining and visualisation of functional genomics data (transcriptomic and proteomic).

Biological data sources are characterised by a very high degree of heterogeneity in terms of the type of data model used, the schema design within a given data model, as well as incompatible formats and nomenclature of values. Further, such data sources frequently make use of large numbers of unstable, inconsistent identifiers for biological entities. Our architecture addresses these two issues by combining two data integration techniques supporting both data heterogeneity and inconsistent identifiers.

The database community has done much work on integration of data from heterogeneous data sources. Examples of significant applications to biological data sources include DiscoveryLink [8], K2/Kleisli [12] and Tambis [5]. In practice, the most widely used system is Sequence Retrieval System (SRS) [30]. A recent survey is provided by [17].

B. Ludäscher and L. Raschid (Eds.): DILS 2005, LNBI 3615, pp. 191–207, 2005.
© Springer-Verlag Berlin Heidelberg 2005

SRS represents one approach to integration: it acts as a portal to data sources exploiting indexes built by the system. It therefore has a more restricted aim than DiscoveryLink, K2/Kleisli and Tambis which are all aimed at supporting higher level query facilities across data sources. DiscoveryLink and Tambis aim to achieve this without users needing to be aware of source data schemas: in our own work we also aim to insulate users in this way.

The two traditional approaches to providing such transparent access are to materialise the integrated data in a warehouse, or alternatively to provide virtual integration with mediator software supporting access to data in the original data sources. Materializing integrated data in a warehouse is usually done on performance grounds: not only is distributed access to remote data sources avoided, but also centralised database query optimisation techniques can be applied to enable complex queries to be supported more efficiently. Maintaining a materialised warehouse to correctly reflect updates in data sources can be complex, however. While access to a virtual warehouse is likely to be less efficient than with a materialised warehouse, it may be the only option if it is not possible to extract data from the underlying data sources, or if the storage overheads of materialisation would be too high.

In our own work we have chosen to exploit the AutoMed data integration toolkit[1] to support the integration of heterogeneous biological data sources. The particular strength of AutoMed for this application area is that it supports bi-directional, extensible transformations from data source schemas to an integrated schema enabling integration both through explicit materialisation in a data warehouse as well as virtual integration of data remaining in the original data resources. The extensibility of AutoMed transformations is also the basis for update of schemas within both the data sources and any materialised warehouse.

AutoMed does not in itself provide a solution for transformations between unstable, inconsistent identifiers. There are a number of significant initiatives within the Life Sciences community to address the problem of inconsistent identifiers. For example, the Life Sciences Identifiers (LSID) initiative [25] is aimed at a standardised scheme for assigning and recognising identifiers for biological entities, while the International Protein Index (IPI) [10] is developing stable identifiers for human, mouse and rat proteomes. Meeting the needs of applications that process and analyze transcriptomics and proteomics data is a particular motivation for such work. Extensive work has also been done on standardisation in more specialised areas, for example the work of the Microarray Gene Expression Data (MGED) Society on MAGE-ML for standardised recording of data related to microarray gene expression experiments [11]. However, the legacy of very large numbers of inconsistent non-standardised identifiers will remain.

Hence, in our work we have combined AutoMed with a clustering approach to associate biological entities independently of their identifiers. In our application of this approach so far, we have used gene sequence clustering to establish associations, but the approach is not limited to sequence-based clustering.

---

[1] See http://www.doc.ic.ac.uk/automed/

The remainder of the paper is organised as follows. Section 2 introduces those features of AutoMed which have been exploited in our work, together with the basis for combining AutoMed with a clustering approach. Section 3 presents our data integration framework. Section 4 reports on our experience applying this framework to the integration of biological data sources in a warehouse being constructed to support the mining and visualisation of functional genomics data. Conclusions and a discussion of ongoing work are given in Section 5.

2 Background

2.1 The AutoMed Toolkit

AutoMed is a heterogeneous data transformation and integration system which offers the capability to handle virtual, materialised and indeed hybrid data integration across multiple data models. AutoMed supports a low-level hypergraph-based data model (HDM), and provides facilities for specifying higher-level modelling languages in terms of this HDM. These specifications are stored within AutoMed's Metadata Repository [1]. In the specific application described in this paper, the problem addressed has been the integration of relational data sources into a relational data warehouse.

AutoMed provides a set of primitive schema transformations that can be applied to schema constructs. In particular, for every construct of a modelling language $\mathcal{M}$ there is an add and a delete primitive transformation which add to/delete from a schema an instance of that construct. For those constructs of $\mathcal{M}$ which have textual names, there is also a rename primitive transformation. For example, in a simple relational model there may be four kinds of modelling construct, Rel, Att, primaryKey and foreignKey.

Instances of modelling constructs within a particular schema are uniquely identified by their *scheme*, enclosed within double chevrons $\langle\!\langle ...\rangle\!\rangle$. AutoMed schemas can be incrementally transformed by applying to them a sequence of primitive transformations, each adding, deleting or renaming just one schema construct (thus, in general, AutoMed schemas may contain constructs of more than one modelling language). Each add or delete transformation is accompanied by a query specifying the extent of the new or deleted construct in terms of the rest of the constructs in the schema. This query is expressed in a functional query language, IQL[2]. AutoMed also provides contract and extend primitive transformations which behave in the same way as add and delete except that they indicate that their accompanying query may only partially specify the extent of the new/removed schema construct. Their query may just be the constant Void, indicating that the extent of the new/removed construct cannot be specified even partially, in which case the query can be omitted.

---

[2] IQL is a comprehensions-based functional query language, and we refer the reader to [18] for details of its syntax, semantics and implementation. Such languages subsume query languages such as SQL and OQL in expressiveness [2].

A sequence of primitive transformations from one schema S_1 to another schema S_2 is termed a transformation *pathway* from S_1 to S_2, denoted by $S_1 \rightarrow S_2$. All source, intermediate, and global schemas, and the pathways between them, are stored in AutoMed's Metadata Repository.

AutoMed has its theoretical foundations in the schema transformation and integration framework described in [22] where it was shown that this approach generalises all the previous notions of 'schema equivalence'. Intuitively, this is because: (a) sequences of the primitive transformations are able to express syntactically any transformation from one schema to another, with first a 'growing' phase which adds missing schema constructs and then a 'shrinking' phase which removes redundant schema constructs; (b) IQL queries are able to express the semantic relationships between a new schema construct and the existing constructs, or between a removed schema construct and the remaining constructs.

The IQL queries present within transformations that add or delete schema constructs mean that each primitive transformation has an automatically derivable *reverse transformation*. In particular, each add/extend transformation is reversed by a delete/contract transformation with the same arguments, while each rename transformation is reversed by swapping its two arguments. [19] discusses how the queries present within these reversible schema transformation pathways can be used to generate view definitions for global schema constructs in terms of source schema constructs. Essentially, this is by means of query unfolding using the queries within delete, contract and rename transformations along the set of reverse pathways from a global schema to a set of source schemas.

AutoMed pathways can be used to express the data cleansing, transformation and integration processes involved in heterogeneous data integration. The queries within transformations also allow the pathways to be used for materialising and incrementally maintaining a materialised global database, and any materialised databases derived from it, in the face of insertions/ deletions/ updates to the data sources. The queries within transformations also allow the pathways to be used for tracing the *lineage* of data in a materialised global database, or any materialised databases derived from it, to the data sources. We refer the reader to [13, 14] for details of these uses of AutoMed pathways.

In any heterogeneous data integration environment, it is possible for either a data source schema or the global database schema to evolve. This schema evolution may be a change in the schema, or a change in the data model in which the schema is expressed, or both. An AutoMed pathway can be used to express the schema evolution in all of these cases. Once the current transformation network has been extended in this way, the actions taken to evolve the rest of the transformation network and schemas, and any materialised derived data, are localised to just those schema constructs that are affected by the evolution. We refer the reader to [23, 24, 15] for details of how this can be achieved in both virtual [23, 24] and materialised [15] integration scenarios. The algorithms used are mainly automatic, except for input of domain or expert human knowledge regarding the semantics of new schema constructs added to a local or global schema which are not semantically equivalent to any existing constructs in the schema.

For our particular application here, the task has been to support the transformation of biological data source schemas into a global warehouse schema. The data source and warehouse schemas were relational, while we have used an XML-based unifying data model for the intermediate schemas. We made this choice in order to allow the use of AutoMed's facilities for automatically transforming and integrating XML data, which are discussed in detail in [28, 29].

The standard schema definition languages for XML are DTD and XML Schema. However, both of these provide grammars to which conforming documents adhere to, and do not summarise the tree structure of the data sources. In our schema transformation setting, schemas of this type are preferable as this facilitates schema traversal, structural comparison between a source and a target schema, and restructuring the source schema(s) that are to be transformed and/or integrated. Moreover, such a schema type means that the queries supplied with AutoMed primitive transformations are essentially path queries, which are easily generated.

The AutoMed toolkit therefore supports a modelling language *XML Data-Source Schema* (XMLDSS) which summarises the tree structure of XML documents, much like DataGuides [16]. XMLDSS schemas consist of four kinds of constructs (see [28] for details of their specification in terms of the HDM): Element, Attribute, PCData and NestList. The last of these are parent-child relationships between two elements e_p and e_c and are identified by a scheme of the form $\langle\langle i, e_p, e_c \rangle\rangle$, where i is the position of e_c within the list of children of e_p in the XMLDSS schema. In an XML document there may be elements with the same name occurring at different positions in the tree. To avoid ambiguity, in XMLDSS schemas we use an identifier of the form elementName\$count for each element, where count is a counter incremented every time the same elementName is encountered in a depth-first traversal of the schema. An XMLDSS schema can be automatically derived from an XML document, as discussed in [28], and it is also possible to automatically derive an XMLDSS schema from a DTD or an XML Schema specification, if available.

2.2   Clustering for Supporting Multiple IDs

While AutoMed is well-suited to the task of supporting transformations of data source schemas into a global warehouse schema, it provides no mechanisms for supporting the equivalence of inconsistent identifiers. Integrating data sources usually results in incomplete matching of related entities in the different data sets, either due to identifier redundancy or due to the use of different reference identifiers. In the case of some biological databases, the percentage of entities that can be matched using a single identifier can be very low. When trying to match proteins from KEGG Gene to the Gene Ontology Gene Products less than 40% match, despite the sources nominally describing the same entities.

Data-based entity clustering provides a general approach to integrating any set of logically related entities and hence supporting multiple identifiers. Under this approach, an appropriate relatedness measure is developed (for example sequence or structure similarity), allowing each entity in the data being integrated to be compared to each of the other entities and a similarity index derived. Once the similarity measure values have been obtained they can be used to organise the entities hierarchically into nested sets. Each level of nesting represents an increasing degree of similarity between the entities contained in the set, allowing each application built on the integrated data source to determine what is an appropriate degree of clustering for that application. In the context of biological data, for example, protein structure is conserved at low levels of sequence similarity compared to function and therefore clusters with lower levels of similarity can be used when structural annotation is desired rather than functional.

Having generated such sets of related entities, information applicable to each set may be extracted and associated with that set. Moreover, an attribute which is only defined for a subset of members may be inferred for remaining members of a set if it is known that the attribute will be shared amongst similar entities.

Use of an appropriate similarity measure and clustering algorithm provides sets of entities that represent the same 'real world' entity that may never have been associated based purely on an identifier mapping. Sets of entities with a lower level of similarity represent entities that are less closely related. While this approach does not allow identification of identical entities, in biological contexts it is often at least, if not more useful to identify similar entities, given the incomplete knowledge about any individual entity.

This type of approach is applicable to many types of data. There is no inherent limitation on the type of clustering or the type or types of similarity measures used to compare entities. For example given a measure of similarity of scientific publications was available, the related articles could be organised into clusters providing links between articles on similar topics. In the simplest case this might be based on keyword matching, but other far more sophisticated approaches are available.

3   Our Data Integration Framework

The architecture of our biological data integration framework is illustrated in Figure 1. There are two principal sources of information for the Global Schema — data sources and cluster data — which are processed in the same way but contain different types of information. Each Data Source is an externally maintained resource that is to be integrated as part of the global database. A data source could be a conventional relational or other structured database, or a semi-structured data source, such as an XML file. Conceptually, a data source describes facts about biological entities. Each Cluster Data resource is constructed from one or more data sources and provides the basis for a generally applicable approach to the integration of data lacking a common reference identifier as discussed in Section 2.2 above. Conceptually, a cluster data resource provides a data-dependent classification of the entities within data sources into related sets.

Each data source is either a structured data source such as a relational database (in which case its associated Schema is a relational schema) or a semi-

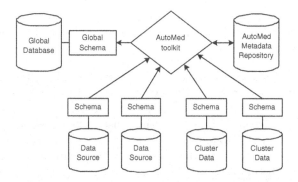

Fig. 1. Architectural Overview of the Data Integration Framework

structured file (in which case it may or may not have an associated schema). In the latter case a schema appropriate for the data source can be generated by the appropriate AutoMed wrapper (see [28, 1] for details of extracting schemas from semi-structured data). Cluster data resources are maintained as relational data with an associated relational schema. The schemas of data source and cluster data resources are processed in the same way, and an arbitrary number of data sources and methods of clustering can be integrated.

Some data sources do not contain a primary key identifier that is persistent between versions of the resource. The lack of a persistent primary key identifier makes the identification of changes between each version difficult. For such data sources a non-volatile, primary key identifier is generated for each entity and added to the data source. Persistent primary key identifiers provide a simple, generic primary key for the higher level tools to use and enables synchronisation of the warehouse with the changing content of the underlying data source.

Wrappers provided by the AutoMed Toolkit automatically generate the AutoMed internal representations of the Schemas and the Global Schema, and store these in the AutoMed Metadata Repository. The AutoMed toolkit is then used to generate the transformation pathways from the Schemas to the Global Schema. These are described in detail with illustration from the example application in Section 4.3 below.

Virtual Integration. After the integration process has been completed, and the transformation pathways from a set of data source schemas to a global schema have been set up, queries formulated with respect to the global schema can be evaluated. Such a query is submitted to AutoMed's Global Query Processor (see [18]) which first reformulates it into a query that can be evaluated over the data sources. This is accomplished by following the reverse transformation pathways from the global schema to the data source schemas in order to generate view definitions of global schema constructs in terms of data source constructs. These view definitions are substituted into the original query, which is then optimised. The query evaluator then interacts with the data source wrappers in submitting to them IQL subqueries which they translate into the local query language for evaluation, returning sub-query results back to the evaluator for any further necessary post-processing and merging.

Materialised Integration. The current version of the BioMap warehouse (see Section 4 below) was materialised using conventional SQL queries on relational sources, before the AutoMed components of our architecture were in place. This approach is labour intensive as the queries must all be manually designed. Upcoming iterations of the warehouse will however be able to benefit from AutoMed's facilities for incrementally maintaining the warehouse. In general, the data sources may be updated by the insertion, deletion or modification of data. Deltas on data sources may result in deltas on cluster data resources also. Both kinds of deltas can be propagated through the AutoMed transformation pathways up to the materialised global database (and to any other materialised databases derived from it). In particular, the queries within add and extend transformation steps can be used to compute a new set of deltas from the current set of deltas, all the way up to the target database (see [14]).

4 Application of the Framework to Gene Family Based Integration

The above architecture has been applied to biological data sources integrated within the BioMap data warehouse. In this section we describe how the architecture has been applied and the results of the work to date.

4.1   The BioMap Warehouse

BioMap is a collaborative project to develop a warehouse integrating protein family, structure, function and pathway/process data with gene expression and other experimental data. The aim is to provide an integrated sequence/structure/function resource that supports analysis, mining and visualisation of functional genomics data (transcriptomic and proteomic). The warehouse is implemented within Oracle, extending techniques developed for the CATH-PFDB database [26] and is designed to serve as a source for data marts which will themselves be constructed using the AutoMed techniques presented in this paper.

Current data sources include the CATH protein structure family database [6], KEGG pathway database [20], Gene3D annotated protein sequence database [21], Gene Ontology [9], EBI Macromolecular structure database (MSD) [4] and ten other resources. Thus far, we have taken CATH, Gene3D, KEGG_Gene, KEGG_Genome, KEGG_Orthology, and also a CLUSTER data source discussed below, representing a significant subset of BioMap data sources describing structural, functional, sequence and ontological information. These contain a diverse set of data structures, formatting conventions and sizes to use for evaluation of our data integration framework.

4.2   The Clustering Approach

There are a variety of methods for classifying biological entities into sets and these methods can be used on the facts within the data warehouse. The facts concerning individual entities within a set will not all derive from precisely the same biological entity, but by choosing an appropriate algorithm to create the sets, the set will contain valuable information about biological entities that are similar (in some way) to each other. One such categorisation method is UniGene [7]. Our categorisation method is based on the PFScape protocol [21] which is in turn based on the TRIBE-MCL algorithm [3]. The PFScape protocol was developed for Gene3D and has been adapted and improved for BioMap. In brief, to construct Gene3D the peptide sequences of more than 120 completed genomes were obtained from the NCBI and from ENSEMBL. An 'all vs all' BLAST was performed using the blastpgp program from the NCBI. The BLAST was performed using a cluster of 50 dual processor machines running GNU/Linux using Sun Grid Engine. An e-value cut off of 0.001 was used. The results were used to create a similarity matrix which was used by TRIBE-MCL to create protein families.

Since then, many more completed genomes have become available, in particular the genome of the Rat. Other genomes have been revised. For the BioMap project an extension of the PFScape protocol has been developed to update the Gene3D families.

The complete genomes of more than 203 Archea, Prokaryotes and Eukaryotes were downloaded from the EBI. For each sequence in Gene3D and the downloaded proteomes an md5 was calculated and an 'all vs all' BLAST performed. The BLAST results were filtered using an 80 percent overlap cutoff to select only the BLAST hits that represented whole chain matches. Each novel sequence was assigned into the best hit family for each of the new sequences, or if no family was identified then a new family was created. Within the protein families multi-linkage clustering was performed based on sequence identity using cluster thresholds of 0.3, 0.35, 0.4, 0.5, 0.6, 0.7, 0.8, 0.9 and 1. The clustering was performed using TCluster, a locally developed program.

To integrate the other data sources, a representative sequence was obtained for each entity in the data source and a md5 calculated. The set of md5s that were not present in the genomic sequences was then obtained. The sequences corresponding to those md5s were then compared to the genomic sequences using BLAST as described above. The entities were then classified in terms of the genomic clusters based on their best hits.

4.3   The Integration Process

The integration process consists of the following steps, of which steps 3 to 6 are explained in more detail below. Steps 1 and 2 are carried out automatically by AutoMed's relational wrapper, as mentioned in Section 3.

1. Automatic generation of the AutoMed relational schemas, $LS_1, \ldots, LS_n$, corresponding to the Data Source and Cluster Data Schemas.

2. Similarly, automatic generation of the AutoMed relational schema, GS, corresponding to the Global schema.
3. Automatic translation of schemas $LS_1, \ldots, LS_n$ and GS into the corresponding XMLDSS schemas $X_1, \ldots, X_n$ and GX.
4. Partial conformance of each schema X_i to GX by means of appropriate rename transformations, to ensure that only semantically equivalent schema constructs share the same name, and that all equivalent schema constructs do share the same name. This results in a set of new schemas $X_1', \ldots, X_n'$.
5. Completing the conformance of each schema X_i' to GX by applying an automatic XMLDSS schema transformation algorithm to each pair of schemas X_i', GX, creating a set of new schemas $X_1'', \ldots, X_n''$.
6. Application of any necessary data cleansing transformations on each X_i'', creating a set of schemas $GX_1, \ldots, GX_n$. As the integration of the schemas up to this point does not involve any reference to the actual data, the data cleansing does not have to be performed prior to this step.

In Steps 4 - 6, the pathways $LS_1 \rightarrow X_1, \ldots, LS_n \rightarrow X_n$ generated by Step 3 are extended with further primitive transformations, leading finally to the schemas $GX_1, \ldots, GX_n$ in Step 6.

Each GX_i is identical to the global XMLDSS schema GX from Step 3. The reverse of the pathway $GS \rightarrow GX$ generated in Step 3 can finally be appended to each GX_i to transform it into the relational global schema GS.

Step 3: Translating AutoMed relational to XMLDSS schemas. To translate a relational schema into an XMLDSS schema we first generate a graph, G, from the relational schema. There is a node in G corresponding to each table in the relational schema. There is an edge from R_1 to R_2 in G if there is a foreign key in R_2 referencing the primary key of R_1. In the given relation schemas there are no cycles in G — in a general setting, we would have to break any cycles at this point. We create a set of trees, T, obtained by traversing G from each node that has no incoming edges, and we convert T into a single tree by adding a generic root. We finally use T to generate the pathway from the relational schema to its corresponding XMLDSS schema. This last phase consists of traversing T and, for each node t encountered, doing the following:

(i) If t is the root, insert a PCData construct into the current schema, and then insert the root itself as an Element construct.
(ii) else:
    (a) insert t as an Element
    (b) insert a NestList construct from the parent of t to t
    (c) find the columns c_i belonging to the table that corresponds to t, and for each c_i: insert c_i as an Element construct; insert a NestList construct from t to c_i; and insert NestList constructs from c_i to PCData.
(iii) For each child of t, t_i', treat t_i' as t in step (i).
(iv) Remove the now redundant relational constructs from the schema.

To illustrate the translation, the top of Figure 2 illustrates a part of the schema of the CLUSTER data source (where ASSIGNMENT_TYPE_ID in ASSIGNMENT_TYPES is referenced by CLUSTER_TYPE in CLUSTER_DATA,

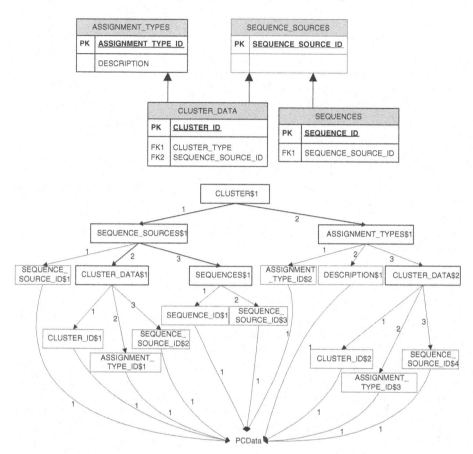

Fig. 2. Top: part of the CLUSTER relational schema. Bottom: corresponding part of the CLUSTER XMLDSS schema

and the rest of the foreign keys have the same name as the primary keys they reference). At the bottom, the XMLDSS schema that corresponds to this relational schema is illustrated. Similarly, Figure 3 illustrates a part of the relational global schema and the corresponding AutoMed XMLDSS schema.

Step 4: Schema Matching. The XMLDSS schema transformation algorithm used in Step 5 of the integration process assumes that if two schema constructs in a local schema and in the global schema, respectively, have the same name, then they refer to the same real-world concept, and if they do not have the same name, they do not. We do not currently support automatic schema matching in our integration process. Thus, after the XMLDSS schemas are produced, and before the application of the schema transformation algorithm in Step 5, the necessary rename transformations must be manually issued on each source XMLDSS schema. These rename transformations effectively simulate a schema

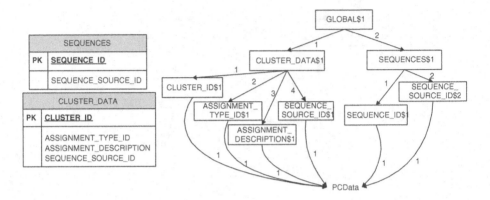

Fig. 3. Left: part of the global relational schema. Right: corresponding part of the XMLDSS schema

matching phase and in our case they have been produced by a domain expert. However, the AutoMed toolkit also offers a tool for performing semi-automatic schema matching and generating the corresponding AutoMed transformation pathways [1]. We also note that this schema matching step does not have to be performed on the XMLDSS schemas, but could instead be performed on the source relational schemas. The only necessity is for this step to be performed before the application of the schema transformation algorithm.

In our running example, the domain expert produced the following rename transformations on the XMLDSS schema in Figure 2:

```
rename(<<CLUSTER$1>>,<<GLOBAL$1>>);
rename(<<DESCRIPTION$1>>,<<ASSIGNMENT_DESCRIPTION$1>>);
rename(<<SEQUENCE_SOURCE_ID$1>>,<<PSEQID>>);
rename(<<SEQUENCE_SOURCE_ID$2>>,<<SEQUENCE_SOURCE_ID$1>>);
rename(<<SEQUENCE_SOURCE_ID$3>>,<<SSEQID>>);
rename(<<SEQUENCE_SOURCE_ID$4>>,<<SEQUENCE_SOURCE_ID$2>>);
rename(<<ASSIGNMENT_TYPE_ID$2>>,<<PASSID>>);
rename(<<ASSIGNMENT_TYPE_ID$3>>,<<ASSIGNMENT_TYPE_ID$2>>)
```

and the following rename transformation on the XMLDSS schema in Figure 3:

```
rename(<<SEQUENCE_SOURCE_ID$2>>,<<SSEQID>>)
```

Step 5: Automatic XMLDSS-based integration. The algorithm for automatically transforming a source XMLDSS schema S into a target XMLDSS schema T has three phases:

Growing phase: Traverse T in a depth-first fashion and for every schema construct encountered that is not present in S, issue an add or extend transformation, resulting in an intermediate schema S_1.

Shrinking phase: Traverse S_1 in a depth-first fashion and for every schema construct encountered that is not present in T, issue a delete or contract transformation, resulting in an intermediate schema S_2.

Renaming phase: Traverse S_2 in a depth-first fashion and issue the necessary rename transformations needed to rename the ordering labels of the NestList constructs in order to create the correct ordering of these constructs, resulting in a final schema S_T syntactically identical to the target XMLDSS schema T.

For reasons of space, we refer the reader to [29] for a detailed description of this algorithm. To illustrate the algorithm, we list below a part of the pathway generated to transform the XMLDSS schema in Figure 2 to the XMLDSS schema in Figure 3, after the earlier rename transformations of Step 4 have first been applied. Here makelist is a built-in IQL function that takes a value v and a number n and produces a list consisting of n copies of v:

```
add(<<0,GLOBAL$1,CLUSTER_DATA$1>>,
        [{v0,v2}|{v0,v1}<-<<GLOBAL$1,SEQUENCE_SOURCES$1>>;
        {v1,v2}<-<<SEQUENCE_SOURCES$1,CLUSTER_DATA$1>>]);
add(<<0,GLOBAL$1,SEQUENCES$1>>,
        [{v0,v2}|{v0,v1}<-<<GLOBAL$1,SEQUENCE_SOURCES$1>>;
        {v1,v2}<-<<SEQUENCE_SOURCES$1,SEQUENCES$1>>]);
extend(<<0,CLUSTER_DATA$1,ASSIGNMENT_DESCRIPTION$1>>,
        [{v1,v2}|{v0,v1}<-<<ASSIGNMENT_TYPES$1,CLUSTER_DATA$2>>;
        {v0,v2}<-<<ASSIGNMENT_TYPES$1,ASSIGNMENT_DESCRIPTION$1>>);
delete(<<1,GLOBAL$1,SEQUENCE_SOURCES$1>>,
        makelist {'GLOBAL$1','SEQUENCE_SOURCES$1'}
                (count <<SEQUENCE_SOURCES$1>>));
delete(<<1,SEQUENCE_SOURCES$1,PSEQID>>,
        makelist {'SEQUENCE_SOURCES$1','PSEQID'}
                (count <<PSEQID>>));
contract (<<1,PSEQID,PCData>>);
contract (<<PSEQID>>);
delete(<<2,SEQUENCE_SOURCES$1,CLUSTER_DATA$1>>,
        makelist {'SEQUENCE_SOURCES$1','CLUSTER_DATA$1'}
                (count <<CLUSTER_DATA$1>>));
delete(<<3,SEQUENCE_SOURCES$1,SEQUENCES$1>>,
        makelist {'SEQUENCE_SOURCES$1','SEQUENCES$1'}
                (count <<SEQUENCES$1>>));
contract(<<SEQUENCE_SOURCES$1>>);
```

The unwanted edges on the RHS of the XMLDSS schema of Figure 2 are deleted/contracted similarly. A series of rename transformations then follows to create a contiguous ordering of edges beneath a parent element.

Step 6: Data cleansing. After the local XMLDSS schemas have been conformed with the global XMLDSS schema, the domain expert can manually issue any further necessary transformations to remove any representational heterogeneities at the data level. AutoMed transformations can express the transformation of data from one format to another in the same way as they can express the transformation of schema structures. For example, consider in our running example attribute DESCRIPTION in relation ASSIGNMENT_TYPES (see Figure 2). The extent of this attribute in the data source consists of mixed case strings. In the CLUSTER XMLDSS schema this attribute is called DESCRIPTION$1. After the partial conformance step (Step 4 in Section 4.3), the attribute has been renamed to ASSIGNMENT_DESCRIPTION$1. To turn the extent of this attribute to uppercase strings before merging with the other data sources in the global schema, the following transformations can be appended to the transformation pathway resulting from the conformance step (Step 5 in Section 4.3):

```
add(<<0,ASSIGNMENT_DESCRIPTION$1,PCData>>,
    [{v0,stringUpper v1} |
     {v0,v1}<-<<1,ASSIGNMENT_DESCRIPTION$1,PCData>>]);
contract(<<1,ASSIGNMENT_DESCRIPTION$1,PCData>>);
rename(<<0,ASSIGNMENT_DESCRIPTION$1,PCData>>,
       <<1,ASSIGNMENT_DESCRIPTION$1,PCData>>)
```

Here stringUpper is a built-in IQL function that converts all the alphabetic characters in a string to upper-case. Several other string-handling functions are supported by IQL e.g. stringLower, stringConcat and stringSplit. The IQL query processor is implemented in such a way that extending it with new built-in functions is straightforward.

In general with AutoMed, these kinds of data cleansing transformations can take place at any stage of the integration process. It is also possible to incorporate materialised correspondences between data values in source and target schemas into data cleansing transformations — this extensional information is treated as another data source.

Implementation and Results. The above integration process was carried out on a Pentium 4 2.8Ghz, with 1Gb RAM and Linux as the operating system. The Gene3D, KEGG_Gene, KEGG_Genome, KEGG_Orthology, CATH and CLUSTER data sources, and the global database are all Oracle databases. The AutoMed repository is stored in a PostgreSQL database, and the AutoMed toolkit itself is written in Java. The integration of each data source took under 15 minutes, resulting in a total running time of about 85 minutes. Many of the algorithms are not yet fully optimised and therefore we expect a major performance improvement as more optimisations are built into the AutoMed toolkit.

5   Conclusions and Future Work

This paper has presented a data integration framework for biological data sources that combines techniques to support the diversity of data models, schemas and formats which are characteristic of biological data together with a clustering approach developed to support distributed biological data sources with inconsistent identification of biological objects.

The work we have described is this paper is currently being extended in a number of areas. First, the approach is being applied to the other data sources noted in Section 4 with further detailed evaluation of the results obtained. The clustering approach is also being extended: while the use of sequence families is described here, other methods of classification could be used including structural and many other approaches. We are currently working on a method that integrates feature and domain recognition (hidden Markov model) approaches to identify attributes of sequences. These attributes (i.e. a structural domain, or a protein active site) can be used to form clades, within which the existing clustering information can be organised. This combination of two clustering approaches will provide the best features of the extremely sensitive, but time consuming scanning approaches with the less sensitive, but much faster simple sequence comparisons.

In the BioMap warehouse we have so far successfully applied the AutoMed-based techniques for the data cleansing, transformation and integration processes as presented in Section 4. We are currently implementing AutoMed-based materialisation and maintenance of the global database, which have been manual processes to date. Use of AutoMed will enable delta changes to be automatically propagated to the global database as well as allowing schema changes to accommodated.

The techniques presented on this paper have not so far been applied to integrating textual data sources such as PubMed abstracts within BioMap. However, work has already been done on extending AutoMed with facilities for integrating unstructured text with structured data [27], and these techniques will be applied to textual biological data sources.

A further collaborative project, ISPIDER, aims to develop Grid-based data integration of biological data resources. The strengths of AutoMed for supporting bi-directional and incrementally constructed transformation pathways are of particular value in a Grid environment, and work is being pursued on developing these techniques and integrating them with existing Web Service and Grid middleware components for service discovery and metadata management.

References

1. M. Boyd, S. Kittivoravitkul, C. Lazanitis, P.J. McBrien, and N. Rizopoulos. AutoMed: A BAV data integration system for heterogeneous data sources. In *Proc. CAiSE'04*, 2004.
2. P. Buneman, L. Libkin, D. Suciu, V. Tannen, and L. Wong. Comprehension syntax. *SIGMOD Record*, 23(1):87–96, 1994.
3. A J Enright, S Van Dongen, and C A Ouzounis. An efficient algorithm for large-scale detection of protein families. *Nucleic Acids Res*, 30(7):1575–84, 2002.
4. A. Golovin *et al.* E-MSD: an integrated data resource for bioinformatics. *Nucleic Acids Res*, 32 Database issue(1362-4962):D211–6, 2004.
5. C.A. Goble *et al.* Transparent access to multiple bioinformatics information sources. *IBM Systems Journal*, 40(20):532–552, 2001.
6. C.A. Orengo *et al.* CATH–a hierarchic classification of protein domain structures. *Structure*, 5(8):1093–108, 1997.
7. D.L. Wheeler *et al.* Database resources of the National Center for Biotechnology: update. *Nucl. Acids Res.*, 32 Database Issue:D35–D40, 2004.

8. L.M. Haas *et al.* DiscoveryLink: A system for integrated access to life sciences data sources. *IBM Systems Journal*, 40(20):489–511, 2001.

9. M. Ashburner *et al.* Gene Ontology: tool for the unification of biology. The Gene Ontology Consortium. *Nature Genet.*, 25(1):25–29, 2000.

10. P.J. Kersey *et al.* The International Protein Index: An integrated database for proteomics experiments. *Proteomics*, 4(7):1985–1988, 2004.

11. P.T. Spellman *et al.* Design and implementation of microarray gene expression markup language (MAGE-ML). *Genome Biology*, 3(9):research0046.1–0046.9–1988, 2002.

12. S. Davidson *et al.* K2/Kleisli and GUS: Experiments in integrated access to genomic data sources. *IBM Systems Journal*, 40(20):512–531, 2001.

13. H. Fan and A. Poulovassilis. Tracing data lineage using schema transformation pathways. In *Knowledge Transformation for the Semantic Web*, volume 95 of *Frontiers in Artificial Intelligence and Applications*, pages 64–79. IOS Press, 2003.

14. H. Fan and A. Poulovassilis. Using AutoMed metadata in data warehousing environments. In *Proc. DOLAP 2003*, pages 86–93, 2003.

15. H. Fan and A. Poulovassilis. Schema evolution in data warehousing environments — a schema transformation-based approach. In *Proc. ER'04*, pages 639–653, 2004.

16. R. Goldman and J. Widom. DataGuides: Enabling Query Formulation and Optimization in Semistructured Databases. In *Proc. VLDB'97*, pages 436–445, 1997.

17. T. Hernandez and S. Kambhampati. Integration of biological sources: Current systems and challenges ahead. *Sigmod Record*, 33(3):51–60, 2004.

18. E. Jasper, A. Poulovassilis, and L. Zamboulis. Processing IQL queries and migrating data in the AutoMed toolkit. Technical Report 20, Automed Project, 2003.

19. E. Jasper, N. Tong, P. McBrien, and A. Poulovassilis. View generation and optimisation in the AutoMed data integration framework. In *Proc. CAiSE Forum at CAiSE'03*, pages 29–32. Univ. of Maribor Press, June 2003.

20. M. Kanehisa, S. Goto, S. Kawashima, Y. Okuno, and M. Hattori. The KEGG resources for deciphering the genome. *Nucl. Acids Res.*, 32 Database Issue:D277–D280, 2004.

21. D. Lee, A. Grant, R. Marsden, and C. Orengo. Identification and distribution of protein families in 120 completed genomes using Gene3D. *Proteins (In Press)*, 2004.

22. P. McBrien and A. Poulovassilis. A formalisation of semantic schema integration. *Inf. Syst.*, 23(5):307–334, 1998.

23. P. McBrien and A. Poulovassilis. Schema evolution in heterogeneous database architectures, a schema transformation approach. In *Proc. CAiSE'02*, pages 484–499, 2002.

24. P. McBrien and A. Poulovassilis. Data integration by bi-directional schema transformation rules. In *Proc. ICDE'03*, pages 227–238, 2003.

25. OMG. Life Sciences Identifiers RFP response. http://www.omg.org/cgi-bin/doc?lifesci/2003-12-02, 2003.

26. A.J. Shepherd, N.J. Martin, R.G. Johnson, P. Kellam, and C.A. Orengo. PFDB: a generic protein family database integrating the CATH domain structure database with sequence based protein family resources. *Bioinformatics*, 18(12):1666–72, 2002.

27. D. Williams and A. Poulovassilis. Combining data integration with natural language technology for the semantic web. In *Proc. Workshop on Human Language Technology for the Semantic Web and Web Services, ISWC'03*, 2003.

28. L. Zamboulis. XML data integration by graph restructuring. In *Proceedings of BNCOD'04*. Springer, July 2004.

29. L. Zamboulis and A. Poulovassilis. Using AutoMed for XML data transformation and integration. In *Proceedings of DIWeb'04*. Springer, June 2004.

30. E.M. Zdobnov, R. Lopez, R. Apweiler, and T. Etzold. The EBI SRS Server - recent developments. *Bioinformatics*, 18(2):368–373, 2002.

Hybrid Integration of
Molecular-Biological Annotation Data

Toralf Kirsten[1], Hong-Hai Do[1], Christine Körner[3], and Erhard Rahm[1,2]

[1] Interdisciplinary Centre for Bioinformatics, University of Leipzig
{kirsten, do}@izbi.uni-leipzig.de
http://www.izbi.de
[2] Dept. of Computer Science, University of Leipzig
rahm@informatik.uni-leipzig.de
http://dbs.uni-leipzig.de
[3] Fraunhofer Institute, St. Augustin
christine.koerner@ais.fraunhofer.de
http://www.ais.fraunshofer.de

Abstract. We present a new approach to integrate annotation data from public sources for the expression analysis of genes and proteins. Expression data is materialized in a data warehouse supporting high performance for data-intensive analysis tasks. On the other hand, annotation data is integrated virtually according to analysis needs. Our virtual integration utilizes the commercial product SRS (Sequence Retrieval System) of LION bioscience. To couple the data warehouse and SRS, we implemented a query mediator exploiting correspondences between molecular-biological objects explicitly captured from public data sources. This hybrid integration approach has been implemented for a large gene expression warehouse and supports functional analysis using annotation data from GeneOntology, Locuslink and Ensembl. The paper motivates the chosen approach, details the integration concept and implementation, and provides results of preliminary performance tests.

1 Introduction

After the complete genomes of various organisms have been sequenced, the focus of genomic research has shifted to studying and comparing the functions of genes and their products. The knowledge about molecular-biological objects, such as genes, proteins, pathways etc., is continuously collected, curated and made available in hundreds of publicly accessible data sources [Ga04]. The high number of the data sources and their heterogeneity renders the integration of molecular-biological annotation data for functional analysis a major challenge in the bioinformatics domain.

To illustrate the data we have to deal with, Figure 1 shows a sample annotation for a gene uniquely identified by accession number 15 in the public source Locuslink [PM02]. The entry comprises different descriptions, which we group into *annotation* and *mapping data*. Annotation data consists of source-specific attributes, such as *Product* and *Alternate Symbols*. In contrast, mapping data refers to inter-related objects in other sources and is typically represented by web links. The objects are identified by

B. Ludäscher and L. Raschid (Eds.): DILS 2005, LNBI 3615, pp. 208–223, 2005.

their source-specific accession ids, for example, gene locus 15 in LocusLink, gene cluster Hs.431417 in UniGene [Wh03], or enzyme 2.3.1.87 in Enzyme [Ba00]. We denote the set of correspondences between objects of two data sources as a *mapping*. Inter-relating objects by mappings allows combining the annotation knowledge from multiple sources for analysis. In the example, the analysis of LocusLink genes can be enriched by annotations of the referenced GeneOntology [As00] or UniGene objects.

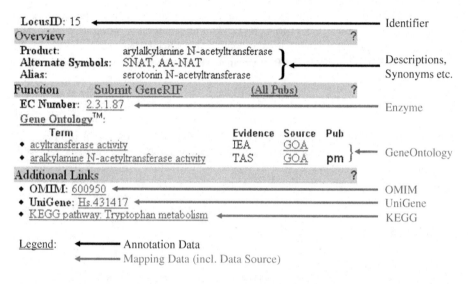

Fig. 1. Annotation and Mapping Data in Locuslink

Establishing and browsing web links represent a first step to integrating different sources, which, due to its simplicity, is widely used. Unfortunately, web links only support interactive analysis for single objects at a time, but not automatic analysis for large sets of objects. Such a set-oriented analysis capability is especially needed for high-throughput expression analysis. In this paper we present a new approach to integrate annotation data from public sources for expression analysis. Large amounts of expression data generated by microarray experiments are physically stored together with experimental descriptions in a data warehouse to support performance-critical analysis tasks. Annotation data, on the other hand, is virtually integrated by a query mediator which utilizes the commercial product SRS (Sequence Retrieval System) to access annotation data of public data sources.

The key aspects of our approach are:

- We combine a materialized and a virtual data integration to exploit their advantages in a new hybrid approach. On the one hand, the data warehouse offers high perform-ance for complex analysis tasks on large amounts of expression data. On the other hand, up-to-date annotation data can be retrieved for analysis when needed.
- Public data sources are uniformly integrated and accessed through the widely accepted SRS tool, which offers wrapper interfaces to a large number of molecular-

biological data sources, including flat files and relational databases. Hence, we avoid the re-implementation of import functions and can easily add sources supported by SRS.

- We explicitly extract mapping data from the data sources and store them in a separate database, the so-called *mapping database*. This separation allows us to determine different join paths between two sources to relate their objects with each other and to pre-compute them for good query performance.
- The approach has been implemented as an extension to the GeWare platform (gene expression warehouse) [KDR03, KDR04] and integrates several public data sources to support the expression analysis. The web interface is accessible under http://www.izbi.de/GEWARE. Performance tests have shown the practicability of our approach.

The rest of the paper is organized as follows. Section 2 discusses related integration approaches. Section 3 describes two main analysis scenarios and their integration requirements. Section 4 gives an overview of our integration concept. Section 5 and 6 describe the central components and their function in more detail, namely the mapping database and the query mediator, respectively. Section 7 presents the results of selected performance tests. Section 8 concludes the paper.

2   Related Work

An overview of representative approaches used for data integration in the bioinformatics domain is given by [LC03], [HK04] and [St03]. Previous solutions mostly follow either a materialized or a virtual integration approach. The former approach physically stores the data in a central database or data warehouse, which can offer high performance for data-intensive analysis tasks. The latter approach typically uses a mediator to perform data access at run time and provide the most current data. In the following we discuss some of these approaches and how they differ from our approach.

Similar to our approach the mediator-based systems DiscoveryLink [Ha01], Kleisli [CCW03, Wo98] and SRS [EHB03, ZD02] do not pursue a (laborious) semantic integration of all data sources by constructing an application-specific global schema. They use a simple schema comprising of the sources and their attributes, which makes it relatively easy to add new data sources. Currently, Kleisli offers interfaces to more than 60 public sources and SRS provides wrappers to more than 700 data sources. Typically, complete copies of data sources are maintained locally and periodically updated for availability and performance reasons. As the price for flexibility, DiscoveryLink and Kleisli leave the task of semantically integration to the responsibility of the user. In particular, the user has to explicitly specify join conditions in queries to relate objects/data from different sources with each other.

SRS and our approach address this problem by capturing and utilizing existing mappings, i.e. correspondences at instance level. SRS maintains indices on these mappings and thus can achieve high query performance. However, SRS only uses the shortest path between two sources as the join path to relate their objects with each other. This represents a restriction as the user may have another preference. Moreover, alternate paths may yield better results than the shortest path [La04]. Therefore, our approach aims at a more flexible and efficient computation of join operations by a) supporting multiple alternative paths and b) pre-computing joins between the sources to a previously determined central source so that join paths with a maximal length of 2 are possible via the central source.

COLUMBA [Ro04] physically integrates protein annotations from several sources into a local database. Source data is imported mainly in its original schema to reduce the effort required for schema integration and data import as much as possible. The source schemas are connected using a mapping to a previously selected central source, the Protein Data Bank (PDB). We also use this technique to construct our mapping database. In contrast to COLUMBA, which only allows a single mapping between a source and the central source, we support multiple mappings, each of which may be pre-computed using a different join path.

ALADIN (Almost Automatic Data Integration) [LN05] generalizes the COLUMBA approach for integrating different kinds of annotation data. A main extension is in the automatic analysis of instance data to detect object associations and duplications. This work is orthogonal to ours and helps to establish new mappings. We are currently focusing on utilizing existing mappings and their compositions.

Our GenMapper [DR04] tool also follows a physical integration of annotation data by using a generic schema called GAM (Generic Annotation Model). The GAM stores both (intra-) associations between objects of the same source and (inter-) association between objects of different sources. High-level operators are used to generate annotation views for different analysis purposes. However, GAM focuses specifically on mapping data and cannot handle data with complex structures, such as geometric data of protein folding structures and genomic sequences. Our hybrid approach uses GenMapper to pre-compute mappings for different join paths, which are then imported into the mapping database.

3 Integration Requirements for Analysis

Figure 2Figure 2 shows two common analysis scenarios in the bioinformatics domain, namely expression and annotation analysis. Expression analysis detects and compares the gene and protein activity under different circumstances, such as in normal and diseased tissues. The main goal is to identify groups of genes or proteins, showing consistently similar or different expression patterns. For example, genes, which are highly active in tumor cells but not in normal cells, could be responsible for the uncontrolled proliferation of the tumor cells. Analyzing the annotations of those genes can reveal the similarities and differences in their currently known functions and infer new gene functions. On the other side, searching in annotation data allows to generally identify genes or proteins with similar functions. This gene / protein groups can be used as input for expression analysis to get insights about their expression behavior.

The usage of annotation data in these scenarios, leads to the following requirements for the integration task.

- *Flexibility and adaptability:* Public data sources are constantly extended and modified at the instance level and also underlie changes at the schema level. Hence, it is important to access current data. Furthermore, the high number of relevant sources presupposes a flexible solution to easily "plug in" a new source when needed. Both observations motivate a virtual integration of annotation data and favor the utilization of a powerful infrastructure such as SRS.
- *Inter-source mappings:* Depending on the research focus of the user, different kinds of annotations may be required for different types of objects. This presupposes the ability to flexibly associate annotations with objects from different sources. For instance, it should be possible to determine functions, e.g. as expressed in GeneOntology terms, for genes in Locuslink, UniGene, NetAffx etc. In addition, filters, such as exact and pattern matching, and their combinations are necessary to identify interesting objects. Finally, alternative join paths should be supported due to the high degree of interconnectivity between sources.
- *Data quality:* Annotations from different sources may largely vary in data quality, e.g. due to different update frequencies and algorithms to calculate object homology. To support user acceptance it is necessary to document how the data has been integrated, e.g. from which source and using which join paths, so that the user can judge its quality.
- *Performance:* Query performance is obviously of key importance for the user acceptance in interactive analysis. Therefore, the physical integration using a data warehouse is recommended for large amounts of expression data. Mediator-based query processing should also be optimized, especially the execution of resource-intensive join operations to relate objects from different public sources. Hence, advanced techniques, such as indexing or pre-computation and materialization of common join paths should be applied to improve query time.

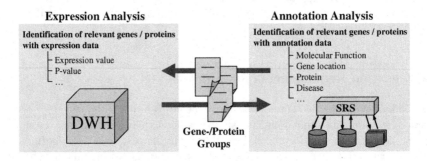

Fig. 2. Different Analysis Scenarios

4 Integration Architecture

4.1 Overview

According to the integration requirements described in the last section, we have designed and implemented a hybrid integration system. Its architecture is illustrated in Figure 3a comprising the following components:

- *GeWare*, a data warehouse supporting expression analysis, is used as integration and test platform for our approach.
- *SRS* is used to query and retrieve annotations from the relevant public sources. Currently the following sources are integrated: the widely used GeneOntology as well as the gene sources Locuslink, Ensembl [Bi04, Po04], UniGene, and the vendor-based source NetAffx [Ch04] providing annotations for the genes of Affymetrix microarrays.
- Our *query mediator* acts as the interface between GeWare and SRS. It transforms user-specified queries into SRS-specific queries which are then forwarded to SRS for execution. Finally, the query mediator combines the results delivered by SRS, performs necessary transformations, and visualizes them on the user web interface.
- The *mapping database* stores pre-computed mappings between the sources. For each source, the mapping database maintains a mapping table storing all correspondences between the source and a pre-selected central source. This star-like schema makes it possible to efficiently perform join operations through the central source.
- The *ADM database* serves administration purposes and stores metadata about the integrated sources, such as their names, attributes and the information about the available mappings (mapping names, and join paths used to compute them). We utilize this metadata to automatically generate the web interface for query formulation.

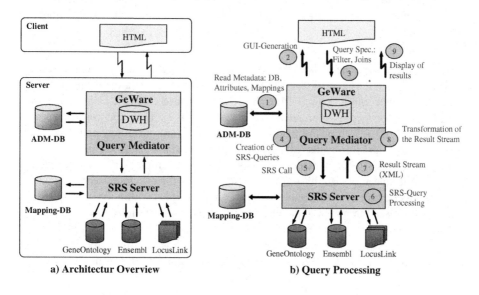

a) **Architectur Overview**   b) **Query Processing**

Fig. 3. Integration approach and corresponding components

The next two subsections describe the interaction between the components in two main processes, the integration of data sources and query processing, respectively. In Section 5 and Section 6, we focus on the issues of metadata management within the mapping and ADM database, and of query processing in the query mediator, respectively.

4.2 Data Source Integration

The comprehensive wrapper library provided by SRS supports numerous data sources available in the bioinformatics domain and allows us to easily add new sources. In particular, we use these wrappers to integrate the flat file-based source Locuslink and two relational databases, Ensembl and GeneOntology. To achieve good performance for interactive queries, we maintain local copies of these data sources for integration in SRS. The ADM database holds metadata about the sources, especially the names of the sources and their attributes.

In our approach, the data sources are organized in a star-like schema supporting efficient join queries. For each object type, one of the sources is chosen as the central source, to which mappings from all other sources of this type are pre-computed. For example, Locuslink is a reference data source for gene annotations. Its identifier, the Locuslink accession, is linked in many other sources and often used for citations in scientific publications. Hence, we choose Locuslink as the central gene source in our current implementation to support gene expression analysis. To construct the mapping database, we import the mappings from Locuslink to all other sources, in particular to UniGene, Ensembl, NetAffx and GeneOntology, which are pre-computed and provided by GenMapper [DR04]. To link a source with the central source, alternative mappings can be computed using different join paths and imported. Each mapping is then registered in the mapping database with the path employed to compute them (see Section 5).

4.3 Query Processing

Figure 3b shows the general workflow of query processing in our system (see Section 6 for more details). The workflow starts with querying metadata about the available sources, attributes and mappings from the ADM database (Step 1). Using this metadata, the web interface is automatically generated (Step 2). Then, the user can formulate the query by selecting the data sources and relevant attributes, and specifying filter conditions and join paths (Step 3). The query mediator interprets the user query and generates a query plan, which consists of one or multiple SRS-specific queries (Step 4). The query plan is passed to the SRS server for execution (Step 5 and 6). While subqueries for selection and projection are performed within the corresponding sources, SRS uses the mapping database to perform join operations. The query result is then returned as one or multiple XML stream (Step 7). The query mediator parses the streams to extract the relevant data (Step 8), which is then prepared in different formats, e.g. HTML for displaying on web browser, and CSV for download (Step 9).

5 Metadata Management

5.1 The Mapping Database

Previous integration systems, such as SRS and GenMapper, determine corresponding objects between two sources using a multi-way join operation along the shortest, automatically determined path connecting them with each other. This approach leads to several problems. First, the shortest path may not always be the best one for joining two particular sources. Other (probably longer) paths may deliver better data, e.g., if the involved sources are updated more frequently than those in the shortest path. Second, the composition of many mappings can lead to performance problems, even for the shortest paths, if they are to be evaluated at run time. One solution to improve query time is to pre-compute and materialize all possible paths in the database. However, this would lead to an enormous amount of mappings and object correspondences (complexity $O(n^2)$ with n sources) which is fairly impractical to manage and update. We address these problems on the one hand by supporting several alternative paths, which can be selected by the users according to their preference or analysis needs. On the other hand, instead of pre-computing join paths between all sources, we identify a central source and pre-compute only the join paths between the remaining sources to the central source, through which the join operations are performed at run-time.

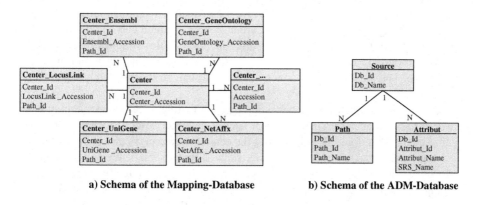

a) Schema of the Mapping-Database   b) Schema of the ADM-Database

Fig. 4. Metadata Management in ADM and Mapping Database

Similarly to COLUMBA, the data sources are connected in a star-like (multidimensional) schema in our approach. In contrast to COLUMBA, we maintain the mappings in a separate database for optimized join processing and support alternative mappings bet-ween a source and the central source. Figure 4a shows the database schema of the mapping database. There is a center table for the central source and a mapping table for each additional data source. All objects of the central source are uniquely identified by the key *Center_ID*. These ids are used as foreign keys in the mapping tables to represent the object relationships at the instance level. Note that a mapping table is used to maintain all mappings of different paths between the respective source and the central source. Each path is identified by a *Path_Id* identifier referring to a specific path which has been used to pre-compute the mappings. Every supported path is described in the ADM database including metadata such as its name and the involved intermediate sources (see Subsection 5.2).

For example, assume we want to relate genes from UniGene with annotations from Ensembl. Neither UniGene nor Ensembl maintain a direct mapping to each other. Hence it is necessary to relate their objects through common objects in other sources.

By analyzing the set of available mappings, we could identify *UniGene-Locuslink-NetAffx-Ensembl* as a possible join path. Without pre-computation, three mappings, each between two neighbor sources in the path, have to be retrieved and successively composed. In our implementation, the mapping table *Center_UniGene* provides a direct mapping *Locuslink-UniGene*. The mapping table *Center_Ensembl* contains the mapping *Locuslink-Ensembl*, which has been previously pre-computed using the path *Locuslink-NetAffx-Ensembl*. Hence, we need only to join these two mappings.

The number of the mappings to be pre-computed and materialized in the mapping database is linear with the number of the sources to be integrated. The support for alternative join paths does not affect the linear complexity (k*n mappings with n sources with k alternative mappings on average per source). New annotation sources can easily be added by creating new mapping tables to hold the corresponding mapping data. This does not affect the run-time complexity because the join operations within the mapping database never involve more than 2 mappings (*source-center-source*). Mapping tables for sources that are no longer required can be removed. Storing mapping data for each source in separate mapping tables simplifies the data update task. In particular, a mapping can be easily updated by deleting it and inserting the new one. The local copies of annotation sources can be independently replaced in SRS by a new version.

The prerequisite to integrate a new source is that there is at least one mapping path between it and the central source or that such a path can be constructed by joining existing paths. Therefore, the selection of the central source plays an important role in this integration approach. Quality criteria, such as update frequency and acceptance by the users, should be considered. Furthermore, if the source already provides direct mappings to many other sources, these mappings can be taken to quickly construct the mapping database. For example, Locuslink and SwissProt represent reference sources for gene and protein annotations, respectively, and maintain a large number of mappings to other (smaller) sources. Hence, they are good candidates for the central source to integrate annotations for gene and protein analysis.

5.2   The ADM Database

Figure 4b shows a portion of the ADM database schema holding metadata about the integrated data sources. The *Source* table records a unique source identifier (*Db_Id*) and the source names. The available attributes of a source are stored in the table *Attribute*, which also contains their SRS-specific names used to translate the user query into a SRS-specific query. All join paths, for which a mapping is materialized in the mapping database, are stored in the *Path* table. The path name concatenates all names of sources that have participated on the join path. Hence, the user can easily differentiate between alternative mappings and identify one for her need. Currently, we import this data partly manually and partly automatically by means of specific database scripts, which extract metadata from the corresponding sources. Subsection 6.2 discusses the process of using this metadata to automatically generate web interfaces for query specification.

6 Query Processing Within the Query Mediator

6.1 Query Types

The query mediator supports two kinds of queries, projection and selection queries, according to the specific requirements of expression and annotation analysis, respectively (see Subsection 3.1):

- *Projection queries* support expression analysis and return a uniform view with user-specified annotation attributes for a given gene group. In a query, the attributes may stem from different sources.
- The goal of *selection queries* is to identify sets of genes showing some common properties. This can be done by applying filter conditions on the corresponding annotation attributes. The gene sets can then be used in expression analysis to compare their expression behavior.

These two query types differ from each other in their input and output data. Projection queries need a gene group as input while selection queries produce a gene group as output. However, they are processed in the same way by associating genes with annotation attributes from the selected sources.

6.2 Query Formulation

The query web interface is generated automatically using the source-specific metadata stored in the ADM database. The user formulates queries on the web interface by selecting relevant attributes (projection queries) and specifying filter conditions (selection queries). Figure 5 shows an example of a selection query to identify all genes, which are located on chromosome *4* and are associated with the biological process *cell migration.*

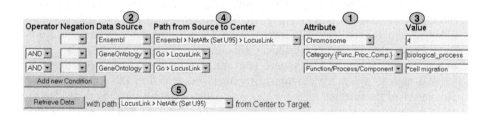

Fig. 5. Query Formulation on the automatically generated web interface

A query may consider attributes (1) stemming from different sources (2). For each attribute, a filter condition (3) can be specified allowing for exact or pattern matching queries. In our example of Figure 5, the asterisk in front of the filter value *"cell migration"* characterizes a similarity search; the other two values are used for exact search. Furthermore, the user has to specify the mapping to connect the source of the selected attribute to the central source by selecting a join path (4). Multiple conditions can be added and combined using the logical operators OR, AND and NOT whereby

OR has the lowest and NOT the highest priority in the query evaluation process. Finally, according to the type of genes to be returned, a mapping between the central source and the target source is to be selected (5).

While SRS only supports to filter attributes of the source from which the data is to be retrieved, our implementation supports the combination of attributes from different sources within a selection query. Moreover, our implementation provides the possibility to combine attributes of different sources (projection) within the same query, which is also currently not directly supported by SRS.

6.3 Generation of Query Plans

From the user specifications on the web interface (see Figure 5), the query mediator generates a SRS-specific query for later execution by the SRS server. This process is performed in three steps, *Block formation* to split the queries into blocks according to the logical operators, *Grouping of source-specific attributes* to determine and group subqueries on attributes belong to the same source to be executed together, and *Assembling SRS query* to generate the final query in SRS-specific syntax and terms. Figure 6 illustrates these steps using the example query from Section 6.2. We discuss the single steps in the following:

1. *Block formation:* First, the filter conditions of a selection query are divided by the logical operator OR into single blocks. Each block contains either one or multiple filter conditions connected with each other by the AND operator. Our query example from Section 6.1 does not contain the OR operator. Hence, there is only one block constructed (see Figure 6, Step 1) holding all three filter conditions. This step is not necessary for projection queries, which do not require filter conditions and build a view for all specified attributes.

2. *Grouping of source-specific attributes:* Within each block obtained from Step 1, the attributes and filter conditions are grouped according to their data source and the mappings to the central source. Each group of attributes and filter conditions concerning the same source and mapping will be valuated together in a subquery. Figure 6, Step 2, shows two identified groups *a* and *b* for the attributes *Category* and *Process* of GeneOntology, and the attribute *Chromosome of Ensembl*, respectively.

3. *Assembling SRS query:* The source and attribute names are replaced by SRS-internal names, which are previously captured and stored in the ADM database. The names of the selected mappings, i.e. the paths, are substituted by their identifiers in the mapping database. For example, Figure 6, Step 3, shows in Line 3 the second and third filter conditions specified on the web interface. The source *GeneOntology* and the attributes *Category* and *Process* are replaced by the internal names *GoTerm*, *typ* and *tna*, respectively. SRS is then invoked by calling its interpreter "*getz*" (Line 1).

From the SRS-specific query in Figure 6, Step 3, we can see, that the objects of *EnsemblGene* and *GoTerm* are first identified by applying the corresponding filters (Lines 2 and 3) and then uniformly mapped to the central identifier *Center_Id* (Line 2) using the mapping ids 1 and 2, respectively. The resulting central identifiers are in turn mapped to the target data source NetAffx using the mapping with id 5, (Line 1). The result of the query consists in a set of NetAffx accessions indicating the corresponding genes.

1. Step: Block formation

Block	Path	Sourcee	Attribute	Filter value
1	Ensembl>NetAffx(Set U95)>LocusLink	Ensembl	Chromosome	4
1	GeneOntology>LocusLink	GeneOntology	Category	biological_process
1	GeneOntology>LocusLink	GeneOntology	Process	*cell migration

2. Step: Grouping of source-specific attributes

Block	Group	Path	Source	Attribute	Filter value
1	a	Ensembl>NetAffx(Set U95)>LocusLink	Ensembl	Chromosome	4
1	b	GeneOntology>LocusLink	GeneOntology	Category	biological_process
1	b	GeneOntology>LocusLink	GeneOntology	Process	*cell migration

3. Step: SRS-Query assembling

```
1 getz -vf "accession" "([Mapping-pid:5]
2   < (Center < ([Mapping-pid:2]<([EnsemblGene-cnm:4]))
3   < ([Mapping-pid:1]<([GoTerm-typ: biological_process] & [GoTerm-tna:*cell migration]))))
```

Fig. 6. Steps for creating the Query Plan

6.4 Extraction and Result Transformation

According to the complexity of the user query specified on the web interface, one or multiple SRS-specific queries are generated and executed. For each such query (e.g. shown in Figure 6, Step 3), SRS returns the result as a XML stream. The stream is then parsed by the query mediator to extract the relevant data. The query mediator then assembles the extracted data of all streams into an internal data structure for later

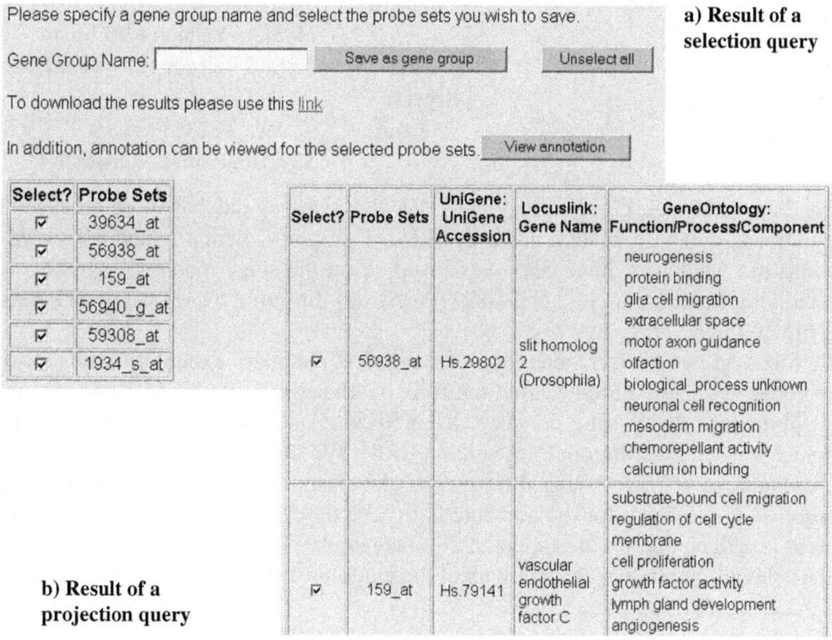

Fig. 7. Results of Projection and Selection Queries

visualization or export. It is also able to perform compensation routines for those functions, which are not yet supported in some DBMS, such as intersection in MySQL, and has not been considered in SRS. A gene group as the result of a selection query can be used as input for a projection query to obtain other annotations for the genes of interest. On the other side, from the result of a projection query, the user can also identify the relevant genes and save them as a new gene group for further queries. The exchange of gene groups between the queries allows us to perform successive refinement for an initially large set of genes.

Figure 7a shows a portion of the result for the example query in Section 6.2. In particular, it contains a set of NetAffx genes which are localized on chromosome *4* and known to have a function in the biological process *cell migration*. The genes are stored in a gene group, for which a projection query is performed to obtain an annotation view as shown in Figure 7b. In particular, the UniGene accession, the Locuslink gene name, and the all functional annotations of GeneOntology are included in the view, based on which the user can further judge the relevance of the genes

7 Performance Analysis

For testing the integration approach and measuring the performance we used an Intel-based platform with the following hard- and software configuration.

Hardware:		**Software:**	
CPU:	4 x Intel Xeon 2.5 GHz	**OS:**	Linux, Fedora 2.4.22
RAM:	8 GB	**DBMS:**	IBM DB2 8.1.0
			MySQL, Version 4.0.17-max
		SRS-Server:	SRS Relational 7.3.1 for Linux
		Java:	Java 2 SUN Platform,
			Standard Edition, Version 1.4.2

The data warehouse GeWare and the ADM and mapping databases are managed by the relational database system DB2 of IBM. The query mediator and all GeWare functions are written in Java. SRS was installed on the same machine together with the locally replicated sources Locuslink (file-based), Ensembl (MySQL) and GeneOntology (MySQL).

We focus on two performance tests investigating the query execution times for different result set sizes. To determine the time overhead induced by SRS, we examine the difference in query time between using SRS to query a relational database and accessing the database directly, i.e. without SRS[1]. We measure the elapsed time of 15 different queries only involving the Ensembl database in MySQL. Each query uses a different filter condition for the attribute *des* (gene description) to return result sets of different size. The queries are repeated 20 times in order to determine the average and standard deviation (shown as error bar) of the elapsed time.

---

[1] To execute a query, SRS in turn creates a query plan consisting of SQL statements to access the corresponding relational database. We use these SQL statements to perform the test in the latter case, i.e. accessing the database directly.

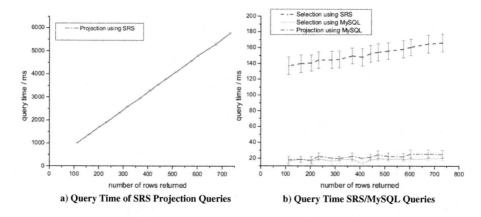

Fig. 8. Performance of projection and selection queries using SRS and MySQL

Due to the large difference in query times, we first show the result for projection queries using SRS in Figure 8a and the remaining results, i.e. for selection queries using SRS and for both selection and projection queries directly accessing MySQL in Figure 8b. Please recall that selection queries only return the accessions of the identified objects, while projection queries return the objects together with the retrieved annotations.

W.r.t to the increasing size of the result set, we observe a significant linear increase in query time for projection queries in SRS (Figure 8a). For selection queries, SRS also requires linear time w.r.t. to the size of the result set (Figure 8b). However, selection queries can be performed much faster than projection queries in SRS. On the other side, we observe almost negligible query time when directly accessing MySQL. For larger result sets, the query time remains almost constant. This leads to the conclusion that SRS produces much time overhead in processing the data obtained from a relational source.

The second test determines the query time for the single steps in the execution of a query involving SRS. For this purpose, we define 11 different queries uniformly involving Ensemble, NetAffx, and the center source Locuslink. They all employ the mappings *Ensembl-Locuslink* and *Locuslink-NetAffx* in order to identify NetAffx genes having a particular pattern in the attribute *des* (gene description) of Ensembl. Figure 9 shows the result of this test. Each query is again repeated 20 times in order to determine the average and standard deviation (shown as error bar) of the elapsed time.

The measured values for each step subsume the elapsed time of all its previous steps. For example, Step 2 performing a mapping between Ensembl and LocusLink subsumes Step 1 to select relevant data from Ensembl. The time of the last step, i.e. Step 4 mapping of the identified Locuslink genes to the required NetAffx genes, represent the entire elapsed time of the query. Overall, the query time increase linear with the amount of the data to be retrieved and is acceptable for even large amount of result data. The first step, selection from Ensembl, performs fastest and the elapsed

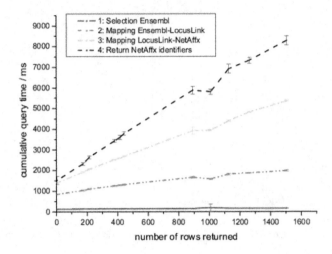

Fig. 9. Query Time of Portions of Selection Queries to Ensemble

time remains relatively constant for different size of the result set. As we can see in Figure 9, querying the mapping database (Step 2-4) to evaluate the specified mappings is more expensive than accessing other sources (Step 1) and thus exhibits high potential for performance optimization. Currently, the mapping database is completely managed and accessed by SRS like other sources. As an alternative, the query mediator may be implemented to directly access the mapping database, so that we obtain more opportunities for tuning.

8 Conclusions

We presented a hybrid approach for the integration of annotation data from public data sources to support expression analysis of genes and proteins. Expression data is physically stored together with diverse experimental descriptions in a data warehouse supporting high performance expression analysis. Up-to-date annotation data is virtually integrated using a mediator and is retrieved on demand according to the analysis needs. The data warehouse and mediator are coupled by means of a query mediator, which exploits existing mappings between the integrated sources for join processing. The mappings are explicitly computed to involve a common central source, through which join operations can be efficiently performed at run time. The use of the a powerful commercial product, SRS of LION bioscience, for the mediator and the generic schema of the database to store the pre-computed mappings allows us to easily integrate a new source or update an existing source. The integration approach has been implemented as an enhancement of our gene expression data warehouse, but is also applicable to other domains, e.g. for protein analysis. The performance evaluation has shown the practicability of our integration approach.

References

[As00]   Ashburner, M. et al.: Gene Ontology: tool for the unification of biology. Nature Genetics 25: 25-29, 2000. http://www.geneontology.org

[Ba00]   Bairoch A.: The ENZYME database in 2000 Nucleic Acids Research 28:304-305 (2000). http://www.expasy.org/enzyme

[Bi04]   Birney, E. et al.: An Overview of Ensembl. Genome Research 14: 925-928, 2004.

[CCW03] Chen, J.; Chung, S.Y.; Wong, L.: The Kleisli Query System as a Backbone for Bioinformatics Data Integration and Analysis. In [LC03]: 147-187.

[Ch04]   Cheng, J. et al.: NetAffx gene ontology mining tool: a visual approach for microarray data analysis. Bioinformatics 20(9), 1462-3, 2004.

[DR04]   Do, H.-H.; Rahm, E.: Flexible Integration of Molecular-biological Annotation Data: The GenMapper Approach. Proc. the 9th Int. Conf. on Extending Database Technology, Heraklion (Greece) 2004. Springer LNCS, 2004.

[EHB03] Etzold, T.; Harris, H.; Beaulah, S.: SRS: An Integration Platform for Databanks and Analysis Tools in Bioinformatics. In [LC03]: 109-145.

[Ga04]   Galperin, M.Y.: The Molecular Biology Database Collection - 2004 update. Nucleic Acids Research 32, Database issue, 2004.

[Ha01]   Haas, L. et al.: DiscoveryLink – A System for Integrated Access to Life Sciences Data Sources. IBM System Journal 40 (2), 2001.

[HK04]   Hernandez, T.; Kambhampati, S.: Integration of Biological Sources: Current Systems and Challenges Ahead. SIGMOD Record 33(3), 2004.

[KDR03] Kirsten, T.; Do, H.-H.; Rahm, E.: A Multidimensional Data Warehouse for Gene Expression Analysis. In: Proc. German Conference on Bioinformatics, Munich 2003.

[KDR04] Kirsten, T.; Do, H.-H.; Rahm, E.: A Data Warehouse for Multidimensional Gene Expression Analysis. Technical Report, IZBI, University of Leipzig, 2004.

[La04]   Lacroix, Z. et al.: Links and Paths through Life Science Data Sources. In [Ra04]: 203 – 211.

[LC03]   Lacroix, Z.; Critchlow T. (Hrsg.): Bioinformatics: Managing Scientific Data. Morgan Kaufmann, 2003.

[LN05]   Leser, U., Naumann, F: (Almost) Hands-Off Information Integration for the Life Sciences. Proc. 2nd Conference on Innovative Data Systems Research (CIDR), 2005

[PM02]   Pruitt, K.D.; Maglott, D.R.: RefSeq and LocusLink: NCBI Gene-centered Resources. Nucleic Acids Research 29 (1), 2001. http://www.ncbi.nlm.nih.gov/projects/LocusLink/

[Po04]   Potter, S.C. et al.: The Ensembl Analysis Pipeline. Genome Research 14: 934-941, 2004.

[Ra04]   Rahm, E. (Ed.): Proceedings 1st Intl. Workshop Data Integration in the Life Sciences (DILS) 2004. LNBI 2994, Springer-Verlag, 2004.

[Ro04]   Rother, K. et al.: COLUMBA: Multidimensional Data Integration of Protein Annotations. In [Ra04]: 156-171.

[St03]   Stein, L.: Integrating Biological Databases. Nature Review Genetics 4(5): 337-345, 2003.

[Wh03]   Wheeler D.L. et al.: Database Resources of the National Center for Biotechnology. Nucleic Acids Research 31: 28-33, 2003. http://www.ncbi.nlm.nih.gov/entrez/query.fcgi?db=unigene

[Wo98]   Wong, L.: Kleisli, a Functional Query System. Journal of Functional Programming, 1 (1): 1-000, 1998.

[Zd02]   Zdobnov, E.M. et al.: The EBI SRS server – recent developments. Bioinformatics 18: 368-373, 2002.

Setup and Annotation of Metabolomic Experiments by Integrating Biological and Mass Spectrometric Metadata

Oliver Fiehn, Gert Wohlgemuth, and Martin Scholz

University of California, Davis Genome Center, GBSF Building
Davis, CA, 95616, USA
ofiehn@ucdavis.edu
http://fiehnlab.ucdavis.edu/

Abstract. Unbiased metabolomic surveys are used for physiological, clinical and genomic studies to infer genotype-phenotype relationships. Long term reusability of metabolomic data needs both correct metabolite annotations and consistent biological classifications. We have developed a system that combines mass spectrometric and biological metadata to achieve this goal. First, an XML-based LIMS system enables entering biological metadata for steering laboratory workflows by generating '*classes*' that reflect experimental designs. After data acquisition, a relational database system (BinBase) is employed for automated metabolite annotation. It consists of a manifold filtering algorithm for matching and generating database objects by utilizing mass spectral metadata such as 'retention index', 'purity', 'signal/noise', and the biological information *class*. Once annotations and quantitations are complete for a specific larger experiment, this information is fed back into the LIMS system to notify supervisors and users. Eventually, qualitative and quantitative results are released to the public for downloads or complex queries.

1 Introduction

Technology advances during the last decade have opened new ways to approach cellular phenotypes. These advances are summarized today as 'omics' platforms which generate quantitative and qualitative data on cellular components such as mRNA transcripts, proteins, or metabolite levels (metabolomics [1]). Metabolomics is a comparatively inexpensive though reliable and informative tool to monitor metabolic states in a variety of different genetic or environmental perturbations. Both for testing and for verifying biological hypotheses, a number of explanatory variables and background information is needed to assist the interpretation (or induction) process. Specifically, there is no way to use data from –omic databases without explaining which biological designs were underlying the experiments. *With other words, data without metadata are junk.* It is a general consensus that scientific experiments and conclusions must be at least explained in such a way that, in

B. Ludäscher and L. Raschid (Eds.): DILS 2005, LNBI 3615, pp. 224–239, 2005.
© Springer-Verlag Berlin Heidelberg 2005

principle, the experiments could be repeated. However, labeling experiments with (biological) metadata is clearly lagging behind descriptions of processes in the data generating technical platforms. It is just now that the metabolomics community has started to develop standards tracking the way from sample to sample processing, data acquisition, data export and normalization to statistics. The ArMet group [2] proposed a generalized framework including various modules to describe a metabolomics experiment. This framework does not detail which (biological or instrumental) metadata are essential to re-use metabolomic experiments for other queries or under other perspectives, and which ontologies need to be used. A related opinion statement on the minimal requirements for a metabolomic experiment (MIAMet) emphasizes the importance for traceable metabolic annotations [3] but does not further embark on biological metadata. A similar trend is seen in the more mature fields of proteomics (the PEDRo standard [4, 5]) and transcriptomics (the MIAME standard [6]). For gene expression experiments, a study-annotator has been developed for describing experimental designs [7]. However, users need to fill 25 forms which relate to 68 tables, and understand and follow pre-defined ontologies that are not authorized by a wide consensus in the biological community.

For metabolomics, an extensive discussion forum is formed by the international working group on *Standard Metabolic Reporting Structures* (SMRS) led by the Imperial College, London, UK [8]. It was summarized in the 2.2 version of the draft document that It should be clear from the previous discussion that the state of biological standardisation for metabonomics experiments is currently non-existent.' [9] The very reason for this inadequacy may be the sheer difficulty to design a comprehensive yet simple schema (and user front end) to capture the ingenuity of experimental designs in biology. We here present pragmatic solution that helps biological researchers defining their experimental design in a coherent and logical metadata structure, with a focus on user friendliness. Together with instrument-related metadata, this design information is used to generate the sample sequence schedule, to define the validity of detected metabolic peaks and to form the basis for statistical treatments of result data. However, we do not envision a direct comparability of the actual data readouts between different experiments: there are no two biological experiments that are totally identical. In fact, it is even difficult to achieve identical results from independent replica setups of experimental designs within a given biological laboratory. The reason for this difficulty in comparability is that there are many fuzzy factors contributing to the actual (metabolic) phenotype of a given individual organism that are hardly controllable in tight manners. Nevertheless, quantitative data outputs will be comparable with respect to trends and magnitudes of control of metabolism, even between laboratories or technology platforms used. In this respect, any information on biological metadata descriptions will enable researcher to (a) carry out own data interpretations and calculations to generate novel hypotheses or (b) combine and compare experiments that share similarities on higher abstraction levels such as àbiotic stress in plant' which would comprise cold, heat, light or nutritional stress.

2 Hierarchical Metadata Defining Biological Experimental *Classes*

We have adopted the general framework laid out by the ArMet group (Architecture for Metabolomics) which consists of nine generic modules [2]:

1. *Admin*: Informal experiment description and contact details.
2. *BiologicalSource*: Genotype and specification of biological source material (BS).
3. *Growth*: Environments in which the biological material developed.
4. *Collection*: Procedures followed for gathering samples BS material.
5. *SampleHandling*: Handling and storage procedures following collection.
6. *SamplePreparation*: Protocols sample preparation prior to data acquisition.
7. *AnalysisSpecificSamplePreparation*: Protocols specific to data acquisition.
8. *InstrumentalAnalysis*: Process description of data acquisition including quality control protocols.
9. *MetabolomeEstimate*: The output of processed data including data processing protocols.

In the implementation period of ArMet, it was found that the accurate description of the biological background of a given sample is the most difficult, but also most important part of the framework. Many steps of modules 4-9 can be easily standardized or described since these are technical procedures that are always performed in a defined manner, at least for a specific routine protocol in a given laboratory. However, the biology experimental part is highly flexible and depends solely on the hypothesis underlying the study. Therefore we decided to use a flexible XML data structure, in order to match a large variety of experimental designs. Given the flexibility and breadth of biological studies, capturing all biological descriptors is technically and intellectually demanding, if not impossible. It is equally difficult to prescribe which of the (potentially very complex) steps of the biological designs are required from the users, and which are just optional. Furthermore, a very in-depth and comprehensive database structure implies that users face highly complex entry forms (and

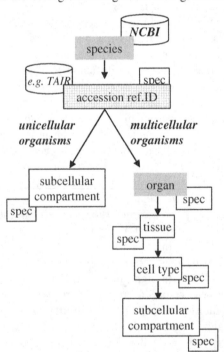

Fig. 1. Description hierarchy for *BioSource*. Use of controlled vocabularies is ensured for specific entries for which authoritative external databases have been assigned (such as NCBI). Others are cross-checked by dictionaries

underlying ontologies) which increases the risk of dummy entries, missing entries or to abstention from populating the database. We have therefore opted for a compromise: we request users to enter the minimal information that would also be required for publishing data in a peer-reviewed scientific journal. In addition we have implemented a structured way to capture this metadata reflecting the underlying biological design. For example, for some relationships and ontologies there are authoritative resources supporting the description of *BioSources* (BS). Besides species names and synonyms, the NCBI database [10](figure 1) supports taxonomic relationships, ultimately up to the top levels śuper kingdom' (arachae, bacteria, eukaryota, viroids and viruses). The underlying taxonomy can be used to distinguish unicellular microorganisms and multicellular (higher) organisms: the latter always consist of distinct órgans' which may further be specified by tissue type, cell type or subcellular compartment that is under study. Microorganisms lack these and can only be further specified by potential subcellular compartments. For setup of an experiment, users can enter more than one species or more than one organ, each of which then may further get specified by additional information. Further authoritative databases are added that help specifying subgroups of species. For example, for the model plant *Arabidopsis thaliana* 831 ecotypes are notified in *t*he Arabidopsis *i*nformation *r*esource TAIR [11], and thousands of well-described Arabidopsis mutant lines, each with a specific ecotype genetic background. All these genetically different Arabidopsis lines are called àccessions' and ar e assigned by a reference identifier in TAIR. As more and more biological communities establish such repositories, these are implemented in our experimental setup designs and made mandatory.

However, even on the level of órgans' there are not many such compulsory lists. For plants, a comprehensive list of organs is given by plantontology.org [12], however, we have not yet identified an accepted standard for naming all mammalian organs, tissues, cell types or eukarytic subcellular compartments: in fact, this is a huge gap in ontology work [13] and frameworks describing relationships between hierarchical levels in biology. In such cases we gradually extend controlled vocabularies by (a) using publicly available lists such as tissue DB [14] that have not yet reached the level of a commonly accepted *de facto* standard and by (b) extending vocabularies used for experimental description in our own database after manual curation. All entries, include strings of flow text descriptions are automatically tested and corrected for spelling by dictionaries and synonym lists.

For a given experiment, all these entities together describe the number of different biological specimen to be tested. It is important to note that each experimental setup necessarily requires description of both *BioSource* and *Growth* condi- tions. It can be expected that metabolic responses on perturbation of growth conditions have at least the same magnitude as effects that are due to genetic changes. This observation is so general that it must be implemented as independent and equally important metadata into a design structure. The resulting biological setup will therefore always span a matrix $M = BioSource_{1n.}$ * $Growth_{1.n}$.

	growth history
BS$_1$	N
BS$_2$	N

Fig. 2. A simple experimental design *BioSource x Growth* for testing biological hypotheses. Each box in the matrices defines a *class* with N ≥6 biological replicates as members

The simplest experimental design that can be devised may compare two different *BioSources* (figure 2), or, alternatively, the same type of *BioSources* under two different *Growth* conditions. There is no *BioSource* that has not been grown in a more or less defined manner. Therefore, the factor 'Growth' is a general property for all biological specimens, however, for some organisms like human patients there is no detailed experimental design. In such cases, generic terms like 'western diet, age' may be used apart from potential treatments (see below) like therapies. This design is equivalent to the well known matrix Genotype *x* Environment' that is used in classical crop breeding. It is important to recognize that each of the different perturbations (*BioSource* or *Growth*) may result in different metabolic states which may be separated into groups or *classes* for statistical analysis of the metabolic levels. For any given experiment, parts of the growth conditions are identical to all *BioSources*. Otherwise, any comparison between the *classes* would be impossible and senseless!These past growth conditions may be described as a growth history G_1. For each species, a minimum set of growth metadata is required whereas other metadata are optional. In plant molecular biology, a single growth history may be defined for which details would be required on sowing and harvest date, harvest time, daylight period, light intensity, humidity, developmental stage, growth medium and type of growth location. Unfortunately, there is no consensus or ontology for this minimum set of background *Growth* metadata'. For a given biological field, experimental descriptors may have been passed on as 'necessary' by journal editors, reviewers and university courses. For example, it is most common to give details on light fluxes in plant biology when explaining the experimental setup in environmental growth chambers. However, it is far less common to say which actual light source was used and the emission spectrum of this, although it is known that plants do react very sensitively on higher or lower levels of red and blue parts of the light spectrum.

In the same way like molecular biologists will vary the genotype (or organ or cell type of a given genotype), physiologists and toxicologists will study variations of *Growth* conditions (including developmental stages) and external environmental impacts such as drugs (treatments'). Each of these growth conditions may again split into different attributes and properties. An example would be 'variation of temperature' in a cold stress experiment in plant physiology, which might utilize high, low, and control temperatures, extending the matrix of BS1 and BS2 (each with three organs) 6 x 3 =18 biological groups or *classes*. It is important to note here that the generation of these *classes* as derived conceptual information from the biology metadata is fed into various other locations within the mass spectral annotation system, most importantly into the data acquisition schedule, the metabolite verification algorithm (see section 3.1) and the statistics workflow. It can easily be imagined that this treatment might be followed in a time dependent manner, which would further increase the matrix (and the complexity of the experimental design). If four time points were included, the overall sample matrix would then be of a dimension of 6 x 3 4 =72 different biological *classes*. In order to perform statistical tests on the resulting metabolomic data, it is wise to use more than six samples per biological *class*, say 10 independent plants. Consequently, 720 samples would be delivered for metabolite analysis: an undertaking that can indeed be carried out in a reasonable time frame and budget in metabolomics, but which would be less feasible for more costly and slower transcriptomic or proteomic experiments (i.e. in case global gene or protein expression levels were to be analyzed). In this respect, metabolomics is different to other –omics techniques because very detailed and structured experimental designs are more likely to be performed with sufficient replicate numbers to carry out statistical tests on the resulting experimental data. In principle, a hierarchical tree of '*Growth*' may be drawn (figure 3).

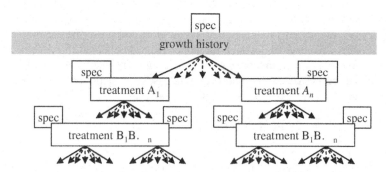

Fig. 3. Flowchart for the description of '*Growth*'. Very complicated experimental designs may be performed, based on the physiological tests that biologists devise. Further specifications (spec) may be entered but are not required

				BS_1 'control'				BS_2 'mutant'			
				blood	liver	heart	kidney	blood	liver	heart	kidney
$drug_1$	$dose_1$	$time_1$		N				N			
		$time_2$		N				N			
		$time_3$		N				N			
		$time_4$		N	N	N	N	N	N	N	N
	$dose_2$	$time_1$		N				N			
		$time_2$		N				N			
		$time_3$		N				N			
		$time_4$		N	N	N	N	N	N	N	N
$drug_2$	$dose_1$	$time_1$		N				N			
		$time_2$		N				N			
		$time_3$		N				N			
		$time_4$		N	N	N	N	N	N	N	N
	$dose_2$	$time_1$		N				N			
		$time_2$		N				N			
		$time_3$		N				N			
		$time_4$		N	N	N	N	N	N	N	N

(left margin label: growth history)

Fig. 4. Pharmacological comparison of two rat strains, four organs, and treatment with two drugs with two different doses which is followed at four time points

This *Growth* design hierarchy is obviously dependent on the underlying metadata from *BioSource*: it does not make sense to require 'light conditions' from a human blood plasma study, and it also is not reasonable to request 'gender' from a plant. However, for a given *BioSource* there is set of growth metadata that is always requested (such as age, sex and other parameters for human samples). The usability of this flowchart for a variety of areas of biological research is exemplified by a pharmacological test setup. The flexible *BioSource x Growth* matrix allows an easy setup of this experiment which may consist of only two rat strains (control and mutant line), on which the effects of two drugs in two different doses is tested on four time points in the blood plasma, and (at end time point 4), also for liver, heart and kidney. Such a pharmacology design is depicted in figure 4. It is important to note that each individual biologist who defines an experiment also defines which metadata are mandatory: in this respect, this metadata layout does not prescribe the biologist what to do but helps scientists to describe the underlying idea behind the design. For both *BioSource* and *Growth*, users may want to add further specific attributes to tables. These cannot be restricted by ontology databases or dictionary comparisons. An example could be þatient ID codes' for clinical samples.

2.1 Technical Implementation of SetupX

We call our system SetupX which sets up experimental design   *classes* and subsequently also manages laboratory workflows and user queries. Although developed for a certain purpose, SetupXarchitecture allows the system to be used in other environments after small adaptations and configurations. A modular structure of this system guarantees that it is reusable, easy to maintain and expandable [15]. All separate functions are offered and used by SetupXin different smaller modules. Communication and interaction between these modules is interceded by the mediator layer. Therefore, different modules can be placed into other environments in short time without requiring major modifications.

Currently there are two access possibilities implemented that allow use of six core modules of SetupXAny external access to   the core modules is shielded by the mediator. One way of external access is the web service module which is based on SOAP (Simple Object Access Protocol) and which allows a platform independent administration and use *via* XML communication. The database is a native XML database that supports storage of metadata in true XML and that also supports the query language XQuery. Native XMLdatabases support data that are not underlying a fixed schema, which is difficult or almost impossible using relational databases.

A swing-user-interface is connected to the web service module for system administration. The second type of external access is the JSP/Servlet module, which generates the standard user-interface for external collaborators and laboratory staff. Part of this user interface is the dynamically generated form for defining biological experiments and *classes*. The six core functions of SetupXinclude user communication and management, interaction with BinBase, generating and writing schedules for the mass spectrometer (based on *class* information), and eventually the definition of the laboratory workflow itself (figure 5).

1. User communication and management
Information stored in BinBase and SetupX must be regarded as confidential. This policy is enforced by defining user authorisations for the different roles. UC Davis users will use their account granted by the campus' Kerberos system. With this account, additional personalized information (e.g. affiliation, address, email, telephone etc) is referenced by SetupXthrough LDAP-directories (Lightweight

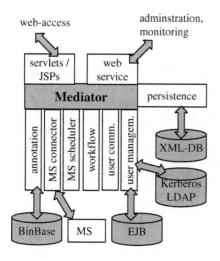

web-access

adminstration, monitoring

persistence

Fig. 5. Modular structure of SetupXand its connected components

Directory Access Protocol). For non-campus users, SetupXneeds to generate an internal authentication. Users need here state once their personal information. Users, and particularly metabolomics staff members, can check the status of laboratory workflows directly by logging in, or are notified by email when predefined workflow parts have been finished or when problems occurred.

2. Interaction with BinBase

Users can request BinBase annotation result files through SetupXwhich activates the BinBase export function by EJB (Enterprise Java Beans) and JMS (Java Messaging Service, a Java interface to Message-Oriented Middleware). BinBase itself requests information about *class* labels of samples using EJBs.

3. Generating and writing schedules for the mass spectrometer

Through the user interfaces, *classes* and the number N of samples per class are entered. SetupX uses this information to generate a run sequence schedule for the mass spectrometer and to communicate this schedule to the instrument in an instrument-specific format. Once the sequence has been started by laboratory staff, an internal scanner is used to grab any information delivered by the mass spectrometer with regards to success or potential failure messages. This information is then fed back into SetupXusing the same   instrument-specific connector.

4. Workflow definition and surveillance

A workflow manager defines the execution sequence of the different modules in order to allow flexible adaptations to new laboratory requirements. In order to make the system independent from the current laboratory workflow definition, a workflow is compiled in a single configuration file. This allows easy update of workflows in case of changes of laboratory protocols or data processing modifications.

5. Persistence and document module

SetupXstores all documents such as experiment description, sample definition etc. as XML files. Consequentially a genuine XML database is used as repository for which XQuery [16] serves as powerful query language. We found XML structure an adequate choice given the fact that the definition of biological *classes* does not allow a unique structure. XML is known as a simple, very flexible text format, which allows the definition of the hierarchy used for the definition of the experiment in an excellent way. Storing this information in a relational database management system would be inappropriate because a large overhead would be generated for mapping this information from XML to the relational structure and back. Speed is not an important aspect for SetupX because no large computational queries are foreseen. Furthermore both the input and export format is XML. Since the database stores the information as unmodified XML the data has never to be mapped.

6. Graphical user interface

One of the main requirements of the new developed LIMS system was that it had to be so user friendly that every user had to be able to use it without any introduction by the staff of the lab and even without reading manuals. The major request was technical – the user interface had to build and generate itself dynamically, because the structure which is represented by the graphical user interface will never, as mentioned above, be fixed to a final set of attributes. We have first explored using a Swing User Interface, similar to the PEDRo-approach. The experimental class structure was defined through an XML schema, and based on this schema the Client-Application created the graphical user interface. However, this schema driven client never matched the requirements of user friendliness and usability, because any fine tuning of *class* definitions and sample specifications were constrained by the technical limitations of the underlying XML schema. Instead we have implemented a server side dynamical created user interface based on Java Server Pages and Java Servlets. This solution is more independent from the experimental design than XML schema. It is therefore possible to add any functionality to this interface that can be implemented in code including functions like real time vocabulary checks or even the adoptions of the user interface to the selected items.

2.2 Experimental Metadata Supporting Other ArMet Modules

Some ArMet modules demand information that is usually stored in classic LIMS implementations such as user logins and user rights. For our case, slight adaptations were needed because many biological experiments are owned by more than one user: it is mandatory in our LIMS implementation to name the principal investigator (usually a faculty member), but also in addition to name the person who was responsible for performing the experiment (who may be research associate or staff). Other modules may be dependent on a given laboratory setup or a given *BioSource*: protocols to prepare samples from plants for metabolomic surveys may be totally inadequate for profiling of human blood plasma.

In order to ensure and monitor long term data quality and reusability, it is good laboratory practice

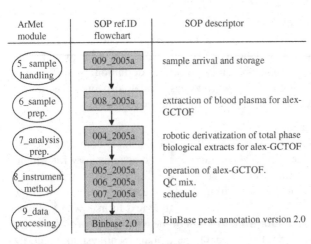

ArMet module	SOP ref.ID flowchart	SOP descriptor
5_ sample handling	009_2005a	sample arrival and storage
6_sample prep.	008_2005a	extraction of blood plasma for alex-GCTOF
7_analysis prep.	004_2005a	robotic derivatization of total phase biological extracts for alex-GCTOF
8_instrument method	005_2005a 006_2005a 007_2005a	operation of alex-GCTOF. QC mix. schedule
9_data processing	Binbase 2.0	BinBase peak annotation version 2.0

Fig. 6. Schematic flow diagram of Standard Operating Procedures (SOPs) in an example for an experiment with *BioSource*: human blood plasma samples

to perform any work by so-called Standard Operating Procedures' (SOPs), both in industrial and semi-industrial analytical environments such as academic core laboratories. Such SOPs include all characteristics needed for direct implementation of sources of metadata into the BinBase system: they include authoritative codes, identifier numbers, clear descriptions of necessary steps and also allowed deviations from protocols. An SOP differs from an academic laboratory protocol in that it must clearly lay out all aspects of a procedure. If a single item of the procedure is changed, it is necessary to state the reason for change, acquire data proving the validity of the change, reinstate permission by the laboratory authority (e.g. the principal investigator) and generate a new SOP number.

Once an SOP is laid out for e.g. sample preparation of a given *BioSource* or data acquisition procedure, it can be made mandatory in a LIMS workflow structure (figure 6). The validity area of SOPs is always clearly defined, but there may be features in the details of SOPs that are shared with external SOPs like the generalized type of the instrument (example in figure 6: a gas chromatography coupled to mass spectrometry) or the type of sample preparation (example in figure 6: cold protein precipitation, silylation). Such higher levels of metadata descriptions yet need to be developed and cannot be made mandatory at present. For example, it is an experience that some analytical instruments are affected by mid-term technical drifts (e.g. in sensititvity). Often, the factors underlying these technical drifts are not well understood and can only be partly controlled. The bottom line of metabolomic experiments is to derive structured information from the acquired data (e.g. by multivariate statistics) and to interpret resulting data clusters by biological metadata. It is obviously of utmost importance that this metabolomic data structure is not affected by non-biological factors such as machine drift. A means to ensure this (apart from instrument quality control) is a randomization of all samples in a sequence, so that each *class* is, on average, affected in the same magnitude as all other *classes*. The easiest way to ensure this is by a random number generator, however, in the laboratory this is almost impossible to put into practice. Therefore, SOP 007_2005a envisions a square root blocking schedule of all replicate samples of each class as compromise between total randomization and laboratory practicability:

$$n_{block} = \sqrt{N}_{class} \tag{1}$$

If a *class* contains a total N=6 biological rep licates, these would be randomized in three blocks of n≥ duplicates over the total instrument run sequence; if a *class* contains 16 biological replicates, these would be blocked into four blocks of four replicates. In summary, the SetupX module generates *classes* via biological metadata and enforces with this information a certain run sequence in the analytical laboratory.

3   Mass Spectral Annotation and Quantitation: BinBase 2.0

All samples are subjected to metabolome data acquisition by automatic liner exchange for gas chromatography/time of flight mass spectrometry (alex-GCTOF). The general output of this instrument is a three dimensional raw data matrix of (time x mass x intensity), which results in 10.8 mio. raw data points for a single sample (415 masses/spectrum x 1300 s x 20 mass spectra/s).

However, biological researchers can only interpret such data matrices if these are transformed into two dimensional data matrices (metabolite x intensity), since metabolite references are found in chemical or biochemical databases like CAS and KEGG and can thus be linked to other important biological objects like proteins and genes. The objective here is therefore to turn (time x mass) information into ͆metabolite' annotations in a routine, but completely unbiased way, and to enable queries in experimental sets of such data matrices.

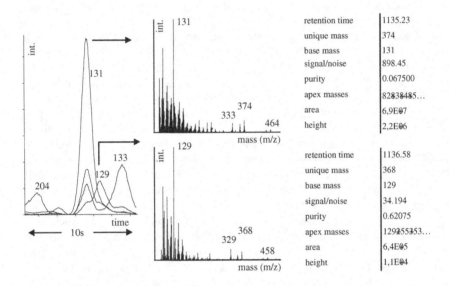

Fig. 7. Deconvolution of raw metabolomic data. Left panel: Overlay of 4 out of 415 measured mass elution profiles (10 s of a total run time 1350 s, profiles for ions m/z 129, 131, 133, 204). Mid panel: Deconvoluted mass spectra of two adjacent, co-eluting peaks with Δtime =1.35 s. Right panel: instrumental metadata labelling these two peaks. Mass spectra and metadata serve as raw data input in Binbase 2.0

It is beyond the scope of this paper to outline theory and concepts of analytical mass spectrometry. It is important to know, though, that in the instrument each metabolite will fragment into more than one mass which will be detected in a finite time frame with an approximately Gaussian intensity time course and identical mass intensity ratios across this ͆elution' time course. This time course is called a ͆peak' with a unique mass/intensity pattern (called ͆mass spectrum'). The peak intensity maxima define the first kind of instrumental metadata, called ͆retention time' (fig. 7). It is unavoidable in metabolomics that peaks overlap (co-elute) since a metabolome of a given sample easily comprises over 1,000 different metabolites. Many mass fragments may be shared between co-eluting peaks. Therefore, the first step of the algorithm is to deconvolute [17] or purify mass spectra from co-eluting peaks, with appropriately assigning the intensity of shared masses to each peak. For this deconvolution we utilize the instrument vendor's software ChromaTOF 2.25. This software detects peaks in an unbiased way and exports one deconvoluted spectrum per peak. In subsequent sections peaks' and spectra' are therefore used as synonyms. After deconvolution, a chromatogram comprises some 400-800 spectra, or a daily output of some 20,000 spectra per day and instrument. BinBase 2.0 then imports these spectra with accompanying metadata such as the unique (model) masses' that best describe the presence of a peak in the local environment. Further instrumental metadata are peak purity' (an estimate of the number, proximity and similarity of co-eluting peaks), signal/noise' (an estimate of peak abundance), apexing masses' (all masses that share maximum intensity with the peak maximum of the unique mass) and other.

3.1 The Filtering Algorithms in BinBase 2.0

Each sample will generate a different number of deconvoluted metadata-labelled spectra. Unfortunately, metabolomic mass spectrometry data sets contain numerous spurious and noisy spectra which need to be detected and deleted prior to annotating and aligning the remaining spectra, and this needs to be performed for multiple samples (n≥1,000) and eventually, multiple of such large experiments. In addition, there may be deconvolution errors reported by ChromaTOF which need to be detected and eliminated. We therefore set out to develop a filtering algorithm that enables metabolite detection and quantification concurrently with automatic extension of metabolic libraries.

The objective of BinBase 2.0 therefore is to three-fold: (a) to annotate all exported spectra to known metabolic peaks that are already compiled as BINs in the database, (b) to automatically add new spectra to the list of BINs and (c) to allow dynamic user queries to export quantitative and qualitative metabolomic information after spectra of all *classes* have been annotated. A BIN is defined as a valid entry in the BinBase that has matched all mass spectral, instrumental and *class* metadata thresholds. In addition to the instrumental metadata, each BIN consists of a set of properties: mass spectrum, retention index (RI), quantification mass, list of unique masses, and a unique identifier number. BINs can be further qualified by super users with *1...n* properties that link further metadata such as metabolite name', ID code referring to external metabolic databases', list of synonyms' or else.

The general algorithm from spectra import to user query export is depicted in figure 8. It starts with importing and storing the .csv data files from all samples of an experiment. The algorithm proceeds by validating all spectra of a sample: check for presence and relative abundance of the unique ion, for presence of all apexing masses in spectrum, for deconvolution error dips, and for the number of spectra per chromatogram that exceed apex intensity thresholds and for the total number of thus detected deconvolution errors. Chromatograms that do not fulfil the latter two criteria will only be used for peak matching, but not for BIN generation. The algorithms then searches spectra of marker compounds that were physically spiked into the samples before data were acquired by using parametrized identification thresholds. With these marker compounds, retention indices (RI) are calculated from retention times to allow retention alignment. This is needed to counteract sample-to-sample retention shifts in the data acquisition procedure. The RI calculation is performed by polynomial regression because absolute and relative retention time shifts markedly differ from linear regressions at early and at late retention times. RIs are never altered or manually adapted for a given data acquisition method, however, they will differ if chromatographic methods are changed. The algorithm then continues by sequentially (*seq.*) selecting all spectra by decreasing intensity (s/n) and testing, whether spectra can be annotated as existing BINs or, if they fail this annotation, if spectra could become new BINs. These decisions work through various filters: first, spectra need to fit into a retention index window, then they need to be labelled with a unique mass that is included in the BIN list of unique masses, afterwards they need to pass a mass spectral similarity filter (sim) that has different thresholds based on the intensity (s/n) and purity of the spectra, and last, spectra need to pass the isomer filter (iso) that selects the best of potentially several matching spectra for a given BIN. The similarity filter currently uses the INCOS algorithm [18], but in principle also other rules could be applied. Spectra that are sorted out in the isomer filter might still be able to match other (neighbouring) BINs and are therefore fed back into the annotation algorithm. Spectra that fail annotation to any existent BIN may generate new BINs. For this, they

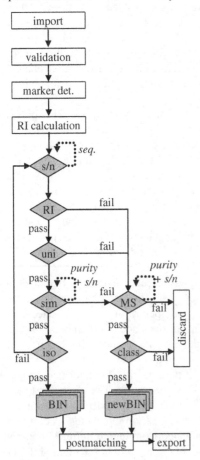

first need to pass mass spectral quality thresholds (MS) that are based on purity and intensity. Thresholds for the MS filter are more draconic than for the similarity filter to ensure that only abundant and pure spectra potentially become new BINs.

Ultimately, a potential new BIN must pass the *class* filter before being validated. This filter demands that a new BIN is detected in at least 80% of all samples of a *class* in order to ensure that this BIN can be supposed to be a genuine metabolic entity and not a spurious contamination. This is also the basic reason why at least *N6* replicates of a given *class* need to be analysed, in order to ensure some level of statistical significance. Once all spectra of all *classes* of a given biological experiment have been annotated, the list of BINs is complete. Then, all spectra are again matched against the BIN list (postmatching) in order to warrant that all BINs (including the new BINs that were generated later in the process) are searched in all samples. Another reason for the postmatching process is that for some samples, spectra may not have passed the (higher) MS thresholds in the BIN generation but would pass the (lower) similarity thresholds in BIN annotation. Therefore, only by final postmatching the eventual result file can be regarded as complete. During the export process, each spectrum is quantified based on intensity of the BIN quantifier mass which is either

Fig. 8. Algorithm for peak annotation and BIN generation. For details, see text

manually set by a super user or (as default value) it uses the unique mass' metadata during BIN generation. Various formats can be used for the final data export, depending on the user's needs. To our notion this is the first published attempt to align and annotate (biological) mass spectra by both instrument-related and biological metadata.

3.2 Technical Implementation of BinBase 2.0

Spectra filtering and BIN databasing is performed in separated modules: it is not advisable to calculate values within a database but use DBs exclusively for queries and data handling. We have employed an SQL 97 conforming database for an efficient data administration and query. The newer SQL 2003 specification was not yet supported by all open source databases. In order to be independent from a specific (supported) database type such as Oracle or SAPDB we have used Java database connectivity (JDBC). It was carefully avoided to program any functions that would be specific to a certain DB type.

BinBase 2.0 predominately consists of $1...n$ table relationships. It is interesting to note that we have implemented the two modules, BinBase and SetupXin two different database structures: for BinBase 2.0, an SQL structure was found to be advantageous due the faster access that is achieved by relational databases with fixed structure, compared to the more flexible but slower XML structure which was used for the (flexible) SetupX system. Furthermore are SQL based systems more mature, offering a wide variety of public or commercial products. For example it is unproblematic to use either Oracle or SAPdb because only minimal adaptations of SQL queries are needed (if programming was done conforming to standards, and if vendor-specific extensions were not used). The largest problems we have encountered were found in storing of all mass spectra. Spectra are imported into BinBase as strings which we first approached to be separated and stored in tables. However, we detected that query times exponentially slowed down with increasing numbers of rows. Therefore spectra are now stored as character large objects' (CLOB) which are dynamically transformed when needed. This procedure has also slowed down performance rates, however, it was found to be still faster than querying tables. The BinBase database itself is configured via XML files, which was found to be a simpler and more flexible solution compared to INI files. Furthermore this configuration offered the possibility to dynamically upload new implementations of the used interfaces *via* Class.forName().

Other components such as SetupXor web interfaces are linked via EJB (Enterprise Java Beans) and JMX(Java Management Extensions). The JMX components enable starting, stopping or querying the status of implemented servers. The EJBs allow querying which samples are being processed or exported during longer sequences. XDoclet was used for generating EJB/JMX configuration files and helper classes. Three servers are implemented: an import server (for importing, matching and BIN generation), a postmatching server (for regular postmatching over the complete database) and a transformation server (for exporting data and file formatting). Currently, plain text, MS Excel and XML is supported. These servers can run independently or together with the EJBs on the JBoss application server.

Finally, front ends have been implemented. A plugin based on Eclipse 3/SWT is used as administrative front end. It includes visualization based on JFreeChart and allows database queries *via* a Hibernate framework. The Hibernate framework supports mapping database documents to objects. Dynamic SWT-tables and visualizations are created from these objects via Java Reflection-API. Therefore, these tables visualize the database contents, for example, all BINs with corresponding metadata. BINs can be modified or manually erased by super users only. A persistence layer is used for user access and user defined queries.

4   Conclusions

This is the first description of a combined system which uses the description of biological experiments to validate metabolic peaks from mass spectra and corresponding mass spectral metadata. Earlier publications have not detailed algorithms how (processed) mass spectrometric peaks are automatically validated and added to a database, but rather focused on database query options [19] or on comparing chromatograms on the base of summing mass spectral intensities [20, 21], instead of alignments of deconvoluted mass spectra and annotation of individual metabolites. The implementation of BinBase 2.0 enables annotating up to 0.5 mio. spectra per day which is far higher than the current production rate of 20,000 spectra/day at the UC Davis Genome Center metabolomics facility. A comparison of manual and automatic validation of such chromatograms will be presented in a bioanalytical journal for the comparison of 1,200 potato tubers from a field trial.

Further improvements will work on parallelization of processes for peak detection and postmatching and on integration of further peak metadata (such as peak tailing factor or profile purity) for automatic flagging of problem cases. SetupX development will consist of further integration of ontologies with a focus on improvements in user friendliness and reducing the time needed for defining each experiment. Ideally, SetupXwould parse the required biological metadata directly from strings that are pasted by users into a single web form, and would only ask for additional information if needed. To this end, however, the abilities of text mining approaches have not been developed far enough yet.

References

1. Fiehn, O.: Metabolomics – the link between genotype and phenotype. Plant Mol. Biol. 48 (2002) 155-171
2. Jenkins, H., Hardy, N., Beckmann, M. *et. al.*.: A proposed framework for the description of plant metabolomics experiments and their results. Nat. Biotechnol. 22 (2004) 1601-1605
3. Bino, R.J., Hall, R.D., Fiehn, O. *et al.*: Potential of metabolomics as a functional genomics tool. Trends Plant Sci. 9 (2004) 418-425
4. Garwood, K., McLaughlin, T., Garwood, C., *et al.*: PEDRo: A database for storing, searching and disseminating experimental proteomics data BMC Genomics 5 (2004) Art. No. 68

5. Jones, A., Hunt, E., Wastling, J.M., Pizarro, A., Stoeckert, C.J.: An object model and database for functional genomics. Bioinformatics 20 (2004) 1583-1590

6. Ball, C.A., Brazma, A. , Causton, H. *et al.*: Submission of Microarray Data to Public Repositories. PLoS Biology 2 (2005) e317, 1276-12773

7. Manduchi, E., Grant, G.R., He, H. *et al.*: RAD and the RAD study-annotator: an approach to collection, organization and exchange of all relevant information for high-throughput gene expression studies. Bioinformatics 20 (2004), 452-459

8. The Standard Metabolic Reporting Structure -An Open Standard for Reporting Metabolic Data . (http://www.smrsgroup.org/ - March 09, 2005)

9. Lindon, J.C. (ed.) Standardisation of Reporting Methods for Metabolic Analyses: A Draft Policy. Document from the Standard Metabolic Reporting Structures (SMRS) Group. *4.5. Summary*, p. 10. (http://www.smrsgroup.org/documents/SMRS_policy_draft_v2.3.pdf - February 01, 2005)

10. Wheeler, D.L., Barrett, T., Benson, D.A. *et al.*: Database resources of the National Center for Biotechnology Information. Nucl. Acids Res. (2005) D39-D45 Sp. Iss. SI

11. Rhee, S.Y., Beavis, W., Berardini, T.Z, *et al.*: The Arabidopsis Information Resource (TAIR): a model organism database providing a centralized, curated gateway to Arabidopsis biology, research materials and community. Nucl. Acids Res. 31 (2003) 224-228

12. Bruskiewich, R., Coe, E.H., Jaiswal, P. *et al*: The Plant OntologyTM Consortium and Plant Ontologies. Comparative and Functional Genomics, 2002, 3(2), 137-142

13. Loranger, S., Higgins, G., Sen, S., Kelly H. The digital human: Towards a unified ontology. Omics 7 (2003), 421-424

14. http://tissuedb.ontology.ims.u-tokyo.ac.jp:8082/tissuedb/ May 16, 2005

15. Erich Gamma et al.: Design patterns : elements of reusable object-oriented software. (Addison- Wesley, Reading, Massachusetts 1995)

16. W3C XQuery 1.0: An XML Query Language. W3C Working Draft. (http://www.w3.org/TR/xquery/ - Feb. 12, 2005)

17. Stein, S.E.: An integrated method for spectrum extraction and compound identification from gas chromatography/mass spectrometry data. J.Am.Soc. Mass Spectrom. 10 (1999) 770–781.

18. McLafferty, F.W., Zhang, M.Y., Stauffer, D.B., Loh, S.Y.: Comparison of algorithms and databases for matching unknown mass spectra. J.Am.Soc. Mass Spectrom. 9 (1998) 92-95

19. Kopka, J., Schauer, N., Krueger, S. *et al.*: GMD@CSB.DB: the Golm Metabolome Database. Bioinformatics 21 (2005), 1635-1638

20. Jonsson, P., Gullberg, J., Nordstrom, A. *et al.*: A strategy for identifying differences in large series of metabolomic samples analyzed by GC/MS. Anal. Chem. 76 (2004), 1738-1745

21. Duran, A.L., Yang, J., Wang, L.J., Sumner, L.W.: Metabolomics spectral formatting, alignment and conversion tools (MSFACTs). Bioinformatics 19 (2003) 2283-2293

Performance-Oriented Privacy-Preserving Data Integration

Raymond K. Pon[1] and Terence Critchlow[2]

[1] UCLA Computer Science Department, Los Angeles, California, USA
rpon@cs.ucla.edu
[2] Lawrence Livermore National Laboratory, Livermore, California, USA
critchlow@llnl.gov

Abstract. Current solutions to integrating private data with public data have provided useful privacy metrics, such as relative information gain, that can be used to evaluate alternative approaches. Unfortunately, they have not addressed critical performance issues, especially when the public database is very large. The use of hashes and noise yields better performance than existing techniques, while still making it difficult for unauthorized entities to distinguish which data items truly exist in the private database. As we show here, the uncertainty introduced by collisions caused by hashing and the injection of noise can be leveraged to perform a privacy-preserving relational join operation between a massive public table and a relatively smaller private one.

1 Introduction

Data is often generated or collected by multiple parties, and the need to integrate the resulting disparate data sources has been identified by the research community [1-6]. Although heterogeneity of the schemas is being addressed, most data integration approaches have not yet efficiently addressed privacy concerns.

Legal and social circumstances have made data privacy a significant issue [7-8], resulting in the need for Hippocratic databases (i.e., database that include privacy as a central concern) [9], particularly in sharing scientific or medical data. Without strong privacy guarantees, scientists often refuse to share data with others for reasons such as subject/patient confidentiality, proprietary/sensitive data restrictions, competition, and potential conflict and disagreement [10]. An application where both data sharing and privacy are important is biomedical research. In this domain research facilities frequently collaborate with each other, sharing experimental data and results. In particular, comparing genome sequences from different species has become an important tool for identifying functions of genes [11]. However, this necessitates integrating different databases. Unfortunately, while there is a significant amount of publicly available data, information provided by most companies, such as proprietary genome sequences, must be kept private.

More concretely, imagine that a scientist wishes to perform a query across a table in his private database (e.g., proprietary genome sequences) and a table in a public data warehouse (e.g., GenBank [12]) in the most efficient manner possible (shown in

B. Ludäscher and L. Raschid (Eds.): DILS 2005, LNBI 3615, pp. 240–256, 2005.
© Springer-Verlag Berlin Heidelberg 2005

Figure 1). Ignoring privacy restrictions, the problem is reduced to a distributed data-base problem that can be solved by shipping the scientist's table to the warehouse and performing the join at the warehouse. However, if the scientist's data set is proprie-tary, it cannot be sent verbatim to the warehouse. The naive solution is for the scien-tist to download the entire public table to his local machine and perform the query there. But to do so would be prohibitively expensive if the public table is very large or the communications link is limited. It would be impossible if the publicly available data cannot be duplicated, for example because of intellectual property constraints.

Assuming that all data sources are abstracted as relational tables and schema recon-ciliation has already been done, the problem can be formalized as the following: table $R = (A, B)$ from a small private database db is to be joined with table $S = (B, C)$ from a large data warehouse dw on column B, yielding the desired table $Goal = R \bowtie_B S$. Table R is private and any party other than the owner of db cannot know the identity of the data items in R. Table S is publicly available and accessible. It is assumed that the system operates in a semi-honest model, where both parties will behave according to their prescribed role in any given protocol. However, there are no restrictions on the use of information that has been learned during the data exchange after the protocol is completed. Thus, from the privacy perspective, dw is treated as an adversary.

Our solution to this problem augments the well-known semi-join framework [13], "hiding" the actual values of the join column of table R by hashing them and includ-ing additional artificial values. The resulting collection is sent to the data warehouse to retrieve a subset of table S that includes data required to answer the original query along with some false positives. Although, this method will not provide for absolute privacy (i.e., the adversary can infer something about the contents of table R), the hash/noise method can guarantee an upper bound on the amount of privacy loss when data is exchanged. By sacrificing a small amount of privacy, this method significantly reduces transmission costs compared to techniques that provide absolute privacy.

1.1 Challenges and Related Work

There are several challenges in privacy-preserving data integration, including: defin-ing privacy, correctness and efficiency. This section provides a short summary of the most relevant work being done by others to meet these challenges, as well as related work on general approaches to privacy preservation. Following this overview, Section 2 describes our privacy metric; Section 3 presents our hash/noise approach; Section 4 outlines a proof of concept implementation and initial experimental results, and; Sec-tion 5 summarizes our work and explores future roads of research.

First, a metric is needed to measure the amount of privacy loss that is incurred when data is exposed. In [14], variable privacy is proposed as a method in which some information can be revealed for some benefit. Privacy loss is likened to a com-munications channel, in which the difference between a priori (i.e., before data has been revealed) and a posteriori (i.e., after data has been revealed) distributions of information measures privacy loss. In [15], the likelihood of what can be inferred about a query posed by the user is used as a measure of privacy loss. In [16] and [17], a metric for measuring the inherent uncertainty of a random variable based on its differential entropy is used as a measure for privacy. These proposed metrics are related to relative information gain, which has also been used in many privacy-preserving applications [18], making it a likely candidate for measuring privacy loss.

Fig. 1. General problem

The second challenge is producing exact and correct answers to queries posed by users. Work in privacy-preserving data mining [19-22] has focused on changing the actual values of data items so that the values of data items are hidden but the distribution of the perturbed data is similar to that of the original data distribution. However, the exact original data values cannot be accurately recovered. While this is acceptable in data mining applications, exact answers are required for data integration.

The third challenge is to perform the private join operation efficiently. It has been shown that to completely guarantee the privacy of the queries, the entire contents of table S should be downloaded [23]. However, in some cases this is not practical and an alternative solution is needed. If the user is willing to sacrifice a small portion of his data privacy, the join operation can be done without retrieving all of table S.

Commutative encryption-based approaches have been proposed to solve the private data integration problem [24-26]. These approaches take advantage of a family of encryption functions in which the order that data items are encrypted by two different keys does not matter. These techniques require the exchange of both parties' encrypted data so that they can mutually encrypt each other's data, making them very expensive. Similarly, oblivious transfer [27-29] allows the user to secretly pose a query and only receive the result of the query and nothing else, but the encryption and transmission of all data items held by dw to the user is required.

There has also been work in private information retrieval schemes [23, 30], which allow a user to retrieve information from a database while maintaining the privacy of his query. In these schemes, table S would be replicated at multiple sites. Given a query, multiple queries are generated and sent to each of site such that no site can learn the actual original query by acting alone. However, users working with sensitive data would be unwilling to trust such a system if no guarantee of enforcement of non-collusion among the sites.

Our hash/noise method takes an approach similar to that of the one discussed in [15], which takes advantage of collisions caused by hashes to introduce uncertainty in the true contents of a private database's table. A hash value is generated for each data item in both tables each time a query is posed. The size of the hash is varied to control the amount of privacy loss, so traditional indexing mechanisms cannot be used to accelerate querying time. A sequential scan of both tables involved in the join is necessary to compute the hash values of all data items in both tables. As a result, the join operation becomes a very expensive operation. There has been work in using Bloom filters to make joins in a distributed database system more efficient and private [31-33]. Similar to the hashing approach, however, Bloom filters would require a sequential scan of both tables to apply a Bloom filter to each of the data items and would not allow the use of traditional indexing mechanism to speed up querying.

In contrast, to these two approaches, our hash/noise method approach uses a set of fixed hashing and artificial hash values (i.e., noise) to control the amount of uncertainty in the identity of the join column values, thereby controlling the level of privacy loss incurred. Because the hashes are known in advance, we can store and index the resulting hash values in the database and would not need to recompute them for each query, enabling indexes to be used to speed up querying. Because the hash functions are known in advance, a dictionary-attack is possible but is partially alleviated by using artificial hash values.

Furthermore, privacy control by hash truncation alone as suggested by [15] is very coarse. For example, suppose that a 16-bit hash does not satisfy a given privacy constraint, so a 15-bit hash was selected instead. However, the 15-bit hash doubles the collision rate of the 16-bit hash, doubling the size of the candidate set for the join result. In contrast, the same 16-bit hash with additional artificial hash values could have satisfied the same privacy constraint and yield fewer records in the candidate set.

2 Privacy Metric

For our work, we use relative information gain [34] as a basis for a metric to measure privacy loss when data is exchanged. Relative information gain is closely related to entropy, which is the amount of uncertainty in a random variable X. If the random variable X can take on a set of finite values $x_1, x_2, \ldots x_n$, then its entropy is defined as:

$$H(X) = -\sum_{i=1}^{n} P(X = x_i) \log_2 P(X = x_i) \tag{1}$$

The conditional entropy $H(X|Y)$ is the amount of uncertainty in X after Y has been observed. Relative information gain, or the fraction of information revealed by Y about X, is defined as:

$$RIG(X;Y) = \frac{H(X) - H(X|Y)}{H(X)} \tag{2}$$

Privacy loss can be thought as the amount of information gained by an adversary about the contents of set of sensitive data items, which in this case are the contents of column B of table R. If dw (i.e., the adversary) has no knowledge about the distribution of column B of table R, then it can only assume that each value that belongs to the domain of B (i.e., U) are equally likely to occur. Let $\tilde{R}$ be a random variable describing the column B values (the only information revealed in a semi-join by db), of a tuple in table R. Absolute privacy loss p_{abs} is defined as the relative information gain on $\tilde{R}$ when any data set N is revealed to dw by db. By doing a simple substitution with equation 2, absolute privacy loss is:

$$p_{abs} = \frac{H(\tilde{R}) - H(\tilde{R}|N)}{H(\tilde{R})} = \frac{\log_2 |U| - H(\tilde{R}|N)}{\log_2 |U|} \tag{3}$$

It is possible that an adversary will make use of any available information to infer the contents of table R, in particular the contents of table S, since it is publicly available. Thus, relative privacy loss is defined as:

$$P_{rel} = \frac{H(\tilde{R} \mid S) - H(\tilde{R} \mid N)}{H(\tilde{R} \mid S)} \tag{4}$$

In this case, the adversary uses the distribution of values in column B of table S as a hint to the possible distribution of values in column B of table R. $H(\tilde{R} \mid S)$ (the uncertainty of the join column values of a tuple in table R given the contents of table S) can be found by directly applying equation 2 on the distribution of values in column B of table S. Because this metric captures the information gained by an adversary with respect to its current knowledge in contrast to absolute privacy loss, it is the metric we have chosen for evaluation of our approach.

3 Privacy-Preserving Distributed Join

Figure 2 outlines our approach to finding $R \bowtie_B S$ when a privacy constraint exists. The first step projects column B from table R and applies a hashing function h to each value in column B, yielding table $h(R)$ with column $h(B)$. Step 2 generates artificial hash values, yielding table n. In step 3, table N is derived from the union of n and $h(R)$. Table N is then shipped to the data warehouse in step 4. At the data warehouse in step 5, table S and N are joined on column $h(B)$, yielding table F. Table F is a set of tuples from dw that contain the final result of the join operation and which is shipped to db in step 6. The final result, *Goal*, is found by filtering out the false positives in F by joining tables R and F.

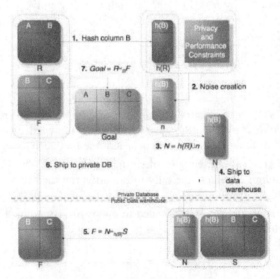

Fig. 2. Privacy-preserving distributed join

3.1 Privacy Constraint Satisfaction

Different hash functions yield different collision rates. Hash functions with large ranges tend to yield low collision rates; whereas, hash functions with smaller ranges tend to yield high collision rates. A hash function h with a high collision rate introduces large amounts of uncertainty about x when $h(x)$ is known. This uncertainty is used to mask the true identity of a join column value in table R. Hash functions also hide clusters of data by hashing clustered values to uniformly-distributed hashed values. A hash function with a high collision rate has the side effect of "compressing" the values of column B from table R since a single hash value can be used to represent multiple actual values. However, if the collision rate is too high, many false positives will occur in F due to the high number of collisions, yielding unnecessary transmission costs. Thus, it is important to use a hashing function that provides an acceptable level of performance while providing enough uncertainty to meet the privacy constraint.

It is computationally expensive to dynamically compute the hash values resulting from a new hash function with a different size each time a query is posed on a large data warehouse table. Furthermore, dynamic generation of values prevents indexing mechanism from being used to during the join operation in step 5. Our approach is to predefine a set of m hash functions $h_1, h_2, ..., h_m$ of different sizes. The result of each of these hash functions to column S on table S are stored explicitly (in m different columns) and indexed.

When the user performs a join on his private table R and the public table S, the privacy loss incurred with respect to the contents of table S is constrained to not exceed p_{rel}. In other words:

$$p_{rel} \geq \frac{H(\tilde{R} \mid S) - H(\tilde{R} \mid N)}{H(\tilde{R} \mid S)} \tag{5}$$

Assuming a uniformly-distributing hash function, the number of real values that hash to the same hash value is estimated to be $\frac{|U|}{|H|}$, where $|U|$ is the size of the domain of possible values for column B (the universe) and $|H|$ is the range size of hash function h. H is the set of possible values in the range of h. For any given hash value, $\frac{|U|}{|H|}$ possible values could have been used as input into the hash function and could have belonged to table R. For a set of $|N|$ hash values, there is a total of $|N|\frac{|U|}{|H|}$ possible values that data items in column B of table R can take on with equal probability. Thus, $H(\tilde{R} \mid N)$ is estimated as:

$$H(\tilde{R} \mid N) = \log_2\left(|N|\frac{|U|}{|H|}\right) \tag{6}$$

By combining equations 5 and 6, the constraint on $|N|$ for a given p_{rel} is found to be:

$$|N| \geq \frac{|H|}{|U|} 2^{(1-p_{rel})H(\tilde{R}|\tilde{S})} \qquad (7)$$

Applying equation 7 to each hash function, the minimum number of hash values $|r_1|, |r_2|, \ldots, |r_m|$ for all m available hash functions on dw can be found.

We can estimate the number of unique hash values generated by hashing each tuple in R with h_i analogously to [15] as:

$$|h_i(R)|_{est} = \left(1 - \left(1 - \frac{1}{|H_i|}\right)^{|R|}\right)|H_i| \qquad (8)$$

Then the actual size of the hash value set N_i that db would send to dw, if hash function h_i is selected, is:

$$|N_i| = \max(r_i, |h_i(R)|_{est}) \qquad (9)$$

Note that $|N_i| \leq |h_i(R)|$, so it may be necessary to add artificial hash values to the set N sent by db to dw in addition to $h_i(R)$. This can be done by randomly selecting $|N_i| - |h_i(R)|$ hash values that belong to the range of h_i. The set of artificial hash values is denoted as n_i, where $N_i = h_i(R) \cup n_i$.

3.2   Cost Estimation

To select the appropriate hash function for the data exchange, the transmission cost normalized with respect to the brute-force method (i.e., downloading table S from dw to db) $cost_i$ can be estimated. It is assumed that transmissions costs will dominate the execution costs of the overall join operation since the system will be operating over a limited communications link and search time is kept low with the use of indices.

If the brute-force method was used, $c_t|S|$ time units are required to transmit $|S|$ records from dw to db where c_t is the cost associated with transmitting a single record returned by dw in bytes. The cost of the hash/noise method can be estimated to be the sum of the cost of transmitting hash values from db to dw and the cost of transmitting the set of candidate tuples F returned by dw to db. The cost of sending the hash values is $c_h|N_i|$ time units for a hash function h_i, where c_h is the cost associated with transmitting a single hash value. The cost of the tuples returned by dw to db after the hash values have been sent is $c_t|F|$. Thus, the transmission cost normalized with respect to the brute-force method is summarized as:

$$cost_i = \frac{c_h|N_i| + c_t|F|}{c_t|S|} = \frac{|F|}{|S|} + \frac{c_h|N_i|}{c_t|S|} \qquad (10)$$

Equation 10 shows that as the cost-ratio c_h/c_t approaches zero, the cost of sending hash values $\frac{c_h|N_i|}{c_t|S|}$ becomes small. In other words, as the size of the tuples returned increases, the cost of sending the hash becomes insignificant. As $|F|$ approaches $|S|$, the performance of the hash/noise method is similar to that of the brute-force method; whereas, when $|F| << |S|$, we see significant performance improvement over the brute method. While $|F|$ is not known until the query has been executed, it can be estimated to be the average number of tuples returned by dw given the characteristics of the hash function and the contents of dw. It is found that on average for a given hash value, the number of values in column B that will collide to the some hash value is $\dfrac{|S|}{|H_i|}$ for a hash function h_i. Consequently, the average number of tuples returned by dw to db is $\dfrac{|S|}{|H_i|}|N_i|$. Thus the normalized transmission cost $cost_i$ for a hash function h_i is estimated to be:

$$cost_i = \frac{c_h\,|N_i|+c_t\,\dfrac{|S|}{|H_i|}|N_i|}{c_t\,|S|} \tag{11}$$

The hash function h_i, with the appropriate N_i found with equation 9 that yields the lowest normalized transmission cost according to equation 11 is selected as the hash function for the data exchange. Clearly, if $cost_i \geq 1$, it would be more advantageous to download S since the cost of doing so is either less than or equal to the cost of our hash-noise approach without any loss of privacy.

4 Implementation and Results

A prototype of this system was implemented in Java using MySQL [35]. Borrowing a technique from [15], eight hash functions were created by simply truncating the result of the MD5 hash [36]. Eight sets of hash values were generated for each B column value by truncating the result of the MD5 hash of a column B value to various bit sizes ranging from 8 to 16 bits. The hash value sets were stored and indexed in dw along with their respective S table. $H(\tilde{R}\,|\,S)$ was computed offline and stored for each S table.

Three sets of data were used for three instances of table S. The first two were each comprised of 2.5 million synthetically generated tuples. The values of column B for table S were generated with a uniform distribution of values from 0 to 99,999 for the first set. The second set's column B values were generated with a Gaussian distribution of values from 0 to 99,999 with a mean of 50,000 and a standard deviation of 1000. The third set of data was the "alignment block in rat chain of chromosome 10" table, taken from the UCSC Genome Browser Project [37]. The genome data set contains approximately 2.4 million records and was biased towards low join column values.

The size of the domain U for the uniformly and Gaussian-distributed join column values was 100,000. There were approximately 123,598 different values for the join column in the genome data set, so the size of domain U for join column values was approximated to be 2^{17}. Unless otherwise specified, the cost-ratio c_h/c_t was ½ (i.e., the cost of transmitting of a hash value is half the cost of transmitting a record from table S).

For each experiment, the R tables were generated randomly. The R tables to be joined with a uniformly or a Gaussian-distributed table S were generated by randomly selecting a value for column B from the range of 0 to 99,999. The R tables to be joined with the genome data were generated by randomly selecting tuples from the "summary information about chain of rat" table (also available from [37]). For each data point plotted, five R tables were randomly generated, each of which was joined with table S using the hash/noise method fives times. The maximum and minimum observed values of each studied parameter were ignored, and the rest were averaged. Timings were taken using a dual processor 1.3 GHz Dell workstation.

4.1   Performance Analysis

In this section, we will study the performance implications (i.e., the size of data sets transmitted and execution time) of different distributions of the private table and different privacy requirements.

4.1.1   Effect of Private Table Distributions

To begin the execution time analysis, we study the effect of different private table distributions on performance by varying the size of table R in relation to the size of the set of possible key values U ($|R|/|U|$) and fixing the required relative privacy loss to not exceed 0.01. Figure 3 shows how execution time varies as $|R|/|U|$ changes. Figure 4 shows how the size of the transmitted sets $|N|$ and $|F|$ varies as $|R|/|U|$ changes. For each of the execution time tests, the transmission cost of transmitting a hash value was equivalent to transmitting a 4-byte integer, and the cost of transmitting a tuple from S was equivalent to transmitting two 4-byte integers.

For a Gaussian distribution and genome data distributions of table S, execution time increases linearly as $|R|/|U|$ increases as do the sizes of N and F. Thus, as expected, the processing (i.e., transmission and computation time) of the two intermediate sets dominate the execution time for these two data distributions.

For a uniform distribution of table S, the execution time behaves as a step function, transitioning when $|R|/|U| = 0.6$. Figure 4 shows that $|N|$ increases along with the execution time curve; whereas, $|F|$ remains relatively constant. While initially surprising, as shown in Figure 8, when $|R|/|U|$ transitions from 0.6 to 0.7, the system

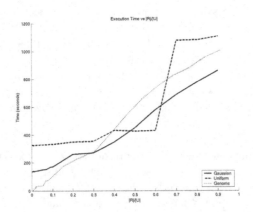

Fig. 3. Execution times for variable $|R|/|U|$. Target $p_{rel} = 0.01$ and $c_h/c_t = 1/2$

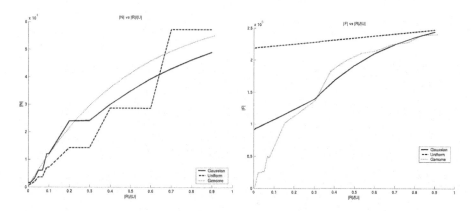

Fig. 4. Set sizes $|N|$ and $|F|$ for variable $|R|/|U|$ Target $p_{rel} = 0.01$ and $c_h/c_t = \frac{1}{2}$

experiences the largest increase in hash size $|H|$, resulting in far fewer collisions; and, consequently many more hash values are sent to dw to meet the privacy constraint. Because, the largest hash size increase occurs at much lower $|R|/|U|$ values for the Gaussian and genome distributions, any sharp increases in execution times are less apparent for those distributions.

Comparing the behavior of the various distributions, the execution time of the distributed join operation is directly related to the size of tables R, N, and F for the Gaussian and genome data distribution. However, for a uniform distribution, the execution time is generally independent of $|R|/|U|$, except when there is a large transition in hash values used, because the transmission of noise and false-positives dominate the cost. From this figure, it can also be seen that the execution times for join operations operating over the genome data distribution are lower than those of the Gaussian distribution, which are usually lower than those of uniform distribution. Less uniform distributions will usually result in better execution times because they are more biased and thus will have less entropy. Uniform distributions have the most entropy of any distribution, requiring either far more hash values or far more false positives to be returned by dw to satisfy the privacy constraint.

4.1.2 Effect of Privacy Requirements

In the second set of execution time analyses, we will study performance implications of different privacy requirements by fixing $|R|/|U|$ to 0.1 and by varying the maximum privacy loss, or the target relative privacy loss p_{rel}, from 0.01 to 0.96, in intervals of 0.05. Figure 5 shows how execution times vary as the target p_{rel} changes. Figure 6 shows how $|N|$ and $|F|$ vary as the target p_{rel} changes in the second graph. Intuitively, as the privacy constraint is relaxed, execution times for both the Gaussian and uniform data distributions decrease since fewer hash values are needed to satisfy the privacy constraint. For any join operation whose target p_{rel} is greater than 0.21, the execution times, $|N|$, and $|F|$ remain constant. In such cases, $|h(R)|$ is large enough to satisfy the privacy constraint without any noise. Thus, there is very little performance gain by increasing the target relative privacy loss greater than 21% for private tables containing only 10% of the total possible keys.

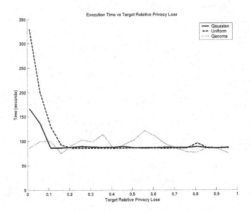

Fig. 5. Execution times for variable target p_{rel}. $|R|/|U| = 0.1$ and $c_h/c_t = \frac{1}{2}$

Figure 5 also shows that the execution time of the genome data set remains relatively constant, with minor variations in execution times due to the randomness of data items in set R and consequently the high randomness of data items in set F. Furthermore, $|N|$ remains constant regardless of the target privacy; and consequently, only the varying sizes of table F contribute to the variation in execution times, which is determined by the random selection of tuples in table R. This is shown in the second graph of Figure 6. The variance in execution times is more than that of the other distributions because the data in the genome data set is much less uniformly distributed than the other two distributions.

In summary, when target p_{rel} is low, there is more variation in execution times for the Gaussian and uniform distributions. When the privacy constraint is relaxed, there is little or no change in execution times.

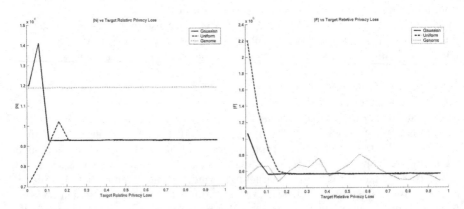

Fig. 6. Set sizes $|N|$ and $|F|$ for variable target p_{rel}. $|R|/|U| = 0.1$ and $c_h/c_t = \frac{1}{2}$

4.2 Absolute Privacy Loss Analysis

Figure 7 shows how absolute privacy loss varies as $|R|$ changes and the target p_{rel} is fixed at 0.01. For the uniform distribution, the absolute privacy loss is kept very low and close to the target p_{rel} of 0.01 since satisfying the relative privacy loss constraint for a uniform distribution is almost identical to satisfying an absolute privacy constraint of the same magnitude. However, for the Gaussian and genome data distributions, the absolute privacy loss differs greatly from the target relative p_{rel}, because far less effort is required to satisfy the relative privacy loss constraint than that required to satisfy an absolute privacy loss constraint of equal magnitude due to less uniformity in these distributions. For non-uniform distributions, achieving low absolute privacy loss would be much more expensive than achieving low relative absolute privacy loss; whereas, the cost for achieving both for a uniform distribution would be relatively the same.

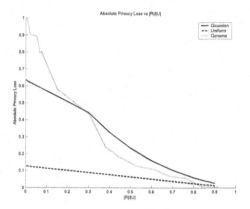

Fig. 7. Varying absolute privacy. Target p_{rel} =0.01 and c_h/c_t =½

Figure 7 also shows that as $|R|/|U|$ increases, absolute privacy loss decreases. In general, as $|R|/|U|$ increases, the data revealed by db to dw increases. As a result, the pool of possible values that an adversary can use to infer the actual values of column B in table R increases as well, resulting in far greater uncertainty about the actual value of a column B value in table R.

4.3 Hash Selection Analysis

In this analysis, we determined the size of the selected hash function that yields the lowest transmission cost increases as $|R|/|U|$ increases, for all distributions. We experimented with hash sizes ranging from 8 to 16 bits because any larger hash sizes, such as 17 bits would yield almost no collisions. It was found that as the uniformity of table S increases, a wider range of hash values is required to account for any variations in sizes of table R provided by a user. Depending on the size of $|R|/|U|$, for the uniform distribution, hash sizes ranging from 10-bits to 16-bits are required. For the Gaussian distribution, hash sizes ranging from 12-bits to 16-bits are required. Finally, for the genome data set, hash sizes ranging from 14-bits to 16-bits are needed.

4.4 Transmission Cost Analysis

In this set of analyses, the transmission costs of the hash/noise method in relation to the brute-force are studied.

The observed normalized transmission cost based on equation 10 using the observed $|F|$ is compared to the estimated norma lized transmission cost based on equation 11. The first graph of Figure 8 shows that the hash/noise method works well when $|R|/|U|$ is very low, and especially well when the distribution of key values in table S is very biased. For uniform distributions of table S and a target p_{rel} of 0.01, the transmission costs of the hash/noise method was 90% or more of the transmission costs of the brute-force method, costing as much as the brute-force method. For a Gaussian-distributed data set, the transmission costs ranged from 35% to 95% of the brute-force method, depending on $|R|/|U|$. For the skewed genome data set, the transmission cost also varied significantly depending on the size of $R|/|U|$.

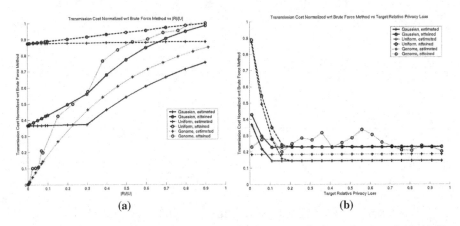

Fig. 8. Varying normalized transmission costs with respect to the brute-force method. (a) Target $p_{rel} = 0.01$ and $c_h/c_t = \frac{1}{2}$. (b) $|R|/|U| = 0.1$ and $c_h/c_t = \frac{1}{2}$

The second graph shows that the transmission cost steeply decreases as the target p_{rel} increases from 0.01 to 0.2 for both Gaussian and uniform distributions. For any target p_{rel} greater than 0.2, transmission costs are 25% of that of the brute-force method, for all distributions. The general behavior of steeply decreasing and flattening out was predicted by the estimated normalized transmission cost curves, but the actual transmission costs were not accurately estimated. For the less uniform genome data, the transmission costs remain relatively constant with an average of 25% of that of the brute-force method, for all target relative p_{rel} values and when $|R|/|U|$ is 0.1. Like for the other distributions, the general behavior of the observed transmission cost curve was predicted by the estimated transmission cost curves, but the actual transmission costs were poorly predicted.

Figure 9 compares the attained normalized transmission cost of the hash/noise method with the cost of simple semi-joins (i.e., no privacy constraints enforced). The graph shows that $|R|/|U|$ is directly proportional to what the cost of the semi-join would be. The graph summarizes how much more the hash/noise method costs to satisfy a maximum relative privacy loss of 0.01 in comparison to a semi-join, which provides for no privacy. Using the hash/noise method, it is very expensive to achieve a maximum relative privacy loss of 0.01 when the distribution of the column B values

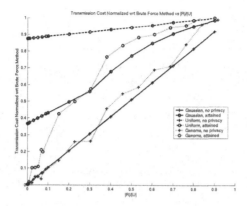

Fig. 9. Attained normalized transmission costs of join with privacy constraints and join without privacy constraints. Target $p_{rel} = 0.01$ and $c_k/c_t = \frac{1}{2}$. Cost of transmitting the key of a record from db is half the cost of transmitting a tuple from dw

of the S table is uniform. In contrast, when the S table is non-uniform, there is much less additional cost for the added privacy that the hash/noise method provides.

5   Conclusion and Future Work

A practical solution to the private date integration problem must maintain privacy while remaining efficient. Based on the metric of relative information gain, we have presented an efficient approach to performing joins between a relatively small database and a large, public data repository. By making use of predefined hash functions and noise injection to satisfy the privacy constraints, traditional indexing mechanisms can be used. Thus, the total cost of a distributed join is dominated by transmission costs rather than by search and computational costs.

Based on our preliminary results, several future research directions can be pursued. Our current cost estimation uses the average number of collisions to estimate the number of tuples to be returned by dw, which works well for uniformly-distributed data but poorly for non-uniformly distributed data. In future work, additional features such as the distribution of table S could be incorporated into the estimate. This work can also be expanded to infinite domains (e.g., people's names), specifically to develop a privacy loss metric relevant to these domains. Additionally, our method only protects the privacy of data over a single query; and, it may be possible for adversaries to make inferences over multiple queries. Perhaps, some caching can be used to avoid exposing the same private data set more than once. Finally, the presented hash/noise technique only works for the equijoin operation. There may be a need to develop methods to protect the privacy of data that are processed by general joins.

Our promising initial results show the merit of using hashing and noise injection to solve the problem of efficiently integrating small amounts private data with large amounts of public data. In comparison to other related approaches, the hash/noise technique does not assume non-collusion, does not require downloading the entire data warehouse table, leverages existing indexing mechanisms, and provides for finer-grain control of privacy than simple hashing.

Acknowledgements

This work (UCRL-CONF-206647) was performed under the auspices of the U.S. Department of Energy by the University of California Lawrence Livermore National Laboratory under contract no. W-7405-Eng-48. The authors would like to thank Tina Eliassi-Rad, David Buttler, and Roderick Son for their valuable input.

References

1. S. Phillippi and J. Kohler, "Using XML Technology for the Ontology-Based Semantic Integration of Life Science Databases," IEEE Transactions on Information Technology in Biomedicine, vol. 8, no. 2, pp. 154-160, June 2004.
2. Tomasic, L. Raschid, and P. Valduriez, "Scaling Access to Heterogeneous Data Sources with DISCO," IEEE Transactions on Knowledge and Data Engineering, vol. 16, no. 5, pp. 808-823, Sept/Oct 1998.
3. S.B. Davidson, et al, "Transforming and Integrating Biomedical Data using Kleisli: A Perspective," ACM SIGBIO Newsletter, vol. 19, no. 2, pp. 8-13, 1999.
4. ZLacroix, O. Boucelma, and M. Essid, "The Biological Integration System," in Proceedings of WIDM '03, pp. 45-49, New Orleans, LA, Nov. 7-8, 2003.
5. M. Alvarez, et al, "FINDER: A Mediator System for Structured and Semi-Structured Data Integration," in Proceedings of the 13th International Workshop on Database and Expert Systems Applications (DEXA'02), pp. 847, Aix-en-Provence, France, Sept. 2-6, 2002.
6. L.M. Haas, et al "DiscoveryLink: A System for Integrated Access to Life Sciences Data Sources," IBM Systems Journal, vol. 40, no. 2, pp. 489-511, 2001.
7. B. Thuraisingham, "Data Mining, National Security, Privacy and Civil Liberties," ACM Special Interest Group on Knowledge Discovery in Data and Data Mining (SIGKDD) Explorations Newsletter, vol. 4, no. 2, pp. 1-5, June 2002.
8. M.S. Olivier, "Database Privacy: Balancing Confidentiality, Integrity and Availability," ACM Special Interest Group on Knowledge Discovery in Data and Data Mining (SIGKDD) Explorations Newsletter, vol. 4, no. 2, pp. 20-27, June 2002.
9. R. Agrawal et al, " Hippocratic Databases," in Proceedings of the 28th Very Large Databases (VLDB) Conference, Hong Kong, China, 2002.
10. T.D. Sterling and J.J. Weinkam, "Sharing Scientific Data," Communications of the ACM, vol. 33, no. 8, pp. 113-119, Aug. 1990.
11. F. S. Collins, E. D. Green, A. E. Guttmacher, and M. S. Guyer, "A vision for the future of genomics research,"Nature, vol. 422, no. 6934, pp. 835-847, 2003.
12. NCBI, GenBank,'[Online] Available: http://www.ncbi.nlm.nih.gov/Genbank/index.html, 2004.
13. P.A. Bernstein and D.W. Chiu, "Using Semi-Joins to Solve Relational Queries," Journal of the ACM, vol. 28, no. 1, pp. 25-40, Jan. 1981.
14. P.L. Vora, "Towards a Theory of Variable Privacy," in review, May 7, 2003.
15. G. Schadow, S.J. Grannis, and C.J. McDonald, "Privacy-Preserving Distributed Queries for a Clinical Case Research Network," in Proceedings of IEEE International Conference on Data Mining Workshop on Privacy, Security, and Data Mining, Maebashi City, Japan, 2002.
16. D. Agrawal and C.C. Aggarwal, "On the Design and Quantification of Privacy Preserving Data Mining Algorithms," in Proceedings of Principles of Database Systems (PODS) 2001, pp. 247-255, Santa Barbara, CA, 2001.

17. C. Clifton, M. Kantarcioglu, and J. Vaidya, "Defining Privacy for Data Mining," in Proceedings of the National Science Foundation Workshop on Next Generation Data Mining, Nov. 1-3, 2002, Baltimore, MD.

18. C. Clifton, et al, "Privacy-Preserving Data Integration and Sharing," in Proceedings of Data Mining and Knowledge Discovery (DMKD) '04, Paris, France, June 13, 2004.

19. J. Vaidya and C. Clifton, "Privacy Preserving Association Rule Mining in Vertically Partitioned Data," in Proceedings of ACM Special Interest Group on Knowledge Discovery in Data and Data Mining (SIGKDD) International Conference on Knowledge Discovery and Data Mining (KDD '02), Edmonton, Alberta, Canada, 2002.

20. S. Agrawal, V. Krishnan, and J. Haritsa, "On Addressing Efficiency Concerns in Privacy-Preserving Data Mining," in Proceedings of the International Conference on Database Systems for Advanced Applications (DAFSAA) 2004, pp. 113-114, Jeju Island, Korea, Mar. 17-19, 2004.

21. W. Du and Z.Zhan,"Using Randomized Response Techniques for Privacy-Preserving Data Mining," in Proceedings of ACM Special Interest Group on Knowledge Discovery in Data and Data Mining (SIGKDD) International Conference on Knowledge Discovery and Data Mining (KDD '03), Aug. 24-27, 2003.

22. R. Agrawal and R. Srikant, "Privacy-Preserving Data Mining," in Proceedings of the 2000 ACM International Conference on Management of Data, pp. 439-450, Dallas, TX,2000.

23. B. Chor et al, "Private Information Retrieval," Journal of the ACM, pp. 965-982, vol. 45, no. 6, Nov. 1998.

24. R. Agrawal, A. Evfimievski, and R. Srikant, "Information Sharing Across Private Databases," in Proceedings of the Special Interest Group on Management of Data (SIGMOD) 2003, pp. 86-97, San Diego, CA, June 9-12, 2003.

25. M. Kantarcioglu and C. Clifton, "Assuring Privacy when Big Brother is Watching," in Proceedings of Data Mining and Knowledge Discovery (DMKD) '03, San Diego, CA, June 13, 2004.

26. C. Clifton, et al, "Tools for Privacy Preserving Distributed Data Mining, ACM Special Interest Group on Knowledge Discovery in Data and Data Mining (SIGKDD) Explorations Newsletter, vol. 4, no 2, pp. 28-34, Dec. 2002.

27. M. Naor and B. Pinkas, "Efficient Oblivious Transfer Protocols," in Proceedings of Society of Industrial and Applied Mathematics (SIAM) Symposium on Discrete Algorithms, Washington, DC, Jan. 7-9, 2001.

28. M. Bellare and S. Micali, "Non-Interactive Oblivious Transfer and Applications," in Proceedings on Advances in Cryptology, pp. 547-557, Santa Barbara, CA, 1989.

29. M.J. Freedman, K. Nissim, and B. Pinkas, "Efficient Private Matching and Set Intersection," in Proceedings of Eurocrpyt 2004, Interlaken, Switzerland, May 2-6, 2004.

30. Y. Gertner et al, "Protecting Data Privacy in Private Information Retrieval Schemes," in Proceedings of the 13th Annual ACM Symposium on Theory of Computing, pp. 151-160, Dallas, TX1998.

31. J.K. Mullin, "Optimal Semijoins for Distributed Database Systems," IEEE Transactions on Software Engineering, vol. 16, no. 5, pp. 558-560, May 1990.

32. J.M. Morrissey and W.K. Osborn, "Distributed Query Optimization Using Reduction Filters," in Proceedings of IEEE Canadian Conference on Electrical and Computer Engineering, vol. 2, pp. 707-710, May 24-28 1998.

33. S. Bellovin and W. R. Cheswick, "Privacy-Enhanced Searches Using Encrypted Bloom Filters," in Proceedings of DIMACS/Portia Workshop on Privacy-Preserving Data Mining, Piscataway, NJ, Mar. 15-16, 2004.

34. C.E. Shannon, "A Mathematical Theory of Communication," Bell System Technical Journal, vol. 27, pp. 379-423 and 623-656, July and Oct. 1948.
35. MySQL AB, "MySQL: The World's Most Popular Open Source Database," Aug. 2004; http://dev.mysql.com/
36. A.J. Menezes, P.C. van Oorschot, and S.A. Vanstone, Handbook of Applied Cryptography, CRC Press, 1997, pp. 347.
37. UCSC Genome Bioinformatics, "UCSC Genome Browser Home," Aug. 2004; http://genome.ucsc.edu/

Building a Generic Platform for Medical Screening Applications Based on Domain Specific Modeling and Process Orientation

Stefan Jablonski[1], Rainer Lay[1], Sascha Müller[1], Christian Meiler[1], Matthias Faerber[1], Victor Derhartunian[2], and Georg Michelson[2]

[1] Chair for Database Systems, University of Erlangen-Nuremberg,
Martensstraße 3, 91058 Erlangen, Germany
{Stefan.Jablonski, Rainer.Lay, Sascha.Mueller, Christian.Meiler,
Matthias.Faerber}@informatik.uni-erlangen.de
[2] Ophthalmic Department of the University of Erlangen-Nuremberg,
Schwabachanlage 6, 91054 Erlangen, Germany
Victor.Derhartunian@augen.imed.uni-erlangen.de
Georg.Michelson@augen.imed.uni-erlangen.de

Abstract. The German health care system increasingly encounters an enormous cost pressure. Preventive medicine opens the possibility to avoid cost for the treatment of chronically sick persons and, especially, for the highly expensive hospitalization. Since with screenings, a special discipline of preventive medicine, a large number of persons are to be examined, information technology plays an important role to reduce cost and to increase treatment quality. We introduce a generic process based platform for distributed screenings for the early detection and diagnosis of the glaucoma disease. Thereby, glaucoma is merely one disease pattern which can be covered with the generic process based platform. Methods and concepts for the enactment of different screening processes and the integration of various modalities are in the center of our interest.

1 Introduction

Glaucoma is a prevalent disease of the eyes which is characterized by a damage of the head of the optic nerve. Besides diabetes mellitus it is the second most prevalent cause for blindness in the industrial countries. From an economic perspective glaucoma causes – as a chronic disease – high expenses [1].

Since an early detection of glaucoma allows for better therapy, the aim is to detect the disease in the earliest possible state. To identify people with glaucoma in that early state, screening measures are performed to detect early signs of the glaucoma disease. To support the diagnosis for glaucoma with high specificity, multiple independent indicators for the glaucoma disease have to be measured and assessed [2].

The screening process which we introduce here is divided into two parts. In the first phase, patient data (identification, medical history, etc.) and examination data (images, measures, etc.) are collected. Thereby, a semi-mobile screening setting is used to examine patients by assistant medical technicians on site, e.g. in companies,

B. Ludäscher and L. Raschid (Eds.): DILS 2005, LNBI 3615, pp. 257–265, 2005.

public utilities, clinics. In a second phase, the data are analyzed telemedically by a specialized ophthalmologist, physically and timely separated from the first phase.

1.1   Medical and Economical Requirements

In the context of the "Sonderforschungsbereich 539"[1] at the ophthalmic clinic of the University of Erlangen-Nuremberg a process for the semi-automatic determination of glaucoma risk is developed in an interdisciplinary way by physicians, bio statisticians and computer scientists. A screening process is characterized by a set of modalities and a clinical algorithm which reflects the criteria of medical decisions. This algorithm contains medical, evidence based knowledge which supports the physician by proposing a standardized diagnosis. The integration of modalities decisively determines the cost of the screening; it also influences the process itself and the clinical algorithm used. Flexibility is therefore one of the major issues for a basis platform for screening examinations. Depending on the medical problem to investigate two perspectives of flexibility can be identified:

1. A certain set of modalities of examinations must be integrated into the screening process.
2. The screening process has to implement specific medical, evidence based knowledge. Here, recommendations from medical guidelines and directives must be incorporated.

When a basic platform for screening fulfills these requirements, it can be used easily for screening examinations of other diseases like diabetes mellitus, stroke, etc. [3].

1.2   Derived Technical Requirements

The medical and economical requirements impose major challenges on the technical configuration of a semi-mobile screening setting. The desired simple re-configuration of the screening platform to support different medical questions requires the dynamic loading of different medical processes. This implies the following:

1. Support for different screening processes requires a comprehensive concept to specify the used modalities and the data flow between all participating devices.
2. The generic screening platform must be able to visualize arbitrary multi media patient data.
3. Decisions during an examination must be supported by collected medical knowledge (mostly formulated as medical rules).

To optimize the deployment of medical personnel, the basic screening platform must separate the screening work place from the diagnosis work place. This implies data communication between the former and the latter (and vice versa). Security issues are to be considered here thoroughly: data has to be encrypted; the completeness and consistency of the transferred data must be guaranteed. The diagnosis finally must be made available to the patients, either by putting it into the patient's electronic health record (EHR) or by sending it by mail. Hereby, security issues must be considered as well.

2   Model Based, Process Oriented Approach

Our approach is built on two cornerstones: it is model based and process oriented. The model based aspect divides the specification of a screening process[1] into three phases. In a first general phase, screening examinations are defined, modeled and documented in standardized process model templates (e.g. for glaucoma, diabetes mellitus). In a second phase, these templates are adjusted to the specific setting of modalities that are available for a certain screening project and to the concrete subject that has to be examined. This customization can be done through modification of the underlying screening process models by the medical personnel. In the third phase, the models will be executed. Thus, it is necessary that the models can automatically be transformed into an executable form.

Due to these requirements we decided to deploy the aspect oriented process model introduced in [3]. Its main advantage is its extensibility. This will be used in our application to describe medical facts and issues in an adequate way, so that it is highly illustrative. As shown in [7], aspect orientation allows us to introduce new domain specific modeling constructs like the "evidence based decider". These constructs make it easy for a medical user to understand the content of a process quickly. Further extensions of the basic aspect oriented process model support the integration of the modalities, whereby the complicated data provision task has to be considered. Special constructs for the so called data logistics (Section 3.2) facilitate the necessary data transport from and to the modalities.

However, it has to be mentioned that the extensibility of the aspect oriented process model also comes with some additional costs. Defining extensions, like e.g. the domain specific constructs, is time consuming and laborious. Nevertheless, these will be compensated widely through the benefits of easy readability and technical adequateness.

The following sections discuss the conceptual foundations of our generic platform for screening applications (Section 3) and its enactment (Section 4).

3   Specification of the Screening Platform

This section exposes the conceptual foundation of our approach, i.e. the modeling part of screening applications is analyzed. At first, the essential requirements of our approach are discussed:

- Correct and consistent enactment of screening applications:
  Its basis is the process model describing a screening application. At first, the order of process steps must be determined: decisions to be taken must be described and alternative variants must be prepared. Aspects of quality management have to be considered. Decisions to be taken must be reproducible and connected to execution paths taken and data collected. All in all the process models must follow clinical algorithms [4]. Basing the screening processes on these matured algorithms increases their quality.

---

[1] Although process orientation is the second cornerstone which is motivated next, we anticipate that screening applications are modeled as processes right now.

- Integration of modalities and multi media data:
  Most of the data involved in screening applications are multi media objects. They are in general produced by high-tech modalities. Thus, the connection of these modalities and the integration of created data is one major task. The data are often images; therefore the various standards for image data transfer must be considered [5]. Besides, alphanumerical data are to be integrated as well; these mostly describe the patients themselves and their patient history. Since most of the data are very sensitive, security issues must be taken into account.

- Support of analytical and diagnostic steps:
  Besides the execution of work steps which mostly are connected to modalities and produce data, a second major type exists: those steps which enforce the tight involvement of medical experts. In such work steps typically medical decisions have to be taken. To be able to diagnose a certain situation, the physician must be able to browse through – potentially – all data that are related to the patient. A suitable process model has to offer appropriate means to express the graphical presentation of relevant data.

As explained before, we are pursuing a process oriented approach. In particular, we choose an aspect oriented process model [3], which is enacted by the modeling tool i>ProcessManager (i>PM) [6]. We use it to introduce a domain oriented model extension, i.e. to introduce new modeling constructs which most adequately realize the requirements discussed above. The main purpose of domain specific extensions is to break the tight corset of conventional process modeling elements. The domain specific constructs alleviate the use of a process model which coincides with an increasing acceptance of the applied method. We show in the next sections how the domain specific extension of the i>PM process model facilitate the above requirements.

3.1   Constructs for the Compact Specification of Medical Processes

The first set of modeling constructs reflects special medical situations that have to be specified in a screening process. The constructs introduced are broadly introduced in [7]; in this paper we briefly describe their fundamental properties and discuss their valuable contribution to specify domain specific process models.

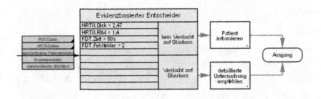

Fig. 1. The Glaucoma screening process (text in German)

Fig. 1 shows a small part of a glaucoma detection and diagnosis process which is taken as reference throughout this paper. The figure depicts a new domain specific modeling construct (big step), the so called "Evidence Based Decider" (EBD). It is designed to automatically generate a proposal for a diagnosis. It works on a set of input data which is shown left from the construct. The principle rules used in the clinical algorithm described by the construct are depicted within the body of the construct. The two alternative decisions proposed can be found at the right edge of the construct. However, finally the physician merely uses this proposal as the only argument, s(he) finally decides about the glaucoma suspicion independently.

From the perspective of our approach, a modeling construct like the EBD bears the following advantages:

- A physician can grasp the modeled situation very rapidly and the compact and customized notation provides him a comprehensive and transparent perspective at the patient's condition.
- All data and the algorithm used to assess the patient's condition are traceable. This is of enormous importance for quality management.

The algorithm used in the construct can be based on evidence, i.e. on published and well assessed clinical algorithms.

Alternatively the depicted situation could be represented with conventional constructs like sequence or decision. This would drastically complicate the process presentation with the consequence that a physician would presumably not accept it and would therefore not accept the whole approach.

To summarize this subsection, the question after the benefit of domain specific extensions is asked. The answer is divided into three parts: firstly, the domain specific constructs allow incorporating domain specific (here medical) knowledge (for example, how risk factors are assessed), and, secondly, guarantee high quality. Thirdly, these specialized constructs are able to represent that knowledge in an (adjusted) compact way. This is one decisive reason why users (here physicians and technical medical assistants) do accept the approach and are willing to cooperate.

3.2 Constructs for the Specification of Data Logistics

In the process model (Fig. 1) data from five modalities are included. As mentioned before, the data produced by these modalities, mostly multi media data like images, must be incorporated into the screening process, i.e. finally in the screening data base. In order to connect the modalities to the screening process, the so called "Data Logistics" (DL) construct is used (broadly introduced in [8]). Again the DL is a domain specific extension of the basic process model of i>PM. However, while the extensions of Section 3.1 are content related, the DL construct is technically motivated. It provides an elegant concept to include modality data into a medical process. To incorporate a special modality with its data, a DL work step must be specified. The enactment of DL constructs is partially introduced in Section 4.3.

3.3 Constructs for the Specification of Data Visualization

Data visualization steps are used for work steps that are associated with diagnosis and related activities (cf. Section 3). The requirement is to enable the physician to browse all relevant data that are somehow related to the decision. Especially in the medical applications that are investigated in this paper, these data are complex images or sequences of images mixed with tables of alphanumerical data.

4   Enactment of a Semi Mobile Screening Platform

This section presents the basic technical concepts for the enactment of the screening processes. Fig. 2 depicts the essential functional components of the screening platform:

- **Modalities:** Modalities are devices that are involved in the screening process.
- **Users:** The medical personnel which are taking care of the screening process.
- **Data Management:** All data are to be managed that is relevant throughout the screening process and which describes the patients.The necessary mapping between generated data and the underlying database is derived from the process model. The link to the electronic patient record (EPR) must be sustained.

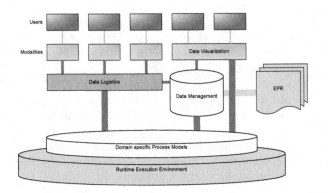

Fig. 2. Architecture of the Screening Platform

- **Domain specific process models:** Extensions of the basic process model that are used to improve the readability of process (cf. Sections 3.1, 3.2, and 3.3).
- **Runtime Execution Environment:** This provides the execution infrastructure for a screening process.
- **Data Logistics**: This is the special data exchange mechanism that takes care of the data transport between the modalities and the other parts of the screening platform in the background (cf. Section 3.2).
- **Data Visualization:** This component takes care of the appropriate presentation of the screening data.
- **Electronic Patient Record (EPR):** All data collected in the context of the screening examinations must potentially be transferred to the patients' EPR.

The following sub-sections discuss the concept and the enactment of the fundamental components of the screening platform that are depicted in Fig. 2. However, technical features of modalities are not analyzed in detail. In Section 4.3, we merely look at these highly sophisticated devices from the viewpoint of data provision.

4.1 Data Management

Data management is responsible to store all data and diagnosis that are produced during a screening process. Very similar to the process meta model, a generic data model is used in our screening platform. Therefore, a repository is taken as a foundation on top of which arbitrary data structures can be defined [9]. This approach allows including screening specific data for a screening process. This is necessary since similar to the examination specific structuring of a screening process (cf. Section 1.1) specific data are to be considered.

4.2 Data Logistics

We refer to [8] and [10] for a detailed discussion of the enactment of DL. Here, we briefly summarize the main aspects of this technique.

Two major tasks must be facilitated for DL: data transformation and data transportation. The latter takes care of moving data between applications and is defined by data access parameters and data quantification. The former is responsible for transforming data in such a way that the sending and the receiving applications can understand them.

4.3 Data Visualization

The data visualization (DV) component is a web based client server application build on the apache Cocoon [11] framework. Its main task is to perform the interaction with the user. The interaction workflow is organized according to the specification in the process model and implemented using Java and Cocoon flowscripts. The DV component extracts and stores patient data (e.g. patient history) and measurements (e.g. images) from the data management component according to the specification described by the data visualization constructs (cf. Section 3.3).

5 Related Work

Preventive screening applications are connected to several fields of medical data management. In this section we outline the intersections of hospital information systems (HIS), workflow management (WFM) and screening platforms.

HIS [12] share some requirements with screening applications, e.g. they gather information from different modalities, organize and process data input or manage EPRs. But in contrary to screening applications which require a very lightweight and quickly adjustable solution, they offer a heavy weight installation that meets the requirements of clinical data management. In other words, the target application area of HIS is very different from the one of our flexible screening platform, whereas the tasks are similar.

WFM [3] is a general concept to control the flow of arbitrary activities. It is based on the specification of a workflow model (i.e. process model) that can be mapped to an execution environment. In this respect WFM is closely related to our approach. HIS are also starting to use workflow technology in a conventional way. But our generic screening platform offers the advantage of application specific solutions, like domain specific modeling constructs, tight integration of modalities, specification of medical scoring algorithms, adaptive data visualization and distributed screening process execution.

6 Conclusion, First Experiences and Outlook

The main objective of the methods presented in this paper is to support screening applications in the context of preventive medicine. Its medical and economical requirements impose challenging requirements on an execution platform for screening processes. Our model based, process oriented approach can cope with these requirements. We have demonstrated how the extensibility of the i>PM process model enables medical domain specific modeling constructs, which facilitate the compact and adequate specification and enactment of screening processes. We have already tested our screening platform in two large applications. In a first application, about 500 employees of a big enterprise in Nuremberg were examined. The second test case was a public examination of about 400 citizens in Erlangen. In these applications, different sets of modalities were used. Diagnoses happened in both cases remotely and after the examinations. These two test applications have proven our concepts perfectly. Using the model based, process oriented specification and execution of the screening applications, the setup of the two applications could be done very efficiently. Minor modifications of the screening processes – caused by experienced gathered through the examinations – could also be performed very well. Altogether, these two test cases were both proofs of concept.

References

1. Sonderforschungsbereich 539: „Glaukome einschließlich Pseudoexfoliationssyndrom" Homepage. http://www.sfb539.forschung.uni-erlangen.de/. retrieved 2004-10-28
2. Michelson G., Groh MJ.: Screening models for glaucoma. Curr. Opin. Ophthalmol. 2001 Apr;12(2):105-11, PMID: 11224716 [PubMed - indexed for MEDLINE]
3. Jablonski S., Bußler C., 1996. Workflow management - modeling concepts, architecture and implementation. London. International Thomson Computer Press, 1996
4. Arbeitsgemeinschaft der Wisenschaftlichen Medizinischen Fachgesellschaften: Homepage, http://awmf.org/, retrieved 2004-11-16
5. Bidgood, W.D., Horii, S.C., Prior, F.W., Van Syckle, D.E.: Understanding and Using DICOM, the Data Interchange Standard for Biomedical Imaging. Journal of the Maerican Medical Informatics Association 1997 4(3): 199-212
6. ProDatO GmbH: i>ProcessManager,http://www .prodato.de/software/processmanager, retrieved 2005-03-09
7. Meiler, C.: Modelling, Planing, and Execution of Clinical Paths. PhD Thesis, University of Erlangen-Nuremberg, 2005
8. Jablonski, S.; Lay, R.; Meiler, C.; Müller, S.: Process Based Data Logistics: A solution for Clinical Integration Problems. In: Rahm, E. (Ed.): First International Workshop on Data Integration in the Life Sciences (DILS 2004), LNBI 2994: Springer Verlag, 2004.
9. Jablonski, S.; Petrov, I.; Meiler, C.; Mayer, U.: Metadata Repositories as Infrastructure for Database Oriented Applications. CAISE 2003 (The 15th Conference on Advanced Information Systems Engineering Velden, Austria), 2003.

10. Jablonski, S.; Lay, R.; Meiler, C.; Müller, S.; Huemmer, W.: Data Logistics as a Means of Integration in Healthcare Applications. The 20th Annual ACM Symposium on Applied Computing, Workshop on Computer Applications in Health Care, Santa Fe, 2005

11. Apache Cocoon Project, Homepage, 2005, http://cocoon.apache.org/, retrieved 2005-03-03

12. Kuhn, K.A.; Giuse, D.A.: From Hospital Information Systems to Health Information Systems - Problems, Challenges, Perspectives. Method Inform Med 2001; 40 (4): 275-287

Automatic Generation of Data Types for Classification of Deep Web Sources*

Anne H.H. Ngu[1],**, David Buttler[2], and Terence Critchlow[2]

[1] Department of Computer Science, Texas State University,
San Marcos, TX 78666
[2] Center for Applied Scientific Computing,
Lawrence Livermore National Laboratory, Livermore, CA 94551

Abstract. A Service Class Description (SCD) is an effective meta-data based approach for discovering Deep Web sources whose data exhibit some regular patterns. However, it is tedious and error prone to create an SCD description manually. Moreover, a manually created SCD is not adaptive to the frequent changes of Web sources. It requires its creator to identify all the possible input and output types of a service *a priori*. In many domains, it is impossible to exhaustively list all the possible input and output data types of a source in advance. In this paper, we describe machine learning approaches for automatic generation of the data types of an SCD. We propose two different approaches for learning data types of a class of Web sources. The Brute-Force Learner is able to generate data types that can achieve high recall, but with low precision. The Clustering-based Learner generates data types that have a high precision rate, but with a lower recall rate. We demonstrate the feasibility of these two learning-based solutions for automatic generation of data types for citation Web sources and presented a quantitative evaluation of these two solutions.

1 Introduction

One of the main impediments to large-scale integration of Deep Web sources is the inability to reconcile the semantic heterogeneity of the sources in an automatic and consistent manner. The problem can be decomposed into homogenization of the input, output, and interaction semantics of a Deep Web source. While there is a large body of research work [4, 9, 5] on homogenizing semantics of the input schema of Web sources, not much work is reported on homogenizing the output and the interaction patterns of Web sources. We proposed a practical, heuristic approach for reconciling the semantic of a class of life science

---

* This work was performed under the auspices of the U.S. Department of Energy by University of California Lawrence Livermore National Laboratory under contract No. W-7405-ENG-48. UCRL-CONF-209719.
** This work was performed while the author was a summer faculty scholar at LLNL.

B. Ludäscher and L. Raschid (Eds.): DILS 2005, LNBI 3615, pp. 266–274, 2005.

Web sources using a Service Class Description (SCD) driven by users application needs. This SCD describes the generic functionalities of services in a particular domain, using example queries, the expected output, and a graph representation (workflow) of how service class members are expected to operate. We are able to classify two-thirds of BLAST sources with 100% accuracy using the manually created SCD in our initial experiments [7].

However, the manual creation of SCD is tedious and error-prone. It requires expert knowledge of both XML and regular expressions. Moreover, any change that affects the input, output, or navigation pattern of the sites, which is not already embedded in the SCD, will affect the accuracy of the classification. It is impossible to be able to anticipate all the possible input and output data types in a class of dynamic web sources. Instead, an adaptive approach where an SCD can be incrementally created and updated is extremely valuable. The automatic creation of SCD enables a scalable and adaptive approach to semantic reconciliation of large numbers of sources in the presence of frequent re-organization, variation in navigation styles and input and output format of underlying sources. Based upon a few known Web sources of interest in a particular domain, a user can interact with those sites by going to a query page, posting a query and retrieving the results. The actions and responses from these chosen sites are used to construct a target SCD that can be fine-tuned with each successive example site.

We propose two different approaches for learning a set of rules that can be used to discover the data types of a class of Web sources. The Brute-Force approach is able to generate regular expressions ranging from the most specific to the most generic patterns for a given tagged example. The generated rules have a high recall rate but low precision. The Clustering-based approach aims to generate regular expressions that best fit the given set of training examples. The generated rules have a high precision rate, but with a lower recall rate than the Brute-Force approach. Both of these approaches allow new rules to be added or revised incrementally with new training examples. Eventually our learner will create a representative set of rules (data types) for a class of Web sources that is tailored to that domain's user information seeking behavior. The main contribution of this paper is the demonstration of the feasibility of these two learning-based solutions for automatic data types generation and a quantitative evaluation of these two solutions.

2   Service Class Description

The service class description provides a mechanism for encapsulating the components that are common to all members of the class and is the means for hiding insignificant differences between individual sources in a particular domain. However, it must also provide enough information to differentiate members of the class from a set of arbitrary Web sources. Service classes are specified by a *service class description*, which uses an XML format and regular expression to define the relevant functionality of a category of Web sources, from an application's perspective. The service class description format supports three categories of information used to define a Deep Web source: *data types, example queries*, and *control flow*.

Data Types are used to describe the input and output of a service class and any data elements that may be required during the course of interacting with a service. The service class data type system is modeled after the XML Schema [3] type system and includes constructs for building atomic and complex types. The `DNASequence` type in Figure 1 is an example of an user-defined type in the nucleotide BLAST service class description. Figure 1 also shows examples of user defined types called `AlignmentSequence` which makes use of `DNASequence` type.

```
<type name="DNASequence" type="string" pattern="[GCATgcat-]+" />
<type name="AlignmentSequence" >
    <element name="AlignmentName" type="string" pattern=".{1,100}:" />
    <element type="whitespace" />
    <element name="m" type="integer" />
    <element type="whitespace" />
    <element name="Sequence" type="DNASequence" />
    <element type="whitespace" />
    <element name="n" type="integer" />
</type>
```

Fig. 1. Sample BLAST service class data type definitions

Control flow graphs are used for enumerating the expected navigational paths used by all members of the service class. A control flow graph consists of a set of states connected by edges. *Examples* contain queries that can be executed against an instance of the service class. Specifically, examples can be used to determine if a site accepts input (data) as required by the service class. In the context of this paper, we only examine the automatic generation of the data types, which are used by both the example queries and the control flows. The automatic generation of example queries and control flow for a specific web site required in a service class description is beyond the scope of this paper.

3   Data Type Learner

The automatic generation of data types for a Service Class Description (SCD) alleviates the tedious and error-prone approach of manual SCD creation. It involves 1) locating the regions in the document that the system is interested in generating the data types for, 2) partitioning the regions into tokens of suitable granularity for data types generation, 3)generating a regular expression (regex) for each data type that balances specificity and generality from the set of annotated examples. We assume that existing techniques such as PageDiff [7], QA-pagelets [2] and Omini region identification [1] can be used to locate regions of interest. We use a simple tokenization mechanism based on whitespace and punctuation marks in our current Data Type Learner for token generation. The core of our Data Type Learner is the automatic generation of regexes (the fundamental data types that makes up SCD definition) based on a set of annotated examples. Our approach to regex generation is similar to WHISK [8] which learns rules in the form of regular expressions from human annotated examples.

The input to the Data Type Learner is a training document, a set of user-annotated examples and a domain specific template type. The output is a set of rules (regex) that can be used to identify the data types specified in the template in an unseen document from the same class. There are two types of atomic rules that can be generated by the Data Type Learner: Matrix rules and Token Set rules. These two types of atomic rules are at the opposite extremes in the spectrum of rule generation. Matrix rules match a fixed number of strictly defined tokens, while Token Set rules match a statistical group of tokens from a set. Tokens in a Matrix rule are strictly ordered, while tokens in a Token Set are unordered. We can create Composite rules from a collection of other rules (such as Matrix rules, Token set rules, and other Composite rules).

Users or domain experts must first describe the generic structure of data in a specific domain that they are interested in before they can use the learner to generate the data types for this domain. This high-level type information is described in a domain specific template type. Figure 2 shows a template for a type meant to describe a citation. The **citation** template states that a citation

```
<type name="citation" type="CompositeRule" >
        <type name="author" type="MatrixRule"/>
        <type name="title" type="MatrixRule"/>
        <type name="venue" type="MatrixRule"/>
        <type name="date" type="MatrixRule"/>
</type>
```

Fig. 2. Template Type for Citation

data type is a composition of author, title, venue and date. For each data type, users can specify the type of rules that can be generated for it. For simplicity in presentation, the example in Figure 2 specifies that Matrix rules need to be generated for all the data types associated with a citation.

A Matrix rule is defined to be a list of rule tokens. Each rule token in a Matrix rule can be a literal string, a semantic class, an user-defined regex type or a regex pattern. The semantic class and user-defined regex type rule tokens are techniques used to inject specific domain knowledge into the Data Type Learner. This will improve the quality of the generated rules. For example, in generating the Matrix rules for a date, knowledge about the valid years, months and days can be incorporated via a semantic class or a regex type. A semantic class is effectively an enumeration of all instances of a type in a specific domain, while regex type is a predefined regular expression for a specific type common to a domain. An example of a predefined regex type for the BLAST domain is DNASequence type shown in Figure 1. The other inputs to a Data Type Learner are the training document and a set of annotated examples from that document. A training document in our case is an html page showing a list of citations. The annotated examples are data that users want to extract.

The Composite rule is defined as an ordered composition of atomic data types or another Composite type. There is no limitation on the number of times a particular atomic data type can appear in the citation. Thus, many matrix rules can be generated for the `author` atomic type. Each matrix rule represents a composition of regex patterns that can be used to identify one form of the atomic type. Figure 3 shows an example of a generated matrix rule for a date instance 2004 `Jan`. A generated matrix rule for the date instance states that the given date can be identified by the regex pattern `\d{4}` followed by a blank space and then by a valid month in the semantic class "month.cls". Being a matrix rule, the ordering is important here.

```
<Rule>
  <matrixRule>
    <TokenList><Text>\d{4}</Text></Tokenlist>
    <TokenList><Text>\p{Blank}</Text></TokenList>
    <TokenList><Text>month.cls</Text></TokenList>
  </matrixRule>
</Rule>
```

Fig. 3. Generated Date Matrix rule

4   Approaches for Learning Data Type Rules

The data type rules generation phase can be divided into three main steps. The first step is the tokenization, the second step is the rule generation, and the third step is the rule measurement. A specific data type example, such as a date, is read in to the rule generator. The example is then broken up into a list of text tokens. For example, the date `May 23,2004` becomes ["May", " ", "23", ",", "2004"]. Then, for each token, a list of rule tokens in the form of regular expression are generated using either the Brute-Force approach or the Clustering-based approach. From the generated set of rule tokens, a filtering mechanism is employed to eliminate rules that resulted in very low precision and add rules with high precision to the final rule set.

4.1   Brute-Force Approach to Rules Generation

The Brute-Force rules generation generates all the candidate regular expressions that can match a given piece of text token. In this approach, regular expression rule tokens are generated from three sources. First, a simple regular expression (regex) that exactly matches the text token is generated. This is analogous

Table 1. Rule Tokens for "May 23,2004" date instance

May		23	,	2004
May	\p{Blank}	23	,	2004
First Name		number	comma	year
Month.cls				
Capitalized				

to creating a literal string for every text token. Second, a list of user-supplied named regex type is searched for generating other regexs that can match the given text token. Finally, a hierarchy of domain independent generic regex types is searched for generating all other potential regexes for the text token. These generic regex types match words in simple ways, such as "Capitalized", "punctuation", or "lower case". Thus, in Brute-Force approach, our date example can be transformed into the ragged matrix as shown in Table 1: Given a ragged matrix, the Brute-Force algorithm creates a list of rules. Potentially, the list of rules equal to $\Pi_{i=1}^{n}|t_i|$, or the total number of combinations of each possible match for each text token in the example. The list of generated rules is then passed to the rule measurement phase. The rule measurement phase removes equivalent rules, consolidates rules when a general rule can be generated to replace two or more specialized rules, and provides a ranking for the generated rules. From the ranked list of rules, obviously bad rules are discarded. Bad rules are those rules that match too many non-examples in the training document.

4.2 Clustering-Based Approach to Rules Generation

The main problem with the Brute-Force approach to rule generation is that an exponential number of rules are being generated as the number of text tokens in an example instance increase. This means the search space is very large when using the generated rules for classification. The Clustering approach is based on the observation that when similar examples are clustered together, the regularity across all the examples can be captured by computing the intersection of the set of regexs that match all the instances in that class. We call the resulting list of regexs from the intersection computation the *maximal* regex for that specific class of examples.

The algorithm for Clustering-based approach to rule generation is shown in Figure 4. It consists of three main steps. The first step clusters the example instances of a specific type with the same number of tokens together. The second step generates the regex patterns for each text token of the example instance of a specific class based on observed characteristics of the example instance as contrasted to using a generic text hierarchy patterns, and the third step computes the *maximal* regex pattern for all instances of a specific class. For each tagged example of a specific type (such as date), we cluster it based on the number of text tokens in that instance. For example, the various instances of dates are clustered into Class3 (which has three text tokens), Class5 (which has five text tokens) and Class7 (which has seven text tokens) in Figure 4. For each cluster

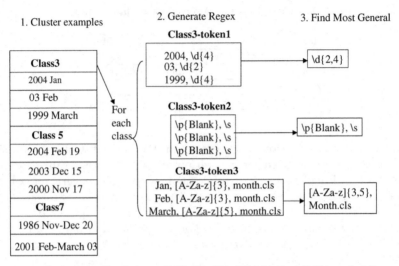

Final rule for Class3 date = [\d{2,4}(\p{Blank}|\s)([A-Za-z]{3,5} |month.cls)]

Fig. 4. A Clustering-based algorithm for rules generation

of example instances of a specific type, we generate regex patterns for each text token of each instance based on a set of rules as described in [6].

5   Experimental Evaluation

In this section, we evaluate the quality of rules generated and the computational trade-off between the two approaches. We use the same training document, annotated examples and template type for both data type learners. The data types that we are interested in generating and measuring are the `author` and `date` in the citation example.

The experiment is conducted in two phases for both types of learners. The first phase is the learning phase where rules for recognizing a citation are generated. The second phase is the classification phase where the generated rules are applied to an unseen citation document. The number of rules generated for each data type by each type of learners is recorded. The time taken to classify the unseen citation document using the generated rules is recorded for each data type learner. These details are shown in Table 2. The larger the number of rules being generated, the longer is the processing time. We use the standard precision and recall to evaluate the effectiveness or the quality of the generated rules. True positive in our results tables are instances that are correctly identified. False negative instances are those that are relevant, but which the generated rules failed to identify them (e.g. those author names which we missed). False positive are instances that are wrongly identified as being relevant. Table 3 shows the result of Cluster Learner while Table 4 shows the result of Brute-Force Learner. A discussion on those results can be found in [6].

Table 2. No of rules generated and processing time for each learner

Data Type Learner	No of rules for Author	No of Rules for Date	Classification Time
Brute-Force Learner	928	126	586 secs
Cluster Learner	1	2	90 secs

Table 3. Results for Cluster Learner

Data Set (Cluster-Learner)	True Positive	False Positive	False Negative	Recall	Precision
Training document (author)	72	3	5	93%	96%
Unseen document(author)	192	7	26	88%	97%
Training document (date)	4	0	0	100%	100%
Unseen document(date)	40	12	1	88%	97%

Table 4. Results for Brute-Force Learner

Data Set (Brute-Force Learner)	True Positive	False Positive	False Negative	Recall	Precision
Training document(author)	75	19	2	97%	79%
Unseen document (author)	208	101	10	95%	67%
Training document(date)	4	15	0	100%	21%
Unseen document (date)	40	70	1	97%	36%

6   Conclusion

We demonstrated in this paper the feasibility of automatically generating data types for service class description. The minor variation in site specific data patterns coupled with common regularity exhibited by the input and output across a class of Web sources lend itself to supervised machine learning technique. We discussed two different approaches to learning the data type rules. The Clustering based approach has a definite advantage over the Brute-Force approach from our initial set of experiments conducted for recognizing citation data types.

The strategies that we employed for learning is fairly simple and preliminary at the moment. We have not yet exploited the ordering and the specific alphabetic letter or digit that can occur in the text token. We have also not explored the orthographic features such as capitalization and position of the text token within an example instance. Incorporating these heuristics will increase the quality of our generated rules. Our learner uses a very simple clustering mechanism based on the number of tokens in the given text instance. More effective clustering that exploits high-level token semantics needs to be investigated.

References

1. David Buttler, Ling Liu, and Calton Pu. A fully automated object extraction system for the world wide web. *Proceedings of IEEE International Conference on Distributed Computing Systems*, April 2001.
2. J. Caverlee, L. Liu, and D. Buttler. Probe, Cluster, and Discover: Focused Extraction of QA-Pagelets from the Deep Web. In *Proceedings of the 20th IEEE International Conference on Data Engineering (ICDE04)*, Boston, USA, 2004.
3. David C. Fallside. XML Schema Part 0: Primer. Technical report, World Wide Web Consortium, http://www.w3.org/TR/xmlschema-0/, 2001.
4. Bin He and Kevin Chen-Chuan Chang. Statistical schema matching across web query interfaces. In *Proceedings of ACM/SIGMOD Conference on Management of Data*, San Diego, CA, 2003. ACM Press.
5. Jayant Madhavan, Philip A. Berstein, and Erhard Rahm. Generic schema matching with cupid. In *Proceedings of the Twenty-seven International Conference on Very Large Databases*, Roma, Italy, 2001. VLDB Endowment.
6. AHH. Ngu, D. Buttler, and T. Critchlow. Automatic Generation of data Types for Classification of Deep Web Sources . Technical Report UCRL-CONF-209719, Lawrence Livermore National Laboratory, 2005.
7. Anne H. Ngu, Daniel Rocco, Terence Critchlow, and David Buttler. Automatic discovery and inferencing of complex Bioinformatics Web Interfaces. Technical Report UCRL-JRNL-201611 (to appear in WWW journal, Springer), Lawrence Livermore National Laboratory, Livermore, CA, 2004.
8. Stephen Soderland. Learning Information Extraction Rules for Semi-structured and Free Text. *Machine Learning*, 1(44):1–44, 1999.
9. Jiying Wang, Ji-Rong Wen, Fred Lochovsky, and Wei-Ying Ma. Instance-based schema matching for web databases by domain-specific query probing. In *Proceedings of the Thirty International Conference on Very Large Databases*, Toronto, Canada, 2004. VLDB Endowment, Toronto, Canada.

BioNavigation: Selecting Optimum Paths Through Biological Resources to Evaluate Ontological Navigational Queries[*]

Zoé Lacroix[1], Kaushal Parekh[1], Maria-Esther Vidal[2],
Marelis Cardenas[2], and Natalia Marquez[2]

[1] Arizona State University, Tempe, AZ 85287, USA
{zoe.lacroix, kaushal}@asu.edu
[2] Universidad Simón Bolívar, Caracas, Venezuela
{mvidal, mcardenas, nmarquez}@ldc.usb.ve

Abstract. In this paper we present the BioNavigation system that allows scientists to express their queries using an ontology representing the conceptual level of scientific classes and labeled relationships. We developed the ESearch algorithm that generates all possible evaluation paths for a given ontological query and ranks them based on source metadata metrics. BioNavigation thus helps the user visualize the resources at a higher ontological level, build queries graphically, and provides valuable guidance on selecting the optimum path through the maze of resources.

1   Introduction

Expressing a scientific protocol, identifying the resources that will be used to implement each of its steps is a tedious task. To help the scientist in the process, two challenges need to be addressed. First, the scientist needs to express a protocol at a conceptual level, independently of any resources available to implement it. Only then, the scientist should identify the resources the most suitable to implement the protocol. Alas, the scientist often expresses ones protocols mixing its conceptual aim with its implementation. Indeed, there are multiple resources (data sources and applications) where to retrieve information about scientific objects and analyze them, and scientists cannot know them all. Each of these resources has its specific data format, data organization, data access, user interface, etc. In addition, although many available resources may look similar, they are different: two similar data sources may offer different coverage, different levels of curation, different characteristics of the scientific objects they provide information about, while two similar applications may generate dramatically different outputs[1]. The wealth of biological resources does not benefit completely the scientists as they typically exploit the few resources they know and

---

[*] This research is partially supported by the National Science Foundation under grants IIS 0223042 and IIS 0222847.

B. Ludäscher and L. Raschid (Eds.): DILS 2005, LNBI 3615, pp. 275–283, 2005.

trust, avoiding the time consuming process of exploring new resources for each of their protocols. Furthermore, the protocols are often driven by the resources known by the scientists to implement them. Instead of selecting the resources best meeting the protocol's needs, the protocol is expressed so that it can exploit the known resources. This may affect significantly the quality and completeness of the dataset collected by the protocol as different selections of resources in the protocol evaluation may generate different datasets [1].

Scientific knowledge involves multiple scientific objects and scientific meaningful relationships among them. This knowledge may be represented with ontologies [2] composed of *concepts*, *relations*, *instances*, and *axioms*. Each concept represents a scientific object (e.g., a gene) that is an abstraction of a set of entities within a domain. Relations represent the interactions between concepts or the properties of a concept. Ontologies aim at modeling scientific information with respect to the understanding of the scientist. This "scientist-friendly" representation of scientific information has proved useful in the past with system such as TAMBIS [3] that used an ontology as a user-interface to access and query multiple integrated databases. In the BioNavigation approach, we aim at exploiting ontologies to provide a scientifically meaningful view of biological resources.

In this paper, we present an extension of the BioNavigation system introduced in [4]. The system provides scientists with the ability to express scientific queries at a conceptual level, and returns scientists evaluation paths composed of physical resources to implement the queries. BioNavigation now exploits an ontology defining a graph of concept classes and multiple labeled edges. The query against the logical graph can be seen as the *design* of the protocol (e.g., "retrieve citations related to a genetic disease"), while the evaluation paths returned by the BioNavigation system are as many possible *implementations* of the protocol (e.g., OMIM (`http://www.ncbi.nlm.nih.gov/omim`) to PubMed (`http://www.ncbi.nlm.nih.gov/entrez`) using the Entrez PubMed Links). Once the system has returned the paths, the scientist may explore the meta-information related to the resources. To better match the scientists needs, the user may select *semantics* that will guide the BioNavigation system in selecting the evaluation path and return them ordered with respect to the semantics. The three semantics are *maximizing the relevance*, *maximizing the number of entries*, and *efficiency*. We use three metrics to compute the probability of each path to validate the semantics.

2   Physical and Logical Map of Resources

Most data sources typically represent a particular type of scientific class. For example, PubMed provides references to published literature, UniProt (`http://www.ebi.ac.uk/uniprot/`) provides information about proteins, etc. There can be several data sources for the same scientific class. For example, one can retrieve 'DNA sequences' from either NCBI Nucleotide or EMBL (`http://www.ebi.ac.uk/embl/`). Data sources also provide links connecting a record to other records in the same data source as well as external data sources in order to provide comprehensive and complete information about the scientific object they represent. Scientists use these links to navigate from one source to another and in the process gathering useful information relevant to the scientific question being studied. Depending on the data source used for obtaining information about a given scientific class and the links followed, scientists may retrieve significantly different information, both in terms of quantity and quality, for the same scientific query [1]. Thus it is important that the scientist should be able to identify and explore all possible paths that can be used to evaluate the query.

We developed the BioNavigation system to allow the scientist to exploit the numerous available data sources and the links between them [4]. It allows the user to generate queries graphically and evaluate them with respect to the above ranking criteria. The main requirement of the interface was to display graphs of sources (nodes) and capabilities (edges) to users in order to interact with and view properties of the sources and capabilities. The BioNavigation interface serves two main purposes: browsing and querying.

The browsing mode allows the user to navigate the conceptual and the physical levels of the resources. The user can select any of the nodes representing the resources in the graph and learn more about its properties including the ontological concept it represents, and in the case of data sources, the URL of the source, the schema of data records, etc. Similarly, the user can click on a capability connecting two physical sources to view its properties. The metadata of the physical resources stored in the physical graph are used to determine optimal paths to evaluate a user query. At the conceptual level, the user can see what scientific class each node represents and their relationships.

We now provide the formal definitions for the two levels of representation, the ontological and the physical, which significantly extends the framework defined in [5]. We also define the mappings that relate the physical resources with concepts and relationships in the ontology. In section 3, we define the query language to formulate the scientists query and used by the ESearch algorithm to identify paths in the physical graph. The logical and physical graphs are respectively defined in definitions 1 and 2. We then introduce, in definition 3, the function ϕ that maps each logical node to the set of its physical implementations.

Definition 1. The Logical Graph $LG = (V_L, E)$ is a directed graph, where:

- V_L is a set of nodes, partitioned into two sets C and A, where, C represents logical classes and A represents logical associations between classes.
- E is a set of directed edges $E \subseteq (C \times A) \cup (A \times C)$ that represents roles played by logical classes in the associations.

Definition 2. The Physical Graph $PG = (V_P, L)$ is a directed graph, where:

- V_P is a set of nodes, partitioned into three subsets, S, AP, and QC, such that, S represents physical data sources, AP represents applications, and QC represents query capabilities.

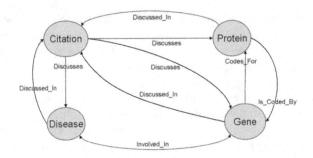

Fig. 1. An Example Ontology of Concepts and Associations

- L is a set of directed edges $L \subseteq V_P \times V_P$ that represents links between sources and applications or query capabilities. If a pair (a, b) belongs to L then, a is a source and b is an application or query capability, or a is an application or query capability and b is a source.

Definition 3. ϕ is a one-to-many mapping from V_L to 2^{V_P} such that it maps (a) a logical class name in C to a set of physical data sources in S and (b) a logical association in A to a set of applications or query capabilities in $2^{AP \cup QC}$. Elements in $\phi(v)$ represent the physical implementations of a logical node v.

Figure 1 provides a small example for a conceptual ontology involving the scientific classes, disease, gene, citation, and protein, and their labeled associations or relationships. The querying mode allows the user to graphically build a regular expression. Consider a scientist interested to *'retrieve citations related to a particular disease'*. An evaluation path for this query could consist of initiating the retrieval process from a particular source that provides information on diseases and then through the links it offers, obtain related citations. One such path could be exploiting the NCBI PubMed Link from OMIM to PubMed. Hence, at the conceptual level the path would be 'd in c' formed from the class 'disease' or 'd', the class 'citation' or 'c', and the association 'discussed in' or 'in'. The user also might want to include in his path any possible intermediate nodes in addition to the direct path.

3   Query Language

We now formally define the language that will be used to express the queries over the logical concepts in set V_L. We use the following notations:

- v is either a class or a logical association in V_L i.e., $v \in V_L$
- $v < AnnotList >$ is an annotated class or association where $< AnnotList >$ is a list of expressions of the form: $OP < PhysicalImpName >$ where OP is either $\neq$ or $=$, and $< PhysicalImpName >$ corresponds to a data source, application or query capability in V_P such that $< PhysicalImpName >$ belongs to $\phi(v)$.

- ε_c is a term representing any possible class in C, similarly, ε_a represents any possible association in A, and ε represents the path $\varepsilon_a\,\varepsilon_c$.

Definition 4. The query language $L(RE)$ over the logical concepts in V_L is defined by the regular expression, $L(RE) = X\,(\varepsilon + Y\,X)^*$, where, $X = \varepsilon_c \mid c \mid c < AnnotList >$ and $Y = \varepsilon_a \mid a \mid a < AnnotList >$

Thus any conceptual level query starts with a logical concept and ends with a logical concept. Two concepts are always connected through a logical association. The term ε allows users to express queries such as '$c_1\,\varepsilon^*\,\varepsilon_a\,c_2$', which means that the path between classes c_1 and c_2 could be of any length and consist of any possible intermediate class and association. A BNF grammar generating the regular expressions is shown in Figure 2. Given the regular expression RE, our

```
<RE>:= <cTerm><Y>
<cTerm>:= <EpsilonC> | <ClassName><SourceAnnotation>
<Y>:= <Epsilon><Y> | <aTerm><cTerm><Y> | empty
<aTerm>:= <EpsilonA> | <AssociationName><LinkAnnotation>
<SourceAnnotation>:= empty | "[" <SourceList>"]"
<SourceList>:=<AnnotatedSource> | <AnnotatedSource> "," <SourceList>
<AnnotatedSource>:=<OP><SourceName>
<LinkAnnotation>:= empty | "[" <LinksList>"]"
<LinkList>:=<AnnotatedLink> | <AnnotatedLink> "," <LinkList>
<AnnotatedLink>:=<OP><LinkName>
<LinkName>:= <ApplicationName> | <QueryCapName>
<OP>:="!=" | "="
```

Fig. 2. BNF grammar of regular expressions

optimization algorithm will identify the set of physical paths in PG that corresponds to the physical implementations of expressions of the language induced by RE, $L(RE)$. The following definition formalizes the paths that are physical implementations of an expression in $L(RE)$.

Definition 5. α is a one-to-many mapping from an expression $e \in L(RE)$ into a set of paths in PG corresponding to the physical implementation of e.

- If e is ε_c, then $\alpha(e) = S$.
- If e is ε_a, then $\alpha(e) = AP \cup QC$.
- If e is a logical concept $l \in V_L$, then $\alpha(e)=\phi(l)$.
- If $e = l < AnnotList >$, where $l \in V_L$ and $< AnnotList >$ is partitioned into $< AnnotListInc >$ and $< AnnotListExc >$, where the former corresponds to the list of sources that must be considered and the latter sources that must be excluded, then, $\alpha(e) = \phi(l)\cap < AnnotListInc > - < AnnotListExc >$
- If $e = e_1e_2$ then,
  $\alpha(e_1e_2) = \{w_1w_2|w_1 \in \alpha(e_1), w_2 \in \alpha(e_2), edge(last(w_1), first(w_2)) \in L\}$,
  where *last* and *first* are functions that respectively map a path with its last and first elements and L is the set of edges in PG (definition 2).

4   Searching the Space of Paths

A path $p = (s_1, a_1, s_2, \ldots, s_{n-1}, a_{n-1}, s_n)$ in PG is defined as a list of sources s_i and applications $a_i \in V_P$. A regular expression r over the alphabet V_L expresses a retrieval query Q_r. The result of Q_r is the set of paths p in PG that interpret r, i.e., the set of paths in PG that correspond to physical implementations of the paths in LG that respect the regular expression Q_r.

A naive method for evaluating a query Q_r is to traverse *all* paths in PG, and to determine if they interpret r. The time complexity of the naive evaluation is exponential in the size of PG because PG has an exponential number of paths. A similar problem was addressed in [6] where it was shown that for (any) graph and regular expression, determining whether a particular edge occurs in a path that satisfies the regular expression and is in the answer is NP complete.

4.1   Assigning Metrics to Physical Paths

The result of a query Q_r is a list of paths that represents the different ways in which the user can navigate through the data sources in order to evaluate Q_r. It becomes important to assign ranks to these paths so that the user can easily select the most suitable one. We use three metrics for ranking the paths:

1. *Path Cardinality* is the number of instances of paths of the result. For a path of length 1 between two sources S1 and S2, it is the number of pairs (e1, e2) of entries e1 of S1 linked to an entry e2 of S2.
2. *Target Object Cardinality* is the number of distinct objects retrieved from the final data source.
3. *Evaluation Cost* is the cost of the evaluation plan, which involves both the local processing cost and remote network access delays.

These three metrics are meaningful to the scientists as the path cardinality computes the probability there exists a path between two sources, the target object cardinality estimates the number of retrieved entries, whereas the evaluation cost guides the scientists to the selection of an efficient evaluation path. These metrics for each path are estimated based on the properties, described in Definition 7, of the links that exist between the data sources in S using the methods introduced in [5] and [7]. The following definitions describe the mappings from the logical associations between two classes in LG to the links between two data sources in PG and their properties.

Definition 6. γ is a one-to-many mapping from a pair of logical associations in $E \times E$ to pairs of physical links in $L \times L$. If a pair (pl_1, pl_2) belongs to $\gamma((la_1, la_2))$, then the following holds:

- $la_1 = (c_1, a)$ and $la_2 = (a, c_2)$ are logical links, where c_1 and c_2 are logical classes, and a is a logical association between c_1 and c_2.
- $pl_1 = (s_1, ap)$ and $pl_2 = (ap, s_2)$ are physical links where s_1 and s_2 are sources and ap is an application or a query capability.
- $s_1 \in \phi(c_1)$, $s_2 \in \phi(c_2)$ and $ap \in \phi(a)$.

Definition 7. The function δ maps a logical association between two logical classes to the set of physical link implementations and the values of the properties link cardinality, domain participation and image participation for each link. Thus, δ is a one-to-many mapping from a pair of edges in $E \times E$ to 5-tuples in $L \times L \times \mathbb{R} \times \mathbb{R} \times \mathbb{R}$. If a 5-tuple (pl_1, pl_2, lc, dp, ip) belongs to $\delta((la_1, la_2))$, then the following holds:

- $la_1 = (c_1, a)$ and $la_2 = (a, c_2)$ are logical links, c_1 and c_2 are logical classes and, a is a logical association between c_1 and c_2.
- $pl_1 = (s_1, ap)$ and $pl_2 = (ap, s_2)$ are physical links and, s_1 and s_2 are sources and ap is an application or a query capability.
- $(pl_1, pl_2) \in \gamma((la_1, la_2))$.
- lc: represents the link cardinality and corresponds to the number of links from all data objects of source s_1 pointing to data objects of s_2.
- dp: represents the domain participation and corresponds the number of objects in s_1 having at least one outgoing link to an object in s_2.
- ip: represents the image participation and corresponds to the number of data objects in s_2 that have at least one incoming link from objects in s_1.

4.2   The ESearch Algorithm

ESearch is an extension of the algorithm presented in [5], to evaluate queries expressed as ontologies such as the one defined in Section 3. *ESearch* is based on an annotated deterministic finite state automaton (DFA) that recognizes a regular expression or query Q_r and the physical implementations that must be excluded from the final result. We refer these regular expressions as annotated regular expressions. The algorithm performs an exhaustive breadth-first search of all paths in PG that respect the regular expression.

Suppose DFA is the automaton that recognizes the annotated regular expression or query Q_r. The annotated DFA is represented by a set of transitions, where a transition is a triple $t=(i, f, e, excImpl)$, where, i represents the initial state of t, f represents the final state of t and, e corresponds to the label of t, note that $e \in V_L$, i.e., e belongs to the set of logical classes and associations between classes. The state i (resp. f) may be a *start state* (resp. *end state*) of DFA. Finally, *excImpl* is a set of physical implementations of e that will not be considered in the physical paths.

The exhaustive algorithm *ESearch* , comprises two phases: (a) *build path* and (b) *print path*. In phase *build path*, for each visited transition $t=(i, f, e, excImpl)$, the algorithm identifies *all* the physical implementations $s_i \in \phi(e)$ that do belong to *excImpl*. If i is not a *start state* of the DFA, then, for each s_i, the algorithm computes a set $s_i.previousImp$. To do so, it considers all the sources or applications that were selected in transition t^p previous to t, and selects the subset of implementations in V_P that are adjacent to s_i in PG; these elements are included in $s_i.previousImp$ in conjunction with the transition used to traverse the node. In phase *print path*, the algorithm starts from the set of implementations corresponding to the final transition, whose final state is an *end state* of the

DFA. For each s_i, it uses the set $s_i.previousImp$ to construct a path. The path terminates in one of the implementations visited by the start transition whose initial state is a *start state* of the DFA. We note that *print path* may commence as soon as *ESearch* visits the first transition for which f is a final state of DFA. In our implementation, we do not consider such potential parallelism between these two phases.

The *ESearch* Algorithm runs in polynomial time in the size of the graph, if the graph is cycle free, all paths are cycle free and, every two nodes in PG are connected by a unique path. Each node (source) in the graph implements only one entity, so a node is visited at most once in each transition (each level of the breadth-first search). Similarly, each node is visited at most once in each iteration of *print path*. An annotated physical graph is produced during the *build path* phase of *ESearch*. For each transition in the DFA, each implementation s_i matching the transition is annotated with implementations in $s_i.previousImp$. If d is the maximum number of nodes that can precede a node in the annotated physical graph, i.e., the cardinality of *previousImp*, and b is the maximum length of (cycle free) paths satisfying the annotated regular expression, then $O(d^b)$ is an upper bound for *ESearch*.

5   Conclusions

BioNavigation can enhance existing mediation approaches by providing scientists with the ability to browse through available integrated resources and to access their properties. The wildcard ε^* allows users to identify alternate paths that may be exploited to evaluate the queries while the annotations aid specifying the resources they may require to be used (or not be used) in the process. The ESearch algorithm designed and implemented for BioNavigation allows efficient search in the space of all possible evaluation paths. Moreover three scientifically meaningful metrics provide scientists the paths that best meet their needs. In the future we will combine the BioNavigation system with SemanticBio [8] that allows users to express and execute scientific workflows with an ontology and Web Services.

References

1. Lacroix, Z., Edupuganti, V.: How biological source capabilities may affect the data collection process. In: Computational Systems Bioinformatics Conference, IEEE Computer Society (2004) 596–597
2. Stevens, R., Goble, C.A., Bechhofer, S.: Ontology-based knowledge representation for bioinformatics. Briefings in Bioinformatics 1 (2000) 398–416
3. Baker, P.G., Brass, A., Bechhofer, S., Goble, C., Paton, N., Stevens, R.: TAMBIS - transparent access to multiple bioinformatics information sources. In: Intelligent Systems for Molecular Biology (ISMB), AAAI Press (1998) 25–43
4. Lacroix, Z., Morris, T., Parekh, K., Raschid, L., Vidal, M.E.: Exploiting multiple paths to express scientific queries. In: Scientific and Statistical Database Management (SSDBM), IEEE Computer Society (2004) 357–360

5. Lacroix, Z., Raschid, L., Vidal, M.E.: Efficient techniques to explore and rank paths in life science data sources. [9] 187–202
6. Mendelzon, A.O., Wood, P.T.: Finding regular simple paths in graph databases. In Apers, P.M.G., Wiederhold, G., eds.: Very Large Data Bases (VLDB), Morgan Kaufmann (1989) 185–193
7. Lacroix, Z., Murthy, H., Naumann, F., Raschid, L.: Links and paths through life science data sources. [9] 203–211
8. Lacroix, Z., Ménager, H.: SemanticBio: Building conceptual scientific workflows over web services. In: Data Integration in the Life Sciences (DILS). Lecture Notes in Bioinformatics, Springer (2005)
9. Rahm, E., ed.: Data Integration in the Life Sciences (DILS). Volume 2994 of Lecture Notes in Computer Science, Lecture Notes in Bioinformatics., Springer (2004)

Support for BioIndexing in BLASTgres

Ruey-Lung Hsiao, D. Stott Parker, and Hung-chih Yang

Computer Science Department, University of California,
Los Angeles, CA 90095-1596, USA
{rlhsiao, stott, hcyang}@cs.ucla.edu

Abstract. The ability to perform genome-wide and cross-genome data analyses can dramatically reduce the time required for new biological discoveries. This raises important issues in bioinformatics database research involving data representations and data integration. Essential biodatatypes (biological datatypes such as sequence locations) and tools (such as the popular BLAST sequence alignment tools) are not supported in traditional database systems, which has forced researchers to represent biological knowledge counterintuitively, and implement codes for data operations. This paper introducesBLASTGRES, an extension of the POSTGRESQL database system, that provides indexable biodatatypes and joinable BLAST alignment.

1   Introduction

Today's vast, distributed biological knowledge is interconnected through many kinds of information about sequences. This information can be indexed, permitting efficient navigational access through browser interfaces and automated traversal of sequence annotations. We are developing *BioIndexing* as a conceptual infrastructure for representing and managing biological knowledge with indexing constructs. This infrastructure has been realized in BLASTGRES, an extension of the POSTGRESQL database system for bioinformatics.

This paper focuses on two aspects of BioIndexing that are implemented in BLASTGRES: indexable *biodatatypes* (biological datatypes) and the ability to join information from external resources. A sequence location biodatatype can relate these two aspects. Sequence locations pervade existing biological information, and define a central biodatatype. They are essential to the sequence alignments produced by tools like BLAST[1, 2], which today provides one of the primary indexing mechanisms for this information. Furthermore, these aspects of BioIndexing are naturally combined within a modern database system, giving a flexible infrastructure for connecting and managing biological information.

BLASTGRES is an implementation that provides these capabilities. We chose to develop BLASTGRES as an extension of the POSTGRESQL [3] database system for several reasons, including in particular that POSTGRESQL: 1) is an open source software package with high performance and stability; 2) facilitates the introduction of user-defined biodatatypes; 3) provides GiST (Generalized Search

B. Ludäscher and L. Raschid (Eds.): DILS 2005, LNBI 3615, pp. 284–287, 2005.

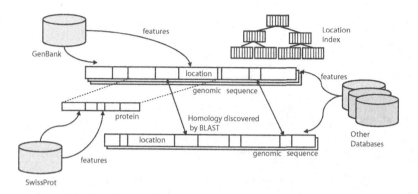

Fig. 1. Much biological knowledge is described in terms of the features associated with locations in sequences and relationships between locations in sequences. BLASTgres supports indexable location biodatatype and BLAST joins between locations. These two capabilities permit efficient large-scale management of sequence-related feature knowledge

Tree) indexing [4]. GiST indexing is particularly important for the location bio-datatype because BLASTGRES needs to support query predicates other than equality test.

2   BLAST Support in BLASTgres

Similarity search is one of the most heavily-used tools in computational and comparative biology. Among available similarity searching tools, the BLAST family is arguably the most popular. BLAST can be viewed as a kind of index; given a sequence of interest, BLAST finds related sequences. Many biological relationships can be represented and related by joining BLAST results.

The integration of BLAST into database management systems enjoys several benefits: 1) today the application of BLAST results in relating biological sequences is often ad hoc. Special codes are needed to interact with a BLAST server, parse BLAST results and integrate the results with other biological data. 2) currently, BLAST provides only a limited set of controls over its result; advanced filtering and query mechanisms (such as sorted by E-values or grouping by a set of attributes) are sometimes required. 3) additional annotational information can be automatically integrated into BLAST results.

In BLASTGRES, BLAST support is achieved by defining a set of user-defined functions that return BLAST results as a table. Biodatatypes are also transformed automatically in this process. Annotational information, such as species, and description for a particular sequences can also be added automatically by invoking the proper function calls. In this way, BLAST results can be easily integrated with other biological information in the system. Besides, annotational data can be attached to BLAST results automatically in order to provide detailed knowledge about these sequences.

Here are some examples:

```
-- blast a given sequence with BLAST default parameters
SELECT * FROM blast('ACTTGATGGTACGTAGTCCGTATAGGCTTAGEACTGGTATCGA', 'blastn', 'nr');
-- blast sequence file against local database
SELECT * FROM local_blast('unknown_proteins.fasta', 'blastp', 'nr');

-- number of hits from different species
SELECT COUNT(*), species FROM annotated_blast( 'AF101044', 'blastn' )
GROUP BY species;

-- retrieve sequences that code for SNRPN proteins in species other than homo sapiens
SELECT * FROM annotated_blast( 'AF101044', 'blastn' )
WHERE descriptions LIKE '%SNRPN%' AND species <> 'Homo sapiens';

-- blast a subsequence in the 2nd exon of Mμs musculus H2-DMB1 gene with stricter condition
SELECT subject_location, length FROM blast('NM_010387.2[265..546][30..60]', 'blastn')
WHERE evalue < 1E-5 AND bitscore > 800;
```

3   The Location Biodatatype and Indexing

Without proper database abstractions, users have to develop specialized codes to handle location operations. Traditional relational databases are generally not equipped to support locations as an abstraction, or permit more powerful querying of locations. A variety of potential issues would arise. For instance, inconsistent representations and interpretations lead to difficulties in data sharing and exchange. Moreover, traditional query processors are sometimes incapable of generating efficient execution plans for complex conjunctive (or disjunctive) normal forms and handling inequality relationship between inter-dependent attributes. Severe performance penalties can result.

Internally, a location is represented as an identifier for the sequence, an integer interval $[lower, upper]$, and the strand in which this sequence resides (only for DNA sequences). The interval range is represented as a closed interval with positions starting at 1, following the convention of most biological databases.

Essential operations and functions for the location biodatatype are supported. More than 30 interval operations are defined (such as Contains, Inside, Equal, Overlaps, Left, Starts_by, Finishes_by, Over_left, etc). Coordinate transformation and slicing operations are also supported. New reference coordinates can be specified by defining aliases to existing locations and slicing can be specified to denote a sub-range within a location. Optimization information (e.g., regarding ordering, commutativity or negation) is also provided to permit optimization of important operations like merge-join, hash-join or general theta-join.

Here is an example of location representation and manipulation:

```
CREATE TABLE alt_splice_homology_map AS
SELECT o.*, d.location, range_start(d.query)+(o.location-range_start(d.hit))/3
FROM alt_splice_exon_obs o, alt_splice_homology d
WHERE    o.location @ d.location  -- contained
       AND d.e_value < 0.01  GROUP BY o

SELECT o.*, f.type, f.location
FROM alt_splice_homology_map o, swiss_feature f
WHERE    o.location &< f.location  -- left overlap
```

In addition, BLASTGRES supports location indexing. Supports for indexing schemes in traditional relational database systems is limited and inflexible. They are only limited to a few well-known index structures and can be used only for a limited set of native datatypes for (in)equality and range queries. In order to support a wide variety of queries, we implemented location indexing under the GiST architecture. GiST is extensible both in datatypes that it can index and query predicates that it supports.

Each BLASTGRES search key contains two ranges: (1) $[id\_lower, id\_upper]$, a pair of integers representing the range of identifiers present in the subtrees, and (2) $[key\_lower, key\_upper]$, a minimal bounding interval that covers the range of location intervals in the subtrees. Common interval predicates, such as Left, Right, Overlaps, Contains, Equal, etc, are supported in our GiST index implementation. Tree search is handled by comparing the minimal bounding intervals with the query according to the GiST search algorithms.

4   Conclusion

The location concept is fundamental to biological knowledge representation since biological features are generally attached to locations and locations are also the bases for maps, alignments and other complex relationships. Naive representation of the location concept, without formal development of a biodatatype, is error-prone and easily leads to inconsistencies in operational definition and poor query performance. In addition, advanced database queries and analysis functions could be used for BLAST queries.

We implemented BLASTGRES, an extension of POSTGRESQL as an illustration of the these concepts. High performance is achieved by bioindexing — the combination of indexing and efficient implementation of the location biodatatype. For instance, in one of our sample databases which keeps track of information for 420,251 genomic locations, it takes 825ms to retrieve a random location without index support. With index support, it takes 17ms.

BLASTGRES is freely available from http://www.blastgres.org/

References

1. Altschul, S. F., Gish W., Miller W., Myers E. W., Lipman D. J.: Basic local alignment search tool. Journal of Molecular Biology, **215**, 403–410
2. Altschul, S.F., Madden, T.L., Schffer, A. A., Zhang, J.,Zhang, E., Miller, W.: Gapped BLAST and PSI-BLAST: a New Generation of Protein Database Search Program. Nucleic Acids Research, 3389–3402. http://www.ncbi.nlm.nih.gov/BLAST/.
3. Stonebraker, M., Kemnitz, G.: The Postgres Next-Generation Database Management System. Communications of the ACM **34:10**. 78–92 http://www.postgresql.org/
4. Hellerstein, J. M., Naughton, J. F., Pfeffer A.: Generalized Search Trees for Database Systems. Proceedings of 21th International Conference on Very Large Data Bases, 562–573.

An Environment to Define and Execute *In-Silico* Workflows Using Web Services[*]

Rafael Targino[1], Maria Claudia Cavalcanti[2], and Marta Mattoso[1]

[1] Computer Science Department, COPPE, Federal University of Rio de Janeiro – Brazil
{targino, marta}@cos.ufrj.br
[2] Computer Science Department, IME-RJ, Rio de Janeiro – Brazil
yoko@de9.ime.eb.br

Abstract. Scientific workflows represent an attractive alternative to describe bioinformatics experiments. They give an adequate support to the Execution and Analysis"cycle, relevant to the process of knowledge discovery. Workflows can create an independent and interoperable environment between the scientific applications and databases, when combined with the Web Services technology. Despite the successful use of these technologies in the business scenario, its use in bioinformatics is still incipient. This work presents an integrated environment that aims at the definition and execution of *in silico* experiments through scientific workflows using Web services. A real bioinformatics experiment was implemented in this environment.

1 Introduction

Scientific workflows represent an attractive alternative to describe bioinformatics experiments. These *in-silico* experiments are usually built by manually composing third-party programs with their input and output data in an execution flow. Output data is analyzed and according to the experiment result, parameters are tuned, workflow is re-executed, programs are replaced on the workflow and partial re-executions are made.

Perl script language has been used to help program invocations and composition, as well as data conversions. However Perl scripts lack flexibility and present difficulty in coping with changes. Recently, the scientific community is moving towards Web services technology [2], [3], [8], [10], [11], [12]. Web services were specially conceived to provide interoperation between applications from different platforms. Thus, they are an appropriate solution to support program composition through geographically distributed programs and data exchange. The Web services technology provides the necessary mechanisms to define workflow processes through the composition of basic web services. Currently, there are a number of language proposals for composing web services. IBM, Microsoft and BEA released BPEL4WS - Business Process Execution Language for Web Services [7], which is becoming a *de facto*

---

[*] This work was partially funded by CNPq.

B. Ludäscher and L. Raschid (Eds.): DILS 2005, LNBI 3615, pp. 288–291, 2005.

standard. However, interoperability issues are only part of the problem. An important knowledge that can be extracted from the results of these experiments, such as data provenance, is lost. With the Web services approach, program composition definition can be registered, but results from these experiments, such as, successful parameter tuning, intermediary results and a log of experiments are not automatically registered. Specific scientific workflow languages [2] have been designed on top of Web services technology to support such experiments management.

In [6] we present SRMW, a scientific metadata architecture that adds semantics to Web services. We evaluated a real bioinformatics workflow implemented through Web services. The flexibility of Web services technology was confirmed, particularly for this specific workflow definition, but no support for defining and managing workflows in SRMW.

Based on this experience we have developed a prototype of an environment, named 10+C, which is presented in this work. 10+C offers services for bioinformatics workflow management based on Web services technology, particularly BPEL [7] and its execution engines. We have built a Web portal that supports definition and execution of workflows using a graphic interface that hides technology specificities. In addition to flexibility, 10+C can register experiments executions, partial results and parameters. Also, it provides partial execution and re-execution of workflows to allow fine tuning of parameters by the scientist. Data persistence services in 10+C represent a first step into data provenance. 10+C can work alone or it can be coupled to SRMW architecture [6], to add semantic support. By using metadata according to the SRMW [5] metamodel, we can identify related programs and experiments and help knowledge extraction from stored experiments.

2   The 10+C Environment

The 10+C environment combines Web services technology with scientific workflow management tools aiming at bioinformatics applications. The main services of 10+C are workflow definition and execution. Workflow definition specifies the execution flow of the programs by defining: the involved activities, restrictions, execution order, input/output data acquisition, conversion and formatting, deviations on the main execution flow, and error handling with compensatory services. Workflow execution is responsible for executing all activities defined at the workflow specification, by following the flow sequence, data conversions and transformations.

The name 10+C stands for ten characteristics available: (1) abstract workflow definition; (2) run time programs workflow definition; (3) automatic workflow language specification; (4) workflow execution; (5) workflow partial and complete re-execution; (6) workflow exception handling; (7) program execution logging; (8) workflow execution logging; (9) remote program execution using open standards; and (10) Web portal interface.

A prototype of 10+C was implemented using JSP/Servlet and MySQL on an Apache Tomcat server. The AXIS [4] package is responsible for providing Java classes as Web services and handling SOAP messages. We chose BPEL as the workflow definition language mainly for its tendency on becoming a standard. To execute BPEL we have used BPWS4J 1.0.1 execution engine, freely available by IBM [9].

Currently a new version BPWS4J has been released, but no longer free. We are evaluating the several open source BPEL execution engines [1] for the next version of 10+C. We can take advantage of BPEL and its execution engines while we concentrate in adding semantics and specific support for bioinformatics workflows. Next, we present the four main components of the 10+C architecture, discussing their support on the ten characteristics.

Scientific Applications. This component contains bioinformatics programs published as Web services. These programs can be third-party code available through CGI, CORBA or any interface different from Web service. In this case, a Web service layer has to be built to invoke the program. Some examples are given in 10+C to help the publisher.

Web service programs or workflows must be registered in 10+C through its WSDL document file. 10+C processes this file and stores invocation information, registering its name, and data types for all input/output of the service. This information is used to help workflow definition, and input/output mappings between program flow.

Data Persistence. Data related to program and workflow execution are persisted using Web services. Examples of persistent data are: parameters, input/output data, results and metadata involved during execution. This persistence can be defined by the user, through the identification of the Web service responsible for this registry. Those Web services store data in flat files or in MySQL database.

Workflow Definition. This component allows for the definition of a workflow. It is based on available Web services of bioinformatics applications. It is responsible for the definition of an execution sequence, based on BPEL constructs to connect programs. A simplified interface is available to view a graphic representation of the workflow. We intend to incorporate an open source workflow editor to improve flexibility. To create a new workflow, the user must give a name and description for the workflow. Then a Web service must be chosen among previously published services and its execution flow is defined for each participant of the workflow.

During workflow definition it is also necessary to map data output to the input of the next program. Finally, a workflow definition file is automatically generated (a BPEL document), which must be later deployed in a workflow engine. Once specified, the workflow is published as a Web service, registered at 10+C and can become part of other workflows.

Workflow Execution. This component builds a visual HTML page for data input of the workflow first step and invokes the corresponding service through its interface. This initial page for data input is automatically built according to the mapping specification previously defined. With these input data, a generic *proxy* is invoked, which is a class that composes SOAP messages to send and receive output data.

Workflows are executed at the workflow engine of the application server by invoking each program along the flow with its corresponding input data. Currently we are using the execution engine BPWS4J, by IBM, but any other BPEL engine could be used, such as open source ActiveBPEL [1]. We have evaluated 10+C services using a real bioinformatics workflow, MHOLline [6] that aims at generating 3D models for proteins. During execution, each program data is persisted using Web services to allow for future analysis or workflows re-executions. Alternative services can be executed along the workflow if a failure occurs and compensatory services are planned.

3 Conclusions

The 10+C environment implements an open platform to define and execute bioinformatic workflows. It is based on Web services technology, which has become an industry tendency. Our approach is highly based on open source tools and standard proposals for Web services workflows, thus, taking advantage of open languages with interoperability and platform independence. Furthermore, wrapping legacy programs into Web services has turned bioinformatics workflow steps into public building blocks, which are available for any other application reuse. Finally, despite the limitations of the Web services description language (WSDL), its extensibility allows to add semantics [6], according to the needs of bioinformatics users.

References

1. ActiveBPEL - The Open Source BPEL Engine, available in http://www.activebpel.org/
2. Addis, M., Ferris, J., Greenwood, M., *et. al*: A.: Experiences with eScience Workflow Specification and Enactment in Bioinformatics. Proceedings UK e-Science All Hands Meeting 2003 Editors - Simon J Cox, (2003) 459-466
3. Altintas, I., Berkley, C., Jaeger, E. ,Jones, M. Ludaescher, B. Mock, S.: Kepler: An Extensible System for Design and Execution of Scientific Workflows. 16th Intl. Conference on Scientific and Statistical Database Management (SSDBM), (2004).
4. Axis Project, available in http://ws.apache.org/axis/
5. Cavalcanti, M., Mattoso, M., Campos, *et. al*:: An Architecture for Managing Distributed Scientific Resources, Proc. Scientific and Statistical Database Management, (2002) 47-55
6. Cavalcanti, M., Targino, R., Baiao, F., Rössle, S., Bisch, P. M., Pires, P. F., Campos, M. L., Mattoso, M.: Managing Structural Genomic Workflows using Web Services. DKE Journal, Elsevier. Special Issue Biological Data Management, Vol. 53, N 1, (2005) 45-74
7. Curbera, F., Goland, Y., Andrews, T., *et. al*: Business Process Execution Language for Web Services, http://www.ibm.com/developerworks/library/ws-bpel/ (2003)
8. Goble, C., Wroe, C., Stevens, R., and the myGrid consortium. The myGrid Project: Services, Architecture and Demonstrator. Proceedings UK e-Science All Hands Meeting 2003 Editors - Simon J Cox, (2003) 595-603
9. IBM BPWS4J, available in http://www.alphaworks.ibm.com/tech/bpws4j
10. Kuenzl, J.: Development of a Workflow-based Infrastructure for Managing and Executing Web Services. Universität Stuttgart, Fakultät Informatik, Diplomarbeit Nr.1997 (2002)
11. Stein L.: Creating a Bioinformatics Nation, Nature 417 (2002) 119-120
12. Wilkinson, M. D., Links, M.: BioMOBY: an Open-source Biological Web Services Proposal. Briefings In Bioinformatics, Vol. 3, N. 4, (2002) 331-341

Web Service Mining for Biological Pathway Discovery

George Zheng and Athman Bouguettaya

Department of Computer Science, Virginia Tech
{gzheng, athman}@vt.edu

Abstract. In this paper, we propose a *Web service mining* approach to automatically discover pathways from biological entities and processes modeled as Web services. We present a preliminary experiment using Web service models of entities such as COX and Aspirin to illustrate the effectiveness of this mining approach.

1   Introduction

The distributed, diverse and complex nature of biological information that is currently available calls for automated tools to help discover biological pathways. While query-based automated pathway discovery mechanisms such as natural language processing (NLP) [2, 1] have been explored to target the free-text format used to annotate biological entities, these mechanisms are inherently limited by the annotative text that is good at describing properties and functions of biological entities but cannot be used to enact these functions. The discovery of biological pathways can be made more feasible if the dynamic functions can be both described and enacted. The enactment of these functions or processes, through the use of biological models, allows previously unknown pathways to be identified and, more importantly, verified through simulations. In addition, the effects caused by disturbances in these pathways that were previously difficult to study via static function descriptions can be made easier to identify. In this paper, we propose a *Web service mining* approach to mine for pathways from biological entities and processes that are modeled as Web services. Query-based discovery approaches mentioned earlier can work well if the user clearly knows what to look for. As the amount of biological information continues to accumulate, it is unavoidable that there would be many pathways hidden in it. While we may not sometimes have the specific queries needed to search for them, the discovery of these pathways could turn out to be the key in unraveling many of the mysteries of life. Our mining approach aims at proactively discovering *unexpected* and potentially *interesting* pathways in a bottom up fashion when we don't know exactly what to look for.

2   Mining Web Service Models for Pathways

A Web service is an application whose operations can be *described*, *published*, *discovered*, and *invoked* by other independently developed applications through

B. Ludäscher and L. Raschid (Eds.): DILS 2005, LNBI 3615, pp. 292–295, 2005.

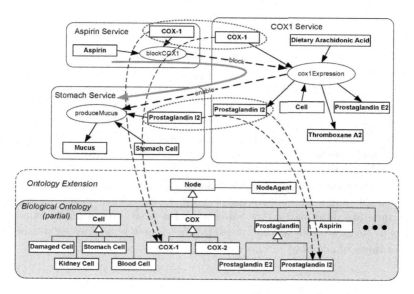

Fig. 1. Operation, Operation Recognition and Pathway

an XML based Web interface. An operation may take several parameters as input, carry out processing tasks using these input parameters, and generate end products as a result of these processing tasks. We refer to the end products as the operation's out parameters. Web service operations can be used to represent functions of a biological entity, such as a gene, a cell, an organ, or a ligand. Fig. 1 depicts three Web services for Aspirin, COX1 and Stomach. An example of a service operation would be cox1Expression of the COX1 Service. Its input parameters include COX1, cell and dietary arachidonic acid, and its output parameters include prostaglandin E2, prostaglandin I2 and thromboxane A2.

Operations from different Web services can recognize one another positively or negatively. When operation A (e.g., cox1Expression) generates some input parameter(s) (e.g., prostaglandin I2) of operation B (e.g., produceMucus), we say that B positively recognizes A. As a result, a *positive bond* is established between the two operations. Similarly, when operation C (e.g., blockCOX1) blocks operation A from consuming some of its input parameters (e.g., COX1), we say that A negatively recognizes C. As a result, a *negative bond* is established between the two operations. A pathway segment is established as a result of a positive or negative bond. Fig. 1 shows an pathway example from blockCOX1 to cox1Expression to produceMucus.

Each of the input and output parameters has a type. The type is defined by a node in a biological ontology that specifies the biological entities and relationships among them. There could be potentially a number of biological ontologies involved in mining. We assume that these ontologies are managed by domain experts. In addition, we assume that developers of biological Web services adhere to such ontologies when defining Web services. The bottom part of Fig. 1 provides an example of how such ontologies can be organized.

We use screening algorithms [4] to screen Web services for potential pathways. When a Web service operation is introduced in our mining process, each of the operation's output parameters will publish to an ontology node it is an instance of. Similarly, each of its input parameters will subscribe to an ontology node it is an instance of. We extend the capability of a conventional ontology by attaching a node agent to every node in the ontology that has been referenced by at least one operation parameter. The node agent keeps track of subscribing and publishing parameters on behalf of a corresponding node. This tracking enables a subscribing parameter (e.g. input parameter prostaglandin I2 to produceMucus) of the node to become essentially notified of the presence of a publishing parameter (e.g. output parameter prostaglandin I2 from cox1Expression). As a result, the operation where the subscribing parameter belongs recognizes the operation where the publishing parameter belongs, and a positive or negative bond is established. As more Web services are introduced, new bonds are established and some of these bonds may extend existing pathways.

3   Experimental Study

We have conducted experiments to assess the effectiveness of the screening algorithms. WSDLs were created with the help of the Systinet Web service plugin to Eclipse 3.0 and used as inputs to our mining process. For each of the Web services discovered, we used the *WSDL2Java* utility provided by Systinet [3] to generate one Java class from its WSDL. This class contains information about operations that are provided by the Web service. Using Java introspection, operations within each of the Java classes can be automatically analyzed to determine their name and information about the input and output parameters. Each operation extracted was first added to an operation library. Our screening algorithms was then used to link these operations. Table 1 contains a list of simplified Web services used in our experiments.

Fig. 2 shows that the mining process identified four pathway trees rooted at *inflammation*, *clot*, *excreteWater*, and *produceMucus*. We only show details of two of the pathway trees in Fig. 2 (a) and (b). Starting from leaf nodes, Fig. 2 (a) shows that operation *blockCOX1* from *AspirinService* blocks the con-

Table 1. Web Services

Web Service	Operation(s)	Input Params	Out Params
COX1Service	cox1Expression	COX1, Cell, DietaryArachidonicAcid	ProstaglandinE2, ProstaglandinI2, ThromboxaneA2
COX2Service	cox2Expression	COX2, DamagedCell, DietaryArachidonicAcid	ProstaglandinE2, ProstaglandinI2
BloodService	clot	BloodCell, LargePlatelet	
AspirinService	blockCOX1	Aspirin	COX1
	blockCOX2	Aspirin	COX2
CelecoxibService	blockCOX2	Celecoxib	COX2
InflammationService	inflammation	ProstaglandinE2, DamagedCell	Swelling, Pain
KidneyService	excreteWater	ProstaglandinE2, KidneyCell	Urine
StomachService	produceMucus	ProstaglandinI2, StomachCell	Mucus
PlateletService	blockPlateletAggregation	ProstaglandinI2	Platelet
	platelet- Aggregation	ThromboxaneA2, Platelet	LargePlatelet

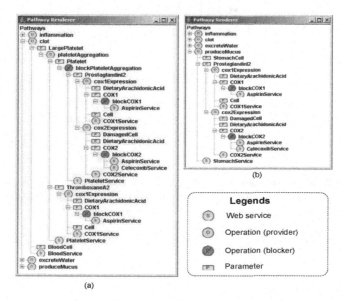

Fig. 2. Pathways Identified via Screening

sumption of *COX1* by operation *cox1Expression* from *COX1Service*. The same *cox1Expression*, in turn, provides *ProstaglandinI2*, which is needed by operation *blockPlateletAggregation* from *PlateletService*. Operation *blockPlateletAggregation* blocks the consumption of *Platelet* by operation *plateletAggregation* from *PlateletService*, which needs both *Platelet* and *ThromboxaneA2* to function. Finally, operation *plateletAggregation* creates *LargePlatelet*, which is needed by operation *clot* from *BloodService*. Operation *blockCOX2* from both *AspirinService* and *CelecoxibService* also blocks the consumption of *COX2* by operation *cox2Expression* from *COX2Service*. Fig. 2 (b) shows the pathway tree that is rooted at operation *produceMucus* from *StomachService*. We can also generate an XML file containing the same content as in Fig. 2. The XML file can be input to tools capable of generating pathway graphs.

References

1. Daming Yao et. al. Pathwayfinder: Paving the way toward automatic pathway extraction. In *Proceedings of the Second Conference on Asia-Pacific Bioinformatics*, volume 29, pages 52 – 62, Dunedin, New Zealand, 2004. Australian Computer Society, Inc.
2. See-Kiong Ng and Marie Wong. Toward routine automatic pathway discovery from on-line scientific text abstracts. volume 10, pages 104 – 112, 1999.
3. Systinet.   Interface   WSDL2Java.   http://www.systinet.com/doc/ssj-55/api/org/idoox/wasp/tools/wsdl2java/WSDL2Java.html.
4. G. Zheng and A. Bouguettaya. Web service modeling for biological processes. In *16th International Conference and Workshop on Database and Expert Systems Applications (DEXA 2005)*, Copenhagen, Denmark, August 2005. To appear.

SemanticBio: Building Conceptual Scientific Workflows over Web Services

Zoé Lacroix and Hervé Ménager

Arizona State University, Tempe, AZ, USA
{zoe.lacroix, herve.menager}@asu.edu
http://bioinfo.eas.asu.edu

Abstract. We introduce here the SemanticBio system [1], which allows expressing scientific protocols as workflows that manipulate scientific objects represented in an ontology. The different tasks are executed using web services, which address many interoperability issues, and are available as interfaces to a variety of life science resources.

1 Introduction

Discovery, in biology as in any other science, is based on scientific reasoning [1]. The basis of this mechanism is that in order to decide whether a hypothesis is valuable, experiments are carried out, and the comparison of the expected results with the actual experimental results leads to supporting it or not. A *scientific protocol* is the description of the coordinated execution of a set of tasks representing the experimental process, and returning a reproduceable set of results.

In the context of modern science, information is increasingly digitalized, as well as experiments, consisting to a greater extent of the collection and analysis of digital datasets. Life science is no exception, and the number of publicly available data sources is growing at a very high rate. In 2005, no less than 719 databases relevant to molecular biology and available on the internet can be listed [2]. This abundance in fact concerns every resource, including data sources (e.g., GenBank) as well as applications (e.g., BLAST). However, scientists do not have the time to learn using them all, and often limit the protocols they express to the few resources they know, hence stating protocols driven by their implementation instead of their scientific aim, with the risk of affecting the value of their endeavour.

We present here the SemanticBio system, that allows scientists to express scientific protocols as workflows at a conceptual level, using ontologies. These conceptual workflows are then translated in a semi-automated process into executable workflows, composed of calls to coordinated web services. In our approach, the emphasis is on the clear separation of the conceptual level, which

---

[1] This research is partially supported by NIH National Library of Medicine grant R03 LM008046-01.

B. Ludäscher and L. Raschid (Eds.): DILS 2005, LNBI 3615, pp. 296–299, 2005.

reflects the protocol *design*, from its *implementation*. The implementation of the protocol, generated according to the user's preferences, can then be *executed* to get the results.

We first describe the requirements of a system aimed at biological data integration and querying. The third section of this paper presents a summary of the SemanticBio architecture. The fourth section introducing an example of a bioinformatics experiment, and a scenario using SemanticBio to carrying it out is followed by the conclusion.

2   Requirements

- *Express queries graphically as data flows* - Scientific protocols are represented as data-flow oriented workflows in a graphical user interface, allowing scientists with few programming skills to design them.
- *Design queries at a conceptual level, using scientific objects* - Scientists design queries using a conceptual model, i.e., an ontology representing scientific objects and the relationships between them.
- *Access a great variety of resources via web services* - Accessing such distributed resources raises many interoperability issues. We believe web services [1] are a good solution to this issues (see also [3]), because they are based on XML, benefiting from the ease of integration this technology provides, and do not require complex firewall parameterization. Furthermore, they are provided by many institutions, such as EBI [2], KEGG [3] or NCBI Entrez [4].
- *Automate the selection and invocation of the required services for each task* - In order to achieve a selection of the resources as automatized as possible, our system provides the user with the available services for each task he defined during the design phase and let him decide which one to use.

3   Characteristics and Architecture of SemanticBio

The system architecture can be divided into different components that are represented in Figure 1.

The user interface is composed of two parts :

- The **web service editor** allows the user to import web services definitions (WSDL files), and describe them semantically, mainly by linking the inputs and outputs of their operations to concepts described in an ontology edited using an **ontology editor**, and the OWL-S [4] language.

---

[1] http://www.w3.org/2002/ws/

[2] http://www.ebi.ac.uk/Tools/webservices/index.html

[3] http://www.genome.jp/kegg/soap/

[4] http://www.ncbi.nlm.nih.gov/entrez/query/static/esoap_help.html

– The **workflow editor** allows users to design a conceptual workflow by dragging components on a graph interface. Once this design phase is achieved, an assistant suggests for each task of the workflow a list of corresponding web services, before to submit and monitor it.

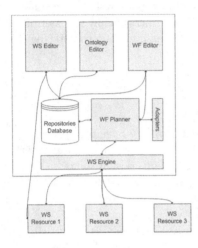

The **repository database** stores all the information used by the system in a native XML database system, more convenient given the format of these data.

The **adapters** are software modules that handle data format transformation between two steps of the workflow.

The **workflow planner** handles the execution of the workflows, coordinating the web services or adapters calls, managing their status, and informing the user of their execution status.

Fig. 1. SemanticBio Architecture

4   Illustrating Example

We consider the following protocol: *"Return all citations that are related to the disease diabetes type 1"* [5]. Its implementation in SemanticBio is a three steps process:

– **Ontology description** - This very simple protocol involves two types of scientific objects: *citation* and *disease*. The protocol evokes a path between those two concepts, that are represented in an ontology.
– **Resources Integration** - To solve this problem, the resources, i.e., the web services implementing the concepts and their relationships, are described in the system, using the **web services editor**. Here, we can use the NCBI EUtils web service. For instance, describing the semantics of the ELink operation is done through mapping its input and output respectively to the corresponding concepts in the source and destination database, in our case OMIM and PubMed. This web service accesses all the scientific objects described in the databases available at NCBI Entrez. Therefore, a single operation, such as the EFetch operation, is described as retrieving the information about all those objects, each time specifying a different value for the database input.
– **Workflow construction** - Once the resources are integrated in the system, users can design the workflow using the **workflow editor**. This workflow is composed of two tasks: one that retrieves the relationships between the *genetic disease* and the *citation* object, and one that retrieves the *citation* object associated to each relationship. The first step takes as input an identifier for the disease, and its output is a list of citation identifiers. The second

step's input is an identifier for a *citation*, and its output is the *citation* itself. Once the graph representing the workflow has been built, the invocation of the implementation assistant offers the user a web service for each of the steps he designed, such as the ones we described earlier.

- **Execution** - The invocation of the execution of this workflow triggers its submission to the **query planner**, which handles the coordination of the calls to web services, notifying the workflow editor of the execution status. In case the output of a task is not be in the same format as the input of the following, an adapter is called to handle this data transformation.

5   Conclusion

The SemanticBio system, currently under development at Arizona State University, addresses the complex issues raised by the integration of biological data with a meta-data driven approach, based on conceptual information. It is aimed at being *scientist-friendly*, making the expression of scientific protocols as easy as possible. Finally, it is built using common formats in order to facilitate its integration with existing resources and its interfacing with other systems. The SemanticBio system relies on an ontology, a collection of available web services, and their semantic description. The collection of this information allows to make extensive reuse and sharing of this knowledge in the life sciences community. The SemanticBio approach is evaluated with scientific protocols currently designed by the Brain Tumor Unit at the Translational Genomics Research Institute (TGen). Future work includes the integration of the BioNavigation system [6], that helps selecting evaluation paths to execute conceptual queries expressed on an ontology, using available resources.

References

1. Lawson, A.: The Living World - BIO100. McGraw Hill Primis (1999)
2. Galperin, M.Y.: The Molecular Biology Database Collection: 2005 update. Nucleic Acids Res (2005) 5–24 vol. 33 Database Issue.
3. Gao, T.H., Hayes, J.H., Cai, H.: Integrating biological research through web services. IEEE Computer **38** (2005) 26–31
4. Martin, D., Burstein, M., Hobbs, J., Lassila, O., McDermott, D., McIlraith, S., Narayanan, S., Paolucci, M., Parsia, B., Payne, T., Sirin, E., Srinivasan, N., Sycara, K.: OWL-S: Semantic Markup for Web Services. W3C Working Draft (2004) http://www.daml.org/services/owl-s/1.1/overview/.
5. Eckman, B.A., Deutsch, K., Janer, M., Lacroix, Z., Raschid, L.: A Query Language to Support Scientific Discovery. In: Computational Systems Bioinformatics, IEEE Computer Society (2003) 388–391
6. Lacroix, Z., Parekh, K., Vidal, M.E., Cardenas, M., Marquez, N.: BioNavigation: Selecting optimum paths through biological resources to evaluate ontological navigational queries. In: Data Integration in the Life Sciences (DILS). Lecture Notes in Bioinformatics, Springer (2005)

PLATCOM: Current Status and Plan for the Next Stages

Kwangmin Choi[1], Jeong-Hyeon Choi[1], Amit Saple[1], Zhiping Wang[1], Jason Lee[1], and Sun Kim[1,2,*]

[1] School of Informatics, Indiana University,
IN 47408, USA
[2] Center for Genomics and Bioinformatics, Indiana University,
IN 47405, USA

Abstract. We have been developing a web-based system for comparing multiple genomes, PLATCOM, where users can choose genomes of their choice freely and perform analysis of the selected genomes with a suite of computational tools. PLATCOM is built on internal databases such as GenBank, COG, KEGG, and Pairwise Comparison Database (PCDB) that contains all pairwise comparisons (97,034 entries) of protein sequence files (.faa) and whole genome sequence files (.fna) of 312 replicons. Since combining multiple tools for sequence analysis requires a significant amount of programming work and knowledge on each tool, we also developed and incorporated high performance sequence data mining tools such as sequence clustering and neighborhood prediction. The next plan includes defining several data types for genome analysis and integrating system modules using data mining tools that operate on the genome data types. PLATCOM is available at `http://platcom.informatics.indiana.edu`.

1 Introduction

The exponential accumulation of genomic sequence data demands systematic analysis of genetic information and requires use of various computational approaches to handle such huge sets of genomic data. Comparative genomics, with such organized data and diverse computational techniques, has become useful not only for finding common features in different genomes, but also for understanding the evolutionary process and mechanism among multiple genomes.

Comparison of multiple genomes is a challenging task partially because combining multiple tools for sequence analysis requires a significant amount of programming work and knowledge on each tool and partially because it handles a huge amount of data. Another problem is the subjectivity of how to select multiple genomes. For example, there are 1,313,400 ($= \binom{200}{3}$) possible selections of three genomes out of 200 completely sequenced genomes. The inconsistency of

---

* Corresponding author.

B. Ludäscher and L. Raschid (Eds.): DILS 2005, LNBI 3615, pp. 300–304, 2005.

input data from existing sources and the effective presentation of analysis results also raise problems.

Considering all these issues, it is not possible to perform multiple genome comparison on the web by simply using sequence analysis tools in an *ad hoc* fashion. SEALS [1], The SEED [2], DAS [3] were developed to achieve such a goal, but they are developed for service on the web.

2 PLATCOM: A Computational Environment for Comparative Genomics on the Web

We have been developing a genome comparison system PLATCOM, which is available at `http://platcom.informatics.indiana.edu`. PLATCOM is designed to be a high performance genome analysis system on the web which is easy to use and easy to maintain and update. These design principles may conflict with other desirable system features such as information richness and a sophisticated user interface. Instead, PLATCOM aims at a flexible, extensible, scalable, and reconfigurable system with emphasis on high-performance data mining. Although PLATCOM does not store or maintain any information on sequences, information on sequences can be obtained via URLs or connectivity tools to other information rich databases.

2.1 Overall System Architecture

PLATCOM consists of four main components; databases, sequence analysis tools, genome analysis modules, and a user interface.

The whole system is built on internal databases, which consist of GenBank, SwissProt, COG, KEGG, and Pairwise Comparison Database (PCDB). PCDB is designed to incorporate new genomes automatically so that PLATCOM can evolve as new genomes become available. FASTA and BLASTZ are used to compute all pairwise comparisons (97,034 entries) of protein sequence files (.faa) and whole genome sequence files (.fna) of 312 replicons. Multiple genome comparisons usually take too much time to compute, but the pre-computed PCDB makes it possible to complete genome analysis very fast even on the web. In general, PLATCOM runs several hundred times faster than a system without PCDB when several genomes are compared. In addition to sequence data, PLATCOM will include more data types such as gene expression data. More importantly, PCDB makes it possible to achieve one of the most important design goals, which is to allow users to select any subset of genomes to be compared freely. We also designed PCDB in a way that the update, introduction of new genomes to PLATCOM, can be performed almost automatically [6].

Sequence analysis tools include widely used high performance sequence analysis tools such as FASTA, BLAST, BLASTZ, HMMER, GIBBS, and MEME. We have also developed high performance data mining tools of our own (see Section 2.2). With the databases and sequence analysis tools, genomes can be compared. There are currently six modules: genome plot, conserved gene neighborhood navigation, metabolic pathways, comparative sequence clustering analysis, putative gene fusion events detection, and multiple genome alignment. A set of genomes selected by users is submitted with parameter settings via a web interface.

2.2   Multi-step Sequence Analysis with Scalable Data Mining Tools

Data mining techniques are useful in combining many sequence analysis tools and databases that can be utilized for genome annotation since data mining tools encapsulate multiple sequence analysis tasks in a single step. Thus well-defined data mining concept and tools can make genome comparison much easier. It is also important that the data mining tools for genome comparison should be scalable. We have been developing such scalable tools: a sequence clustering algorithm BAG [5], a genome sequence alignment tool GAME [7], an algorithm for mining correlated gene sets [9], and a multiple genome sequence alignment algorithm by clustering local matches [8].

To summarize the analysis result, we have developed visualization tools for genome plot, multi-domain, gene-genome matching table, and genome alignment. Since our ultimate goal is to make PLATCOM a flexible system in that users can combine multiple computational tasks freely, it is also important to make visualization modules independent of particular computational tasks. We designed the interface of the visualization modules to use genomes as context so that output from different computational tasks can use the same visualization module.

3   Plan for the Next Stages

We have been using high performance sequence analysis tools to "simplify" sequence analysis tasks. For example, the BAG clustering tool can generate a set of sequence clusters in a single operation, rather than performing many sequence data searches using FASTA or BLAST and then combining the search results. However, our goal is to provide a web-based environment for genome comparison. To achieve this goal, many sequence analysis and data mining tools should be combined freely. Our approach is to introduce several data types for genome analysis so that sequence analysis and data mining tools can be combined using these data types. In this way, a series of sequence analysis tasks can be viewed as a composition of mathematical functions. For example, $\mathcal{F}(\mathcal{H}(x), y)$ can be seen as a two step sequence analysis tasks, $\mathcal{H}$ followed by $\mathcal{F}$, where the functional composition is performed on the same data type for the co-domain of $\mathcal{H}$ and the domain of $\mathcal{F}$. To make the user interfaces easy to use, we introduce only a few data types.

Almost all sequence analysis and data mining tools can be viewed as functions on the domain of "a set of sequences" and "a set of genomes". Thus we introduce two data types, $\mathcal{S}$ for a set of sequences and $\mathcal{G}$ for a set of genomes. To allow users to select sequences and genomes, we introduce selection functions, $I_{\mathcal{S}} : \mathcal{S} \rightarrow \mathcal{S}'$ to select a set $\mathcal{S}'$ of sequences from $\mathcal{S}$, and $I_{\mathcal{G}} : \mathcal{G} \rightarrow \mathcal{G}'$ to select a set $\mathcal{G}'$ of genomes from $\mathcal{G}$. I_S and $I_{\mathcal{G}}$ are implemented as a web page where a set of sequences or genomes are listed and then users select subsets by clicking checkboxes.

We briefly illustrate this concept using an example of combining two existing modules in PLATCOM. GenomePlot(G_i, G_j) computes and plots gene matches in G_i and G_j, and GeneClusterSearch($S, \mathcal{G}$) searches a set $\mathcal{G}$ of genomes for matches of a given set S of sequences. These two modules can be combined as follows:

1. GenomePlot($I_{\mathcal{G}}$ (all genomes in PLATCOM)) generates a set of gene matches MG.
2. MCGS(MG) [9] computes a set of gene clusters, $\{GC_1, \ldots, GC_k\}$.
3. For any GC_i, users can perform GeneClusterSearch($I_S(GC_i)$, $I_{\mathcal{G}}(all\ genomes)$)) where users select a set of genes from GC_i via $I_S(GC_i)$ and searches its occurrences in a set of selected genomes via $I_{\mathcal{G}}(all\ genomes)$).

We are currently working on a complete implementation of this concept in order to provide a flexible genome comparison environment on the web.

Acknowledgments

This work partially supported by INGEN (Indiana Genomics Initiatives) and NSF CAREER Award DBI-0237901.

References

1. Walker, D.R, and Koonin, E.V. : SEALS: A System for Easy Analysis of Lots of Sequences. Intel. Sys. for Mol. Biol. **5** (1997) 333–339
2. Overbeek, R., Disz, T., and Stevens, R. : The SEED: A Peer-to-Peer Environment for Genome Annotation. Comm. of the ACM **47** (2004) 47–50
3. Dowel, R.D., Jokerst, R.M., Eddy, S.R., Stein, L. : The Distributed Annotation System. BMC Bioinfo. (2001) 2:7
4. Choi, K., Ma, Y., Choi, J.-H., and Kim, S. : PLATCOM: A Platform for Computational Comparative Genomics. Bioinfo. (2005)
5. Kim, S. : Graph theoretic sequence clustering algorithms and their applications to genome comparison, Chapter 4 in Comp. Biol. and Genome Informatics, World Scientific. (2003)
6. Ma, Y., Bramley, R., and Kim, S. : A Data Management Architecture for Computational Biology. Indiana University Computer Science Technical Report **607** (2005)
7. Choi, J.-H., Cho, H.-G., and Kim, S. : A Simple and Efficient Alignment Method for Microbial Whole Genomes Using Maximal Exact Match Filtering. Comp. Biol. and Chem. (2005) (To appear).
8. Choi, J.-H., Choi, K., Cho, H.-G., and Kim, S. : Multiple genome alignment by clustering pairwise matches. Proceedings of the 2nd RECOMB Comparative Genomics Satellite Workshop, Lecture Notes in Bioinformatics, Bertinoro, Italy, **3388** (2005). 30–41, Springer-Verlag, Berlin.

9. Kim S., Choi, J.-H., Yang, J. : Gene Teams with Relaxed Proximity Constraint, IEEE Computational Systems Bioinformatics (CSB'05), (2005), San Francisco, CA (To appear)
10. Schwartz, S., Kent, W. J., Smit, A., Zhang, Z., Baertsch, R., Hardison, R. C., Haussler, D., and Miller, W. : Human-mouse alignments with BLASTZ. Genome Res. **13** (2003) 103–107

SOAP API for Integrating Biological Interaction Databases

Seong Joon Yoo[1], Min Kyung Kim[2], Ho Il Lee[3], and Hyun Seok Park[2,3]

[1] School of Computer Engineering, Sejong University, 98 Gunja, Gwangjin,
Seoul, Korea 143-747
sjyoo@sejong.ac.kr
[2] Department of Computer Science and Engineering, Ewha University,
11-1 Daehyun-dong, Seodaemun-gu, Seoul, 120-750, Korea
{minkykim, neo}@ewha.ac.kr
[3] Macrogen Inc., Bioinformatics Institute, 60-24, Gasan-dong,
Geumcheon-gu, Seoul, Korea
headil@macrogen.com

Abstract. This paper presents design and implementation of SOAP API with which bioinformaticians can integrate the biological interaction datbases. While designing a web service based integration framework, it is not easy but important to define API for biological SOAP servers. Therefore, we propose in this paper a web service API especially for the interaction databases: BIND, DIP and MINT. The three databases are mirrored in our local computers on which we have implemented a prototype of SOAP servers for the interaction databases.

1 Introduction

While interaction database is an emerging field of biological research, no database supports SOAP servers yet. Therefore, it is natural to focus on devising a data model to enable interaction databases to provide for web services. It is necessary to define objects and their methods for SOAP servers of the interaction databases. We define the objects by considering the characteristics of each database. Prior to designing the SOAP server objects for each database, this section describes and analyzes characteristics of three most popularly used interaction databases: BIND, DIP and MINT.

BIND(**B**iomolecular **I**nteraction **N**etwork **D**atabase) [1], a research project of Samuel Lunenfeld lab, in Mount Sinai hospital, handles the interaction and pathway information of biomolecules. DIP(**D**atabase of **I**nteracting **P**roteins) [2] is a database that collects and provides the information which is experimentally determined about protein-protein interaction. MINT(**M**olecular **INT**eraction database) [3] is a RDBMS that stores the interaction formation between biomocules.

2 Related Work

Web services have been deployed in a few biological databases including myGRID[4], BioMoby[5], and KEGG[6]. myGRID is designed for data or service

B. Ludäscher and L. Raschid (Eds.): DILS 2005, LNBI 3615, pp. 305 – 308, 2005.

providers who want to build applications for biologist. None of these databases, however, provide web services for interaction information.

The DDBJ SOAP server provides services for the search and analysis of sequence databases. The myGrid is a middleware for *in silico* experiments in biology. It aims at a rather general purpose system by providing several web services including NCBI BLAST, WU BLAST, SRS, etc. The scope of BioMoby is service description, discovery, transaction and simple input/output object type.

KEGG contains data useful for developing bioinformatics technologies such as comparison, reconstruction and design of the metabolic process targeting at the research of the functional genomics. Its data has been partially open through web service technologies since 2003.

There are four objects for KEGG's web service: SSDB(Sequence Similarity DataBase), PATHWAY, GENES and KEGG. SSDB, PATHWAY and GENES objects are the web service objects for SSDB, PATHWAY, GENES databases of KEGG, and KEGG object is the web service object that provides the information about KEGG database, such as version of KEGG database.

3 The Design of the Web Service API for the Interaction Databases

The BIND database has three API classes: BindInteractionIF, BindPathwayIF, and BindComplexIF. The DIP interaction database has defined a single class: DIPInteractionIF. The MINT interaction database provides one class: MINTInteractionIF.

Fig. 1 illustrates the classes and their methods for the API of the BIND SOAP server. The SOAP servers of the MINT and the DIP databases also provide web service API with which application programmers may implement bioinformatics systems accessing interaction databases. The MINT database SOAP server supports the MintInteractionIF class and the DIP database SOAP server supports DipInteractionIF class respectively. Each class has a set of methods as listed in Fig. 2. MintInteractionIF object is used to provide the Interaction data. The meanings of the methods are defined in a similar way as the meaning is defined for the methods of the BIND database.

The BindInteractionIF is a class that is used for accessing Interaction data. This has 18 methods that support querying complex, compound, DNA, protein and RNA objects participating in a specified interaction. The BindInteractionIF class includes four types of methods: methods returning all objects participating in specified interactions, methods for finding information of specified interactions, methods finding interactions by id or names, and methods finding objects by gene ID or names. Bind_InteractionIF also allows querying the information for interaction nodes and edges. Users can query interaction information by gene name or gene ID's.

Bind_ComplexIF is a class for accessing the Complex data. This class allows users to query type, ID, PubMed Information, and interaction ID of complexes by name, ID or gene ID. This class has defined eight methods for supporting the above mentioned queries.

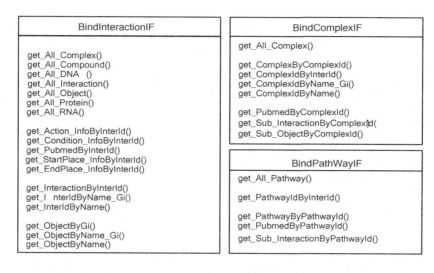

Fig. 1. API and its Methods for the BIND database

MINTInteractionIF	DIPInteractionIF
getMintInteractionByInterId() getMintInteractionBySpId() getMintInteractionBySpIds() getMintInteractDomainByInterId() getMintPubmedByInterId() getMintExperimentByInterId() getCommentByInterId() getNegationByInterId() getInteractionTypeByInterId() getOrganismBySpId() getShortNameBySpId() getMintInteractorFeatureByInterId()	getDipInteractionByNodesId() getDipInteractionByEdgeId() getDipInteractionByNodeId() getDipNodeByNodeId() getNodeNameByNodeId() getSwissprotIdByNodeId() getNcbiIdByNodeId() getPirIdByNodeId() getDescriptionByNodeId() getOrganismByNodeId() getTaxonByNodeId() getNodeIdByNodeName() getNodeIdBySwissprotId() getNodeIdByNcbiId() getNodeIdByPirId() getPubmedIdByEdgeId() getExperimentByEdgeId() getClassByEdgeId() getEdgeFeatureIdByEdgeId() getEdgeFeatureByEdgeId() getEdgeFeatureByEdgeFeatureId() get_StartPlace_InfoByInterId()

Fig. 2. Methods of the Web Service API for the MINT and DIP databases

Bind_PathwayIF is a class for accessing Pathway data. This class can access pathway data by specifying pathway ID. It also supports querying pathway data including a specified interaction. Finally, users can query PubMed information of a pathway. This class has eight methods.

We have developed a prototype of the SOAP servers and databases in our local systems. Since original sites of the above three interaction databases do not provide their own SOAP servers, we needed to build our own databases by copying the original data in non-relational database format and transforming these into relational database format. We have used three servers with Intel Pentium4 1.9 CPU, Memory 512MB, and the software environment of Windows 2000 Professional, JWSDP 1.3, and Java 1.4.

4   Conclusion

This paper proposes a practically useful API for SOAP servers of interaction databases. Bioinformaticians may build their own client software that accesses data from those three interaction databases through these objects on the SOAP servers. Once the source databases implement the proposed API on their own databases and SOAP servers, worldwide users can access the interaction databases more easily. Users can use this system with the interaction of gene ontology for semantic integration of interaction databases. We are working on this issue.

Acknowledgment

Min Kyung Kim is supported by the BK21 program of Korea Research Foundation.

References

1. Bader, G.D. and Hogue, C.W.: BIND-a data specification for storing and describing biomolecular interactions, molecular complexes and pathways, (2000) Bioinformatics. 16, 465-477.
2. Salwinski, L., Miller, C.S., Smith, A.J., Pettit, F.K., Bowie, J.U., and Eisenberg, D.: The Database of Interacting Proteins: 2004 update. (2004) Nucleic Acids Res. 32(Database issue), D449-451.
3. Znzoni, A., Montecchi-Palazzi, L., Quondam, M., Ausiello, G., Helmer-Citterich, M, Cesareni, G. (2002) MINT: a Molecular INTeraction database. FEBS Lett. 513, 135-140.
4. Stevens RD, Robinson AJ, and Goble CA.:myGrid: personalised bioinformatics on the information grid, (2003). Bioinformatics. 19, i302-i304.
5. Wilkinson MD and Links M.,:BioMOBY: an open source biological web services proposal, (2002) Brief Bioinform. 3, 331-341.
6. Shuichi Kawashima, Toshiaki Katayama, Yoko Sato and Minoru Kanehisa,:A Web Service Using SOAP, WSDL to Access the KEGG System, (2003) Genome Informatics. 14, 673-674.

Collaborative Curation of Data from Bio-medical Texts and Abstracts and Its integration

Chitta Baral, Hasan Davulcu, Mutsumi Nakamura, Prabhdeep Singh,
Luis Tari, and Lian Yu

Department of Computer Science and Engineering,
Arizona State University,
PO Box 878809, Tempe, AZ 85287-8809, USA
{chitta, hdavulcu, mutsumi, prabhdeep, luis.tari,
lianyu}@asu.edu

Abstract. We propose an inexpensive and scalable approach for curation that takes advantage of automatic information extraction methods as a starting point, and is based on the premise that if there are a lot of articles, then there must be a lot of readers and authors of these articles. Thus we provide a mechanism by which the readers of the articles can participate and collaborate in the curation of information.

1 Introduction

Besides the data that exists in various public and private databases, there is a much larger and ever increasing amount of information buried in existing biomedical articles. It is beyond human ability to read the various relevant articles and recall relevant findings of these articles for further research. Therefore, it becomes clear that the findings in these articles have to be culled and stored in a database such that the data can be integrated with other existing databases. The sheer volume of the articles and their constant growth makes it prohibitively expensive to employ *(and monetarily compensate)* human curators to read through the articles and cull the necessary knowledge/data buried in them. Nevertheless, such human curation (see for example [1,3-7,21]) has been tried for specific domains. Due to the issue of cost, many of the curated databases are proprietary with limited coverage.

In recent years an alternative approach of using automatic text extraction systems [2,8-20] has been proposed. Although good progress has been made in this area, the systems are not fool-proof. They at times infer incorrect information or miss out important information. Moreover, most existing systems focus on simpler data forms, such as identifying gene or protein names, simple interactions without context. Sometimes such simplicity may lead to inconsistency.

In this paper we propose a solution to the problem of curating information from the large and growing body of biomedical texts and abstracts. We propose a methodology where the community collaboratively contributes to the curation process. We use automatic information extraction methods as a starting point, and promote mass collaboration with the premise that if there are a lot of articles, then there must be a lot of readers and authors of these articles.

B. Ludäscher and L. Raschid (Eds.): DILS 2005, LNBI 3615, pp. 309–312, 2005.
© Springer-Verlag Berlin Heidelberg 2005

2   CBioC System Architecture

The two main components of our CBioC system are (i) the CBioC interface and (ii) the CBioC database. The user interacts with the CBioC system through the CBioC interface. When a user views a PubMed article, the CBioC interface is automatically invoked to display all the extracted interaction data relevant to the article. The user curates the extracted interaction data through voting. Depending on the access level, an user can also enter or modify data.

The CBioC interface has many subcomponents such as the automatic invocation component, the user and access management component, and the voting and other interactions component. Two auxiliary components of the system are (a) a suite of automated text extraction systems and (b) a data exchange system. The text extraction systems are used to automatically extract data from texts and abstracts and the data exchange system is used to download relevant data from existing databases (such as [7,9-13]) and convert them to our format. This is illustrated in Figure 1 below.

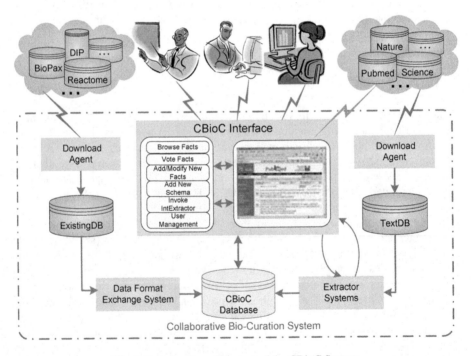

Fig. 1. Functional architecture of the CBioC System

We now illustrate the use of the CBioC system which also further illuminates on the architecture of the CBioC system.

Installation and Invocation: An important goal of ours is to make it easier for a researcher to participate in the collaborative curation. For that a researcher has to download our system and install it in her computer. Once the system is installed it watches the researcher's access of the web through Internet Explorer windows. Whenever the researcher accesses a web page from where she can access an article or an abstract, the CBioC system is invoked and an interaction frame is created, as shown below in Figure 2.

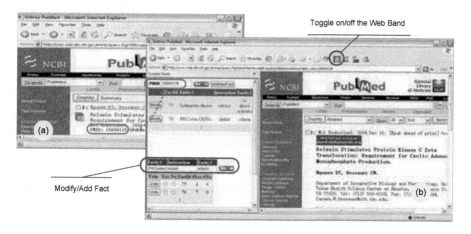

Fig. 2. Automatic triggering of CBioC interaction frame

System Implementation: From the implementation angle, the CBioC system consists of three main parts: (i) Web forms and connection to database; (ii) WebBand and Browser Helper components, and (iii) Connector to Interaction Extractor, and is currently implemented for Internet Explorer in the client side and Linux-MySQL-Php on the server side. This is illustrated in Figure 3 below.

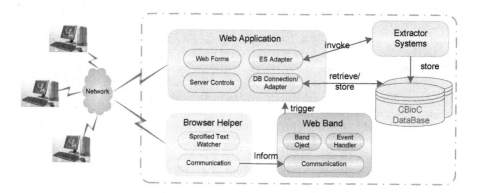

Fig. 3. Implementation Architecture of CBioC System

Acknowledgement. Deepthi, Toufeeq, Haishan, Luping, Ning, Drs. Kambhampati, Das, Berens, Fukuda, Bittner, and Gaasterland and, NSF grant 0412000.

References

[1] Bader, G.D., Donaldson, I., Wolting, C., Ouellette, B.F., Pawson, T., and Hogue, C.W. (2001) BIND-The biomolecular interaction Netwoek database. *Nucleic Ac. R.* **29**, 242-245.

[2] Rzhetsky, A. et al. (2004) Geneways: a system for extracting, analyzing, visualizing, and integrating molecular pathway data. *Journal of Biomedical Informatics* **27**, 43-53.

[3] Stein, Lincoln (2002), Creating a bioinformatics nation, *Nature*, **417**, 119-120.

[4] Xenarios, I. and Eisenberg, D. (2001) Protein interacting databases. *Current Opinion in Biotechnology.* **12**, 334-339.

[5]   KEGG: Kyoto Encyclopedia of Genes and Genomes, http://www.genome.jp/kegg/

[6]   BIND: Interaction Network Database, http://www.bind.ca

[7] HPRD: Human Protein Reference Database, http://www.hprd.org/

[8] Fukuda, K., Tamura, A., Tsunoda, T., and Takagi, T. (1998) Toward information extraction: identifying protein names from biological papers. *PSB 1998*, 707-718

[9] Tanabe, L. and Wilbur, W.J. (2002) Tagging gene and protein names in biomedical text. *Bioinformatics.* 2002 Aug;18(8):1124-1132.

[10] Blaschke, C., Andrade, M.A., Ouzounis, C., and Valencia, A. (1999) Automatic extraction of biological information from scientific text: protein-protein interactions. *Proceedings of the International Conference on Intelligent System Molecular Biology.* 1999, 60-67.

[11] Ono, T., Hishigaki, H., Tanigami, A., and Takagi, T. (2001) Automated extraction of information on protein-protein interactions from the biological literature. *Bioinformatics. 2001* Feb;17(2):155-561.

[12] Novichkova, S., Egorov, S., and Daraselia, N. (2003) MedScan, a natural language processing engine for MEDLINE abstracts. *Bioinformatics. 2003* September. **19(13)**, 1699-1706

[13] Friedman, C., Kra, P., Yu, H., Krauthammer, M., Rzhetsky, A. (2001) GENIES: a natural-language processing system for the extraction of molecular pathways from journal articles. *Bioinformatics.* 2001, **17** Suppl 1:S74-82.

[14] Rzhetsky, A., Iossifov, I., Koike, T., Krauthammer, M., Kra, P., Morris, M., Yu, H., Duboue, P.A., Weng, W., Wilbur, W.J., Hatzivassiloglou, V., and Friedman, C. (2004) GeneWays: a system for extracting, analyzing, visualizing, and integrating molecular pathway data. *J Biomed Inform.* 2004 February, **37(1)**, 43-53.

[15] Corney, D.P., Buxton, B.F., Langdon, W.B., and Jones, D.T. (2004) BioRAT: extracting biological information from full-length papers. *Bioinformatics.* 2004 November 22, **20(17)**, 3206-3213. Epub 2004 Nov 22.

[16] Temkin, J.M. and Gilder, M.R. (2003) Extraction of protein interaction information from unstructured text using a context-free grammar. *Bioinf.* 2003 Nov 1, **19(16)**, 2046-2053.

[17] Chiang, J.H., Yu, H.C., and Hsu, H.J. (2004) GIS: a biomedical text-mining system for gene information discovery. *Bioinformatics.* 2004 Jan 1, **20(1)**, 120-121.

[18] Craven, M. and Kumlien, J. (1999) Constructing biological knowledge bases by extracting information from text sources. *Proceedings of International Conference on Intelligent System Molecular Biology.* 1999, 77-86.

[19] Bunescu, R., Ge, R., Kate, R.K., Marcotte, E.M., Mooney, R.J., Ramani, A.K., and Wong, Y.W. (2004) Comparative Experiments on Learning Information Extractors for Proteins and their Interactions. *Journal Artificial Intelligence in Medicine* 2004.

[20] Ding, J., Berleant, D., X, J., and Fulmer, A. (2003) Extracting biochemical interactions from MEDLINE using a link grammar parser. *In Proceedings of the 15th IEEE International Conference on Tools with Artificial Intelligence (ICTAI'03),* 467. IEEE Computer Society, 2003.

[21] www.biocurator.org

Towards an Ontology Based Visual Query System

Serguei Krivov and Ferdinando Villa

Gund Institute for Ecological Economics, University of Vermont,
590 Main, Burlington, VT., 05405
{skrivov, fvilla}@uvm.edu

Abstract. This paper discusses an ontology based visual query system of Ecosystem Services Database. A new visual query languge for OWL is proposed.

1 Introduction

Large Data Integration projects such as TAMBIS [1], KIND [2], SEEK [3], SE-WASIE [4] use ontologies to provide integrated logical views of heterogeneous data bases, while the queries and the views of the data sources are defined and executed at the level of ontologies. For such applications user friendly ontology–based interfaces are essential. While several ontology editing and visualization frameworks are available [5], few of them support the Semantic Web endorsed ontology language OWL [6]. Visual tools that support query formulation are also appearing [7], however no graphical query language for OWL ontologies has been developed so far. This paper discusses an ontology based visual query system and its application within a web-accessible database. A new graphical query language for OWL is proposed.

2 Ecosystem Services Database and GrOWL

Ecosystem services are the benefits people derive from ecosystems. Quantification of the economic value of ecosystem services has become an important vehicle for assuring social recognition and acceptance of public management of ecosystems [8]. The Ecosystem Services Database (ESD) [9],[10] allows users to compare ecosystem service values across the geographic regions, for different biomes and verify all the components that went into their formulation. Among the novel features of the ESD are the use of knowledge maps and an ontology based visual query system based on GrOWL software. Fig. 1 shows the usage of a biome ontology in query formulation.

The GrOWL visualization model is an accurate mapping of the underlying Description Logics (DL) semantics of OWL ontologies. We presume here that the reader is familiar with DL notations and DL semantics of OWL [11]. Fig.2 describes the mapping of DL class constructors, mapping of ABox expressions

B. Ludäscher and L. Raschid (Eds.): DILS 2005, LNBI 3615, pp. 313–316, 2005.

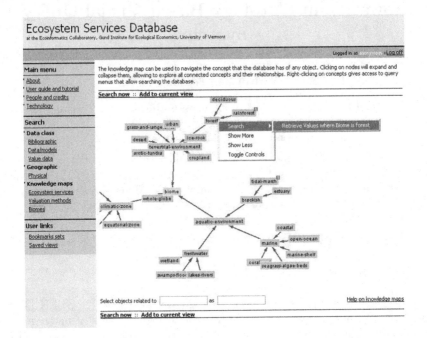

Fig. 1. GrOWL Applet is apart of ESD interface. The knowledge maps provides assistance in formulating ESD queries

Definition of class C	The diagram G(C)	Base node BN(C)
Named Class C	C	C
Intersection $C_1 \sqcap C_2$	BN(C1) ⬌ ⊓ ⬌ BN(C2)	⊓
Union $C_1 \sqcup C_2$	BN(C1) ⬌ ⊔ ⬌ BN(C2)	⊔
Complement $\neg C_1$	¬ ⟶ BN(C1)	¬
Enumeration $\{o_1, o_2\}$	E ↙ ↘ o2 o1	E
Exist Restriction $\exists R.C_1$	∃:R ⟶ BN(C1)	∃:R
For all Restriction $\forall R.C_1$	∀:R ⟶ BN(C1)	∀:R
Number Restriction $\geq nR$	Eg. ≥5:R	≥5:R
Number Restriction $\leq n\,R$	Eg. ≤7:R	≤7:R
Value Restriction $R{:}o$	R ⟶ o	R

Fig. 2. Recursive mapping of DL class constructors

and the queries is illustrated by Fig. 3 . On Fig.2 the diagram $G(C)$ represents the definition of respective class C , while the base node $BN(C)$ represents class C itself.

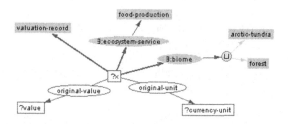

Fig. 3. A simple graphical query

3   Graphical Query Language for OWL

In the realm of DL a conjunctive query condition is a conjunction of query atoms $x{:}C$ and $(x, y){:} R$, where x and y are variables or individuals C is a concept expression and R is a role. We introduce two types of variables: "select" variables prefixed by "?" and "ignore" variables prefixed by "-". With such notations the set of select variables of a query $?x1, \ldots, ?xm$, is defined by the query condition. GrOWL-Query is GrOWL with variables. Queries are GrOWL's ABox diagrams where the variables are allowed in place of individuals. GrOWL-Query diagram in Fig. 3 represents the following query:

$answer(?x, ?value, ?currency\text{-}unit) : -$
   $?x : valuation\text{-}record,$
   $\wedge ?x : \exists ecosystem\text{-}service.food\text{-}production,$
   $\wedge ?x : \exists biome.(forest \sqcup arctic\text{-}tundra),$
   $\wedge (?x, ?currency\text{-}unit) : original\text{-}unit$
   $\wedge (?x, ?value) : original\text{-}value.$

Complex class expressions allow users to effectively describe disjunctive queries (as above) and queries with negations. Detailed description of GrOWL and GrOWL- Query could be found online [12]. GrOWL-Query can be used as a query interface to a DL reasoner. It can also be used as a part of a Data Integration system.

Acknowledgements. *This work has been supported by NSF grants 9982938/ 0243957 and 0225676. GrOWL evolved through the discussions within SEEK community and we owe a lot to every member of the seek-kr-sms discussion group. We are especially grateful to Mark Schildhauer who motivated us to develop a visual representation of OWL and to Rich Williams who made considerable contributions to GrOWL design and implementation.*

References

1. Stevens, R., Baker, P., Bechhofer, S., Ng, G., Jacoby, A., Paton, N., Goble, C., Brass, A.: TAMBIS: Transparent access to multiple bioinformatics information sources. Bioinformatics **16** (2000) 184–5

2. Ludäscher, B., Gupta, A., Martone, M.E.: Model-based mediation with domain maps. In: 17th Intl. Conf. on Data Engineering ICDE, Heidelberg, Germany, IEEE Computer Society (2001)

3. Michener, W.K., Beach, J.H., Jones, M.B., Ludaescher, B., Pennington, D.D., Pereira, R.S., Rajasekar, A., Schildhauer, M.: A knowledge environment for the biodiversity and ecological sciences. (Journal of Intelligent Information Systems (accepted))

4. : SEWASIE- semantic webs and AgentS in integrated economies. (URL: http://www.sewasie.org/)

5. Denny, M.: Ontology building: A survey of editing tools. URL: http://www.xml.com/pub/a/2002/11/06/ontologies.html (2002)

6. Bechhofer, S., Harmelen, F.V., Hendler, J., Horrocks, I., McGuinness, D.L., Patel-Schneider, P.F., Stein, L.A.: OWL web ontology language reference. (M. Dean, and G. Schreiber (eds.) W3C Recommendation 10 February 2004)

7. Catarci, T., Mascio, T.D., Franconi, E., Santucci, G., Tessaris, S.: An ontology based visual tool for query formulation support. In: Proceedings of the 16th European Conference on Artificial Intelligence (ECAI-04), Valencia, Spain. (2004)

8. Costanza, R., D'Arge, R., DeGroot, R., Farber, S., Grasso, M., Hannon, B., Limburg, K., Naeem, S., O'Neill, R., Paruelo, J., Raskin, R., Sutton, P., VanDenBelt, M.: The value of the world's ecosystem services and natural capital. Nature **387** (1997) 254–260

9. Villa, F., M.A.Wilson, DeGroot, R., Farber, S., Costanza, R., Boumans, R.: Design of an integrated knowledge base to support ecosystem services valuation. Ecological Economics, 41 (2002) 445– 456

10. : Ecosystem services database. (URL: http://esd.uvm.edu)

11. Horrocks, I., Patel-Schneider, P.: Reducing owl entailment to description logic satisfiability (2003)

12. Krivov, S.: Graphical query language for OWL (GrOWL). URL: http://ecoinformatics.uvm.edu/dmaps/growl/ (2004)

Data Integration in the Biomedical Informatics Research Network (BIRN)*

Vadim Astakhov, Amarnath Gupta, Simone Santini, and Jeffrey S. Grethe

University of California San Diego, La Jolla, CA 92093, USA
{astakhov, gupta, ssantini}@ncmir.ucsd.edu

Abstract. A goal of the Biomedical Informatics Research Network (BIRN) project sponsored by NCRR/NIH is to develop a multi-institution information management system for Neurosciences, where each participating institution produces a database of their experimental or computationally derived data, and a mediator module performs semantic integration over the databases to enable neuroscientists to perform analyses that could not be executed from any single institution's data. This demonstration paper briefly describes the current capabilities of Metropolis-II, the information integration system for BIRN.

1 Introduction

The goal of the data integration system for the Biomedical Informatics Research Network (BIRN) (www.nbirn.net) is to develop a general-purpose information integration framework which diverse groups of neuroscientists can use for a variety of application problems that arise from different scientific research needs. This framework is designed to support a number of neuroscience research test beds. In the setting of the *mouse BIRN* test bed, a large number of very different information integration applications may need to be designed over a slowly increasing set of very heterogeneous data sources. The data to be integrated range from 3D volumetric data of nerve components, to image feature data of protein distribution in the brain, to genomic data that characterize the anatomical anomalies of different genetically engineered mouse strains and so forth, and there are a number of integrated schemas over different combinations of these sources designed for different study groups. In contrast, the integration requirement of the *human morphometry BIRN* and *human functional imaging BIRN* test beds have a single virtual schema collectively developed by the participating research groups, and an increasing number of research universities are contributing their data to this schema. The data provided by these test beds are mostly deidentified patient records for patients with neurodegenerative diseases, containing, for instance, demographic data, psychological evaluations and medical imaging analyses.

---

* This work is supported by NIH BIRN-CC Award No. 8P41 RR08605-08S1, NIH Human Brain Project Award No. 5RO1DC03192.

B. Ludäscher and L. Raschid (Eds.): DILS 2005, LNBI 3615, pp. 317–320, 2005.

Given this application context, the data integration framework of BIRN consists of a *global-as-view* mediator called Metropolis-II, a number of specialized tools for schema registration, view definition and query building, a number of domain-specific clients, and a set of tree and graph structured ontologies that supply intermediate information such that integrated views can be defined over the sources. Using the external ontologies to integrate information is our way of implementing *semantic integration* [1]. The overall architecture of the system is given in Figure 1.

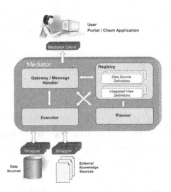

Fig. 1. The general architecture of Metropolis-II

2   Integration Framework

Data Sources. Metropolis-II makes the assumption that a data source is relational that may have a binding pattern for every exported relation. Every schema element *relations, attributes* has a descriptor for keyword search, and a so-called *semantic-type* that can be used to map the element to an ontology [2]. Further, a data source may export a set of functions that are internally treated as relations with the binding pattern $(\bar{b}, f)$ where $\bar{b}$ represents a set of bound arguments and the single f is the free output variable of the function. In Mouse BIRN, for example, specialized functions are used to compare the distributions of proteins in a set of user-specified regions in the brain. Using this model also enables us to treat computational sources such as the R statistical package as a "data source" that contributes only functions and no relations. Integrated views are written using standard data sources as well as these functions. We have also designed source-specific wrappers for sources such as Oracle, Oracle Spatial, and Postgres, where a generic query can be translated into the appropriate flavor of SQL, and functions supported by the specific systems.

Ontological Sources. We use the word ontology here to mean a term-graph whose nodes represent terms from a domain-specific vocabulary, and edges represent relations that also come from an interpreted vocabulary [3]. The nodes and edges are *typed* according to a simple, commonly agreed upon set of type produced by test bed scientists. The most common interpretation is given by rules like the transitivity of *is-a* or *part-of* relations, which can be used, for example, to implement inheritance. However, there are also domain specific rules for relationships such as *volumetric-subpart: brain-region → brain-region* and *measured-by: psych-parameter → cognitive-test* that need special rules of inference. For example, if a brain-region *participates-in* a *brain-function* (like "working memory"), and the brain-function is *measured-by* a *cognitive-test*, then the cognitive-test *functionally-tests* the brain-region. Currently, ontologies are represented as a set of relations reflecting the set of nodes and their properties, a set of edges, and a set of edge properties. Also other operations, including graph functions such as path and descendant finding, and inference functions like finding transitive edges are implemented using an API of functions, as described in the previous paragraph.

View Definition and Query Languages. The query language for Metropolis-II is the union of conjunctive queries, which may contain function terms, as well as the standard aggregate functions. The syntax of the language, expressed in XML, is essentially that of Datalog with aggregate functions [4]; essentially, a query has the form $q(X, F(Y)) : -r_1(X, Z), r_2(Z, Y)$ where $F(Y)$ is the aggregate function operated on sets of Ys and X is a (in reality, a set of) group-by variable. The query planner and execution engine in Metropolis-II translates this expression to

$$q'(X, Y) : -r_1(X, Z), r_2(Z, Y)$$

$$q(X, W) : -F(gb(q'(X, Y)))$$

where the the group-by function gb followed by the aggregate function F is pushed together to the data source whenever possible, and are otherwise evaluated at the mediator. The language also admits nested queries, where inner queries are assigned to intermediate relation variables, that are used by the main query. The view definition language for the system, on the other hand, does not allow aggregates and nested queries at the present time. The language allows only safe negations, where all variables in negated predicates are bound.

Mapping Relations. In the current GAV setting of the mediator, the burden of creating proper integrated views over data sources is on the integration engineer who works with the domain scientists to capture the requirements of the application at hand. This often leads to the pragmatic problem that the relationships between the attributes exported by different sources and those between the data values are, quite often, not obvious. To accommodate for this, the recent version of the system [5] has created additional mapping relations. Currently there are three kinds of mapping relations. The *ontology-map* relation that maps data-values from a source to an ontology term of a known ontology (like the Unified Medical Language System from the National Library of Medicine). The *joinable* relation pairs attributes from different relations if their data type and semantic types match. The *value-map* relation maps a *mediator-supported data value* or a *mediator-supported attribute-value pair* to the equivalent value (resp. attribute-value pair) supported by the source. For example, the mediator may export a demographic attribute called gender with possible values {male, female}, while one source may refer to it as attribute sex with possible values {0, 1}, while another may call it kcr_s57 with the domain {m, f}. The Metropolis-II planner uses a look-up function to make a substitution before dispatching the query plan to the execution engine.

Authentication and Authorization. Access control is a very important aspect of a practical information integration system. For BIRN, this is accomplished in two stages – defining authenticable users, and the implementation of authorization that enables a user to perform only the tasks she is permitted to. The authentication function is handled outside the mediator by a community authorization service. The authorization is handled through an additional access control database that is implemented inside the mediator.

3   The Demonstration

The demonstration will present to the user the information integration system together with the different clients for tasks performed by the submitter of a newly joining source, and integration engineer. These tasks include schema registration, integrated view design, ontology browsing and query design. A number of different query clients are designed for different user groups, and walk through the different stages of query execution in the system. This will include the XML-encoded query language and the view-definition language of the mediator, the plan generated by the system, the communication between the mediator and the different wrappers. As part of this walkthrough, we would also demonstrate how we have used the statistical package R as a computation resource accessed through the mediator. In this process, we will also illustrate the different kinds of data sources and different classes of queries the system can handle.

Acknowledgments. David Little, Maryann Martone, Robin Park, Xufei Qian, Edward Ross, Joshua Tran, Yujun Wang, Wai-Ho Wong, Aylin Yilmaz, Ilya Zaslavsky are the BIRN R&D team. Bertram Ludäscher contributed to the basic research and first version of the system.

References

1. Ludäscher, B., Gupta, A., Martone, M.E.: Model-based mediation with domain maps. In: Proc. 17th Int. Conf. on Data Engineering (ICDE), Washington, DC, USA, IEEE Comp. Soc. (2001) 81–90
2. Gupta, A., Ludäscher, B., Martone, M.E.: Registering scientific information sources for semantic mediation. In: Proc. 21st Int. Conf. on Conceptual Modeling (ER), London, UK, Springer-Verlag (2002) 182–198
3. Gupta, A., Ludäscher, B., Martone, M.E.: Knowledge-based integration of neuroscience data sources. In: Proc. 12th Int. Conf. on Scientific and Statistical Database Management (SSDBM'00), Washington, DC, USA, IEEE Comp. Soc. (2000) 39–52
4. Zaniolo, C., Arni, N., Ong, K.: Negation and aggregates in recursive rules: the ldl++ approach. In: Proc. 3rd Int. Conf. on Deductive and Object-Oriented Databases (DOOD), Springer-Verlag (1993) 204–221
5. Santini, S.: Metropolis MkIII: Prolegomena to design. Technical Report 04-04, Dept. of Neurosciences, Univ. of California San Diego (2004)

Data Integration and Workflow Solutions for Ecology*

William Michener[1], James Beach[2], Shawn Bowers[3], Laura Downey[1],
Matthew Jones[4], Bertram Ludäscher[3], Deana Pennington[1], Arcot Rajasekar[5],
Samantha Romanello[1], Mark Schildhauer[4], Dave Vieglais[2], and Jianting Zhang[1]

[1] LTER Network Office, University of New Mexico
[2] Biodiversity Research Center, University of Kansas
[3] Genome Center & Dept. of Computer Science, University of California, Davis
[4] NCEAS, University of California, Santa Barbara
[5] San Diego Supercomputer Center, University of California, San Diego

1 SEEK: Introduction and Architecture

The Science Environment for Ecological Knowledge[1] (SEEK) is designed to help ecologists overcome data integration and synthesis challenges. The SEEK environment enables ecologists to efficiently capture, organize, and search for data and analytical processes. We describe SEEK and discuss how it can benefit ecological niche modeling in which biodiversity scientists require access and integration of regional and global data as well as significant analytical resources.

SEEK is designed as a three-layer architecture. The *EcoGrid* forms the base layer and provides a uniform and simple programming interface for access to distributed resources such as data, metadata, and workflows. The KEPLER *Scientific Workflow System*[2] forms the topmost layer and provides tools that allow scientists to create and compose scientific workflows (e.g., analytical models), execute them, and archive the results. KEPLER makes extensive use of EcoGrid interfaces. For instance, through the EcoGrid, KEPLER allows scientists to search for and retrieve data and workflows stored across distributed repositories. The *Semantic Mediation System* (SMS) forms the middle layer of the architecture and mediates between heterogeneous resources in the EcoGrid and the analyses and models to be executed in KEPLER. SMS leverages ontologies to facilitate data integration and workflow composition, thereby increasing the scale and complexity of analyses that can be constructed and executed by scientists. Each of these layers is described further below.

The EcoGrid [4] layer forms the underlying cyberinfrastructure within SEEK for enabling remote data and service discovery, data sharing and access, and remote service invocation. EcoGrid services and interfaces are being built using best practices currently available in grid technology (e.g., OGSA/WSRF, SRB, and Condor). EcoGrid provides resource discovery through a registration service. Many data sets accessible through the

* We thank the other members of SEEK, including: Chad Berkley, Dan Higgins, Jessie Kennedy, Ricardo Pereira, Town Peterson, Aimee Stewart, Jing Tao, and Bing Zhu. This work is supported in part by NSF grants ITR 0225674, EF 0225665 and DBI 0129792, DARPA grant N00014-03-1-0900, and the Andrew Mellon Foundation.

[1] seek.ecoinformatics.org

[2] www.kepler-project.org

B. Ludäscher and L. Raschid (Eds.): DILS 2005, LNBI 3615, pp. 321–324, 2005.
© Springer-Verlag Berlin Heidelberg 2005

EcoGrid have Ecological Metadata Language[3] (EML) descriptions that are also used for data discovery, access, and integration.

KEPLER is used to design and execute scientific workflows [6] (see Figure 1). KE-PLER includes components (called *actors*) to access data from the EcoGrid as well as other generic scientific workflow components including R and Matlab modules for statistical analysis. From these components, customized scientific workflows can be built such as the Genetic Algorithm for Ruleset Production (GARP) discussed below. Existing components and workflows can be linked within KEPLER to form a new scientific workflow graph. The inputs and outputs of components are represented using ports, which can have structural types describing the physical representation of data (e.g., double) as well as semantic types describing the conceptual meaning and scientific context of data (e.g., BODYSIZE) [1, 3]. The SMS system uses structural and semantic types to help scientists construct meaningful scientific workflows.

The SMS layer provides ontology-based services to KEPLER including support for data integration, workflow composition, and concept-based searching. The SEEK Knowledge Representation Team (KR) includes ecologists and knowledge engineers who jointly develop and maintain formal ontologies to be used by the SMS. These ontologies cover a number of different areas including measurement, time and space, basic ecological concepts, biodiversity, and unit systems. Also as part of KR/SMS, the SEEK Taxonomic Object Service [5] is being developed to help resolve progressive changes of taxonomic names to sets of taxonomic concepts, providing well defined, authoritative, and (ideally) unambiguous information about the identification of organisms.

2   Use Case: Ecological Niche Modeling

A new and promising paradigm in ecology is the use of ecological niche modeling (ENM) to extrapolate implications of global climate change for biological diversity [7]. Figure 1 shows the KEPLER implementation of an ENM workflow that assesses the implications of climate change for mammals of the Western Hemisphere. Such broad-scale comparative analyses of effects of different climate-change modeling scenarios are difficult to implement due to their computational complexity, which includes data discovery ($> 3,000$ mammal species), data integration (20 climate scenarios), and analytical complexity ($> 180,000$ model runs).

ENM incorporates both spatially explicit point data indicating where a species has been found as well as spatially explicit environmental data such as descriptions of climate, hydrology, and soils. Within KEPLER, scientists can use EcoGrid-based search interfaces to discover occurrence data (e.g., within the DiGIR network) as well as environmental data (e.g., located within Metacat or SRB collections)[4]. These searches leverage metadata about objects to locate relevant items of interest and present them to the user. For ENM, partitioning the relevant taxonomic data into species groupings may be difficult as a result of changes in taxonomic names. The Taxonomic Object Service can be used to help resolve these taxonomic clustering issues.

---

[3] knb.ecoinformatics.org/software/eml/

[4] digir.sourceforge.net/, knb.ecoinformatics.org/software/metacat/, www.sdsc.edu/srb/

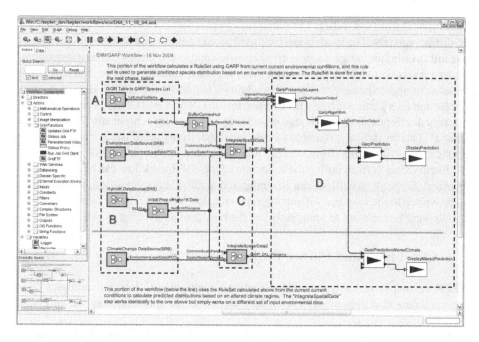

Fig. 1. The ENM workflow in KEPLER with components for: (A) accessing and pre-processing DiGIR species occurrence data; (B) accessing and pre-processing SRB environmental data; (C) integrating occurrence data and environmental layers; and (D) GARP modeling steps

Once relevant data sets are discovered they can be directly imported into KEPLER. Data access components are provided by KEPLER that can use the detailed descriptions of the physical data structure of EML to automate the process of importing data. Thus, using EML-described data sets in a workflow simply involves dragging their associated icons onto the workflow canvas (Figure 1, A and B). KEPLER parses the metadata and exposes output ports that represent each attribute within the data. It also provides a Query-By-Example extension for user-friendly SQL query construction.

SMS provides a generic set of ontology-based languages and tools for storing and exploiting semantic annotations [1, 3], which explicitly link existing data sets and workflow components to ontologies. Through semantic annotations, the mediation layer provides knowledge-based data integration and workflow composition services [2], component and data discovery via concept-based searching, as well as basic services used in workflow modeling such as ensuring that workflows are "semantically" type-safe (based on annotations). In the ENM case, each of the data types must undergo a series of transformations for integration, including re-projection to a common geographic coordinate system, re-scaling to a common resolution, and re-orientation to center the imagery on the same point on the globe. The locations of occurrence points are used to sample the environmental data and create vectors containing many bands of information associated with each occurrence point (Figure 1, C).

The ENM workflow analyzes native distributions of species using a genetic algorithm (GARP; Figure 1, D) written in C++ [8]. The GARP algorithm generates a rule-based model from input data. The workflow is run a number of times to generate a set of distinct models. The models are then used to construct a probabilistic prediction of the full distribution range under current climate conditions, and potential distributions under various climate change scenarios. The workflow consists of more than fifty components in approximately ten nested workflows including the GARP algorithm, grid access and query components, GIS components in GRASS and GDAL, statistical components developed in R, and image processing and viewing components developed in ImageJ. This workflow is reusable by multiple biodiversity scientists in many different applications in its current form, and can readily be modified for additional applications.

Finally, data derived during the execution of the ENM workflow can be saved to the EcoGrid. KEPLER workflows can be configured to allow any output from a component to be written to the EcoGrid with appropriate metadata, completing the "analysis cycle" by allowing future work to seamlessly use the results of an existing workflow.

3   Conclusion and Future Project Directions

SEEK encompasses many cyberinfrastructure tools needed to integrate complex ecological data and enable rapid development and re-use of complex scientific analyses. Nevertheless, many challenges remain. Future work includes: (1) exploration of new ways to leverage and extend the Taxonomic Object Service; (2) use of the Geographical Markup Language to achieve greater interoperability of spatial/GIS data; (3) additional support for semantic annotations in workflow design and execution [3]; (4) native support for scheduling of compute-intensive, distributed scientific workflows; (5) additional geospatial semantics for ontology-based ENM workflow compositions; and (6) usability engineering to improve SEEK tools.

References

1. C. Berkley, S. Bowers, M. Jones, B. Ludäscher, M. Schildhauer, and J. Tao. Incorporating semantics in scientific workflow authoring. In *SSDBM*, 2005.
2. S. Bowers and B. Ludäscher. An ontology-driven framework for data transformation in scientific workflows. In *Intl. Workshop on Data Integration in the Life Sciences*, 2004.
3. S. Bowers and B. Ludäscher. Actor-oriented design of scientific workflows. In *Intl. Conf. on Conceptual Modeling (ER)*, 2005.
4. M. Jones. SEEK EcoGrid: Integrating data and computation resources for ecology. *DataBits: An electronic newsletter for Information Managers*, Spring 2003.
5. J. Kennedy, T. Paterson, and R. Kukla. Scientific names are ambiguous as identifiers for biological taxa: Their context and definition are required for accurate data integration. In *Intl. Workshop on Data Integration in the Life Sciences*, 2005.
6. B. Ludäscher, I. Altintas, C. Berkley, D. Higgins, E. Jaeger-Frank, M. Jones, E. A. Lee, J. Tao, and Y. Zhao. Scientific workflow management and the Kepler system. *Concurrency and Computation: Practic & Experience, Special Issue on Scientific Workflows*, 2005.
7. A. Peterson, M. Ortega-Huerta, J. Bartley, V. Sanchez-Cordero, J. Soberon, R. Buddemeier, and D. Stockwell. Future projections for mexican faunas under global climate change scenarios. *Nature*, 416, April 2002.
8. D. Stockwell and D. Peters. The GARP modelling system: Problems and solutions to automated spatial prediction. *Intl. Journal of Geographical Information Science*, 13, 1999.

Eco-Informatics for Decision Makers Advancing a Research Agenda

Judith Bayard Cushing[1] and Tyrone Wilson[2]

[1] The Evergreen State College, Olympia, WA, USA
judyc@evergreen.edu
[2] National Biological Information Infrastructure, U.S. Geological Survey, Reston, VA, USA
tyrone_wilson@usgs.gov

Larry Brandt, Valerie Gregg, Sylvia Spengler (National Science Foundation)
Alan Borning (University of Washington),
Lois Delcambre (Portland State University),
Geoff Bowker (Santa Clara University), Mike Frame (USGS/NBII),
János Fülöp (Hungarian Academy of Sciences),
Carol Hert (University of Washington-Tacoma),
Eduard Hovy (University of Southern California),
Julia Jones (Oregon State University),
Eric Landis (Natural Resources Information Management),
John L. Schnase (NASA),
Charles Schweik (University of Massachusetts-Amherst),
William Sonntag (EPA)

Abstract. Resource managers often face significant information technology (IT) problems when integrating ecological or environmental information to make decisions. At a workshop sponsored by the NSF and USGS in December 2004, university researchers, natural resource managers, and information managers met to articulate IT problems facing ecology and environmental decision makers. Decision making IT problems were identified in five areas: 1) policy, 2) data presentation, 3) data gaps, 4) tools, and 5) indicators. To alleviate those problems, workshop participants recommended specific informatics research in modeling and simulation, data quality, information integration and ontologies, and social and human aspects. This paper reports the workshop findings, and briefly compares these with research that traditionally falls under the emerging eco-informatics rubric.

1 Introduction

Decision makers at all levels of government and at NGOs who manage natural resources or carry out ecological or environmental policy face significant information technology (IT) problems when integrating ecological or environmental information. Ecology and environmental decision makers work with information providers and data managers, and seek a wide variety of information sources, but little of the data used to produce these sources is collected specifically for the decision making at hand. Thus, the decision maker is faced with data gaps, data presentation

B. Ludäscher and L. Raschid (Eds.): DILS 2005, LNBI 3615, pp. 325–334, 2005.

mismatches, and finding appropriate indicators. These IT issues suggest computer science research needs in information integration, modeling and simulation, data quality, and human-centered issues such as training, technology transfer, best practices for information provision and use, and human-friendly software. While a growing body of research has focused on information technology to help ecology researchers [pcast,bdei], solving IT problems in natural resource management is not simply a matter of adopting the technology developed for other domains, or even applying research completed under the eco-informatics rubric to the decision-makers' problem space. As suggested in discussions of eco-informatics at a Digital Government Conference, the problems are different, and the field of eco-informatics should be extended to include helping decision makers (e.g., policy makers and natural resource managers) utilize data and information more effectively. [dgo.04]

Eco-informatics problems faced by natural resource decision makers require, in addition to new research, sustaining innovation in the public and NGO sectors just as in digital government research [dgo.05].   Researchers must find the right domain problem, distill research that will prove fruitful to a range of stakeholders, find the right agency collaborators, and manage everybody's expectations. [hovy] Researchers in eco-informatics decision-making must also consider combining quantitative with qualitative information and have a basic understanding of decision making.   If computer scientists and social scientists in the academy are not prepared to take on these challenges in addition to demanding research, natural resource eco-informatics will lag behind informatics in other science and policy domains. We base this assessment on the complexity introduced by public policy requirements added to already complex scientific eco-informatics issues.

2   Eco-Informatics Problem Space for Natural Resource Management

Eco-informatics is about both biodiversity-rich conservation managed systems and natural resource protection and human health impacts of environmental, anthropogenic pollutants, on the other.   Rather than sorting out different informatics needs for these two areas, we recognized that the latter presupposes a command of the former, and focused on non-human-health-centered ecological constituencies. Another perspective can be found in Europe, where the research is much broader in nature and includes health and security, as well as ecosystem function. [jensen]

To map the problem space for natural resource management eco-informatics, we note that many organizations have developed IT for natural resource decision makers, and we laid out the eco-informatics problem space collaboratively with personnel from the USGS, NASA, EPA, State Agencies and InterState Consortia. [e.g., guldin, sugarbaker] Resource management informatics is hard, and data and tools form a demand cycle; the more successful one is, the more demand arises. Exemplary projects, as at Ohio State and Oregon State universities show how researchers and coastal policy makers might collaborate, and exemplify the research complexity. Coasts are interaction zones of land, sea, and air, and although they occupy only 3% of sea surface area and 0.5% of ocean volume, about 70% of global fish resources spend time there. Further, about 60% of the world's human population resides close to the coast zone, exploiting it for food, recreation, transport, waste disposal, etc. This brings materials discharge from land to coast, and causes environmental changes through physical, chemical, and biological processes. Enhanced data handling capacity and cooperation among intergovernmental agencies are essential for integrating required multi-source data. [li, wright]   Drawing on these and other experiences, we categorized the problem space for natural resource management eco-informatics into: policy, data presentation, data gaps, tools, and indicators. [context].

2.1 Policy Problems

Policy issues related to ecological and environmental information and decision-making include, but are not limited to, problems that organizations encounter because of policies related to: 1) the provision, production and maintenance of eco-informatics tools and information; 2) the use and possible abuse of tools and information; 3) the cross-organizational sharing (or lack thereof) of tools and information; and 4) the communication (or lack thereof) of environmental management decisions grounded upon eco-informatics-based analysis. From here on, "BDEI" (biodiversity and eco-system informatics) will be used interchangeably with eco-informatics.

1. Provision, production, and maintenance, e.g., data curation or archiving. BDEI tools and information must begin with user needs assessment, but in some cases developers are not doing this. We recommend research into why this is so, how the problem differs from other IT application areas, and how to solve the problem. Though costly and inefficient, data are sometimes collected but only used once, or even not at all. Better systems of metadata and storage retrieval might ensure that what data are collected or generated are used and shared. Whether BDEI tools and information should be treated as public, private or toll goods is another issue. While the tools and information could be considered public goods, there might be important reasons to treat them as a private or toll good in order to collect revenue to absorb some costs.

2. Use and possible abuse of tools or information. Issues here include translating from uncertain scientific models to policy decisions that require a legal burden of proof, and determining whether scientific evidence justifies a policy decision when there is uncertainty in the data. Tools that might be useful for policy analysis are not used in decision making because they a) take too long compared to the political cycle; b) cost too much; c) are based on unrealistic assumptions; or d) are too complex or technical.

3. Cross-organizational data and tool sharing, or lack thereof. We note two levels of organizational cooperation: sharing tools or information, and co-production of tools or information. Organizations generally avoid paying to develop tools if another organization is also involved – a classic free rider problem in collective action theory. Further, organization policies can be barriers to co-production, if employees are not recognized for such efforts. Where organizations are interested in sharing information or models, they may be hindered by inadequate metadata or ontologies that would allow integration with their own data and models. We see both carrot and stick approaches for encouraging data owners to produce and maintain metadata, e.g., employee performance rewards or positive recognition, or mandates with negative consequences for noncompliance.

4. Communication (or lack thereof) of environmental management decisions grounded in informatics-based analysis. Organizations sometimes do a poor job communicating issues discovered through BDEI analysis – an information diffusion problem involving mass communication. Organizations or policy-makers are sometimes caught off-guard by an environmental or ecological problem, and face difficulties addressing the problem because little or no data exist. Further the extant data might exhibit a linear trend when in fact the pattern is more complex. These problems are illustrative of decision-making or policy under circumstances involving uncertainty. [policy]

2.2 Data Presentation

Data Presentation problems arise from complex interactions between user needs and data (e.g.,, metadata, raw data, information, accuracy specifications, methods, documentation, policy). System limitations (e.g., software modalities, hardware availability or costs); and information format further complicate presentation. Needed research includes determining what information is best on which medium, cross-referencing and supporting data across presentations, representing time and change, new media (e.g., 3D, VR), and user task definitions. This problem area can be distilled into two major components: 1) presentation as the mediator between users and their needs, and between task and data or metadata and their characteristics, and 2) the set of research questions and themes that relate to the facilitation of that mediation role. Presentation options must reflect the user experience as well as the nature of the data, with constraints. Presentation types may need to reflect a number of user dimensions: 1) User needs, perhaps conceptualized as tasks or time available, or context, and 2) User characteristics, including preferences, (dis)abilities, and computing capabilities available. On the data side, presentation could reflect the nature and amount of data, metadata available, data and metadata quality, data preparation activities, and policies such as privacy and confidentiality. Presentation instantiations and approaches need to reflect the marriage of the user and the data. In addition, presentation media add their own affordances and issues. Different software modalities may have different suitabilities for different data types, and different hardware media have different costs, availability, and permanence. These components will suggest a range of research questions that will help understand presentation for BDEI decision making. [presentation]

2.3 Data Gaps

Geographic data gaps between biodiversity-rich and conservation-managed land areas adversely impact decision making. These problems stem from lack of the needed data sets or access to them, disjoint data sets that require manipulation to compensate for temporal or spatial gaps, an emphasis on adaptive management which outpaces data reliability, or a paucity of database professionals upon which resource managers can call. Major issues include how to appropriately generalize fine-scale data that will necessarily contain gaps, and decision makers' and policy makers' sensitivity to uncertainty. Next steps to refine this problem area would be to address the original data needs, and define review criteria, such as stable standards for data collection and documentation.

2.4 Tools

Major BDEI tool problems apply across the board to scientific informatics research, namely, how one balances longer term research to advance functionality with supporting users in the short term. Tool problems involve 1) a lack of a tool "clearing-house", i.e., from the developer side getting a tool out to users, and from the user side finding and evaluating tools and determining if a given tool can be applied to other problems or with other data than what it was developed for.   2) new or different data types and data collection methods, 3) lack of user frameworks, product suites, and development standards, 3) lack of tools to support metadata issues, and 4) social science issues of usage, sharing, and adoption.

2.5 Indicators

Indicator problems exist because indicator definition, relevance, and value are neither well-defined nor communicated. Constituents may be uneasy with environmental measures, and data gaps effect reliability and trust that these stakeholders have in indicators. Finally, the inherent complexity of the ecosystem further complicates this issue. Prime examples of the complexity that arises in using indicators include the Death Valley Pupfish and the Washington State Shellfish Bed Closures.

3 Research Issues

Teasing out research issues from the natural resource management problem space was a four-step process. 1) We examined three current research projects to see how interdisciplinary approaches and government partners were involved. 2) In breakout sessions, we articulated research issues, which were 3) in plenary sessions critiqued by a panel of resource managers and researchers.   4) Finally, breakout groups reconvened to refine and prioritize issues, identify strategies for sustaining research, and suggest resource management case histories that exemplified the need for research. [research]

Three NSF Digital Government research case studies, the Forest Portal, UrbanSim, and Understanding Government Statistics. [blm, borning, hert], gave an initial set of research issues to consider. The Forest Portal, an adaptive management tool that harvests information to sustain forests, highlighted the importance of collaboration between federal agencies and academic institutions, and demonstrated the capabilities of using metadata attachments.   UrbanSim demonstrated how ecological models and partnerships contribute to data collection, preparation, and assessment, which in turn would likely lead to realistic policy scenarios and major policy applications. The GovStat project models user access to U.S. government statistical information to better integrate data across agencies.   The project emphasized deploying prototypes to identify research challenges and designing an interface that relies on metadata generated from the web sites. Research issues were categorized into four major areas: 1) modeling and simulation, 2) data quality, 3) data integration and ontologies, and 4) social and human aspects.

3.1 Modeling and Simulation

Modeling and simulation research issues included: coupling diverse models, addressing values in design for diverse stakeholders, incorporating new visualizations for results, representing error and uncertainty when presenting information to decision makers, challenges in handling large data sets, and open source modeling infrastructure.   This group emphasized proposed that an open-source, flexible, reusable modeling infrastructure, along with the social practices that sustain it would allow researchers and decision makers to experiment freely with new models or change existing ones.

3.2 Data Quality

Data quality research issues primarily involve how to determine and communicate uncertainty to decision-makers who use multiple data sources.   Methods are needed to mitigate introducing error when creating and combining data sets, and to associate error with alternative decisions. The question of whether metadata could become an obligatory part of the data set was raised. The overarching research question invoked by this problem is the extent to which uncertainty associated with data quality and synthesis really has an influence on policymaking and plan implementation. Research issues arise also in individual studies and data sharing where diverse data sources are combined. Research is needed to develop methods for 1) reducing the introduction of error when datasets are created and combined, measuring and logging error at each stage of the study and 3) characterizing relationships among errors – additive, multiplicative, averaging.

Where data are shared, for example in data harvesters such as the Long-term Ecological Research network's Clim-DB and Hydro-DB, the major issue is the extent to which metadata can become an integral part of the dataset. What happens to metadata when multiple sources are integrated? How can metadata management be automated? How can data standardization help in combining metadata from multiple sources? How could metadata become a part of the data set. The research challenge is how general can the tools that manage data quality become, and whether they be applied to a wide range of ecological datasets. NSF could develop and publish metadata standards across all grants, instead of just for certain programs. Successful metadata efforts include the Federal Geographic Data Committee within the USGS and the LTER information manager standards used in internal reviews of LTER projects.

To determine the influence of uncertainty associated with data synthesis on policymaking and plan implementation, studies could be done of decision-makers perceptions of the value of science findings made from synthesized or integrated data. For example, data harvesters such as Clim-DB and Hydro-DB have generated publications from combined datasets, which are perhaps being used by land managers or decision makers in the Forest Service and NOAA. How is the increase in power associated with data synthesis balanced by the increase in uncertainty associated with the ways in which the errors were combined? How might synthetic studies stand up in courts of law in comparison with other forms of "expert testimony?"

3.3 Information Integration and Ontologies

Information integration involves mechanisms for reliable, transparent and authoritative data combination. Associated research issues include: defining the dimensions of integration, quantifying semantic distance, integrating multiple ontologies, promoting document modeling, evaluating utility of qualitative and quantitative data, tools to support data integration, and evaluating knowledge from non-traditional sources. Ontologies are useful in providing metadata (semantics) over databases, making cross-disciplinary connections, and thesauri. Ontologies on the Grid would help users find data and functionality. Tools to build, verify and deliver ontologies still require considerable research. Other phenomena that require research are understanding gaps and inconsistencies in ontologies, trust and verification of the content of ontologies, and understanding and handling change in the material represented by ontologies in ways that go beyond simple versioning. The semantics of BDEI is critical, and includes defining and operationalizing meanings, data standardization, and semantic services. Transferring knowledge from other domains to BDEI is itself research. Quality control, data access, and collaborative decision making support are also critical.

The reclassification of rainbow trout as salmon in the early 1990s and a subsequently implemented information system had broad-reaching effects; the moral being that no indicator is innocent, and IT systems have social consequences. How can computer scientists and developers be sensitized to the value judgments inherent in data collection, ontology generation, and modeling? Future IT applications should warn scientists and policy makers of impending circumstances. [ontology]

3.4 Social and Human Aspects

Research issues identified in this area included: eco-informatics tool development and information sharing among decision makers (e.g., measuring success, determining appropriate institutional designs and incentives or disincentives), human-computer interaction (human/tools interface), impact on management practices, education and training, and user requirements and system design. Advancing the eco-informatics agenda hinges on both new technologies and new understandings of how information infrastructures inter-relate between individuals, organizations, communities, disciplines, information resources, and tools. Consider State Agency Official "Jane Doe" prioritizing parcels for conservation. She is interested in forecasting land use change over a region to identify habitat parcels most threatened by human encroachment. Ideally, Jane would like policy-relevant modeling to identify the "development fringe", but she cannot develop that on her own. Because others, whom Jane might not even know about, may be well on their way to doing this, tools to facilitate the investigation would include library management systems, and newer, innovative collaboration tools and computer-based land use change models. We considered this scenario as it would play out now and in ten years if the recommended research were successful. A second scenario involving the Death Valley National Park Devils Hole Pupfish illustrated an immediate need for tools to integrate information over time and across agencies, evaluate legacy data, identify indicators, visualize alternative actions, and model current ecological conditions. [social].

4 Conclusions, Broader Impacts, and Recommendations

One metaphor useful in understanding our natural resource management vision is a fictitious, ideal decision making tool, dubbed Yoda. Yoda sees decision makers as those who choose among alternatives, and what they do as integrating information – via sharable data structures, compatible software tools, human collaboration, and understanding outcomes. Theirs is an awesome task that involves ontologies, semantic distances, data quality assessment, etc., and many complex steps. [tosta].

The sheer number, breadth, and complexity of problems and potential solutions suggested in this report dictate that it will take decades to solve the problems – all while species and ecosystems disappear at an increasing rate, and natural resources are depleted. Thus, we need to prioritize the critical informatics problems – ask where problems intersect across agencies and environments to find the greatest synergies, which of those with the greatest intellectual merit could be solved with focused R&D, and where public and private funds could be leveraged. A follow-on workshop of resource managers, eco-informatics professionals and computer scientists, itself followed by online surveys, auctions or futures markets could accomplish this. Because problems are both technical and sociological, a few well-chosen broad projects in those areas could serve other more focused research.

Two unanticipated issues emerged as we followed our agenda: 1) feedback loops and 2) the nature of decision making. If aresource managers become more effective, the effect on managed system is not negligible. We also saw that understanding decision making is critical for doing the work. Environmental issues are complex with considerable uncertainties, but in the political and policy arena many decisions are placed in a "yes or no" context. Thus, one workshop product is a decision making primer. [decision].

Communication enables collaboration (human centeredness), trustworthiness (ontologies), and data sharing (data integration). Social science is characterized by indigenous local and community knowledge plus the ethics of decision making (data integration), and user needs (the futures market). Ontologies, coupling diverse models and how second and third generation metadata can be used to define data quality are particularly important. One real challenge of this area is the difficulty involved in pursuing research in one of these areas without at least some understanding of the others.

Another challenge involves training computer scientists and social scientists to work in eco-informatics and natural resource management. A graduate student team considered how researchers might articulate educational impacts, involve students in research, and use research as a teaching tool. The students saw the ethical issues around large data repositories as a particularly fruitful area for teachable moments, and funding interdisciplinary mission-oriented tasks that force addressing local problems as a way to pursue these goals. The students encourage NSF to partner with agencies that support applied student research. [gradStudent] We further encourage the early focus in eco-informatics and decision making to be on ecological and biodiversity issues, as environmental health decision making is even more complex and requires natural resource management as input.

Funding agencies must work together and with principal investigators, information managers and decision makers to sustain and encourage innovation in this area. How would researchers funded by NSF find collaborators so they can best understand resource problems, extract the research issues, and test prototypes? How might research results and prototypes make their way to resource managers deployed in field offices? How might product evaluation be fed back to inform new research? Considerable attention should be paid to assuring a cycle of innovation from research to prototype, to development and commercialization, and finally to deployment and evaluation (and back to research). The differing, non-overlapping missions and reward systems built into each agency make it easy to lose momentum at any of these stages. Longer funding cycles are needed to elicit requirements and integrate these into a research agenda, and then enter into an "agile" software cycle of develop, evaluate, and deploy. One year is barely adequate for the first step (eliciting requirements, understanding the domain, and setting up a collaboration); three years is more adequate to developing and evaluating tools with decision maker collaborators, and we recommend special two-year supplements for deployment (given prior evaluation) would continue a cycle of innovation.

Finally, considerable attention must be paid to constant re-prioritization of the research agenda, and assuring development of tools that promise, through extensibility, applicability to a wide range of problems, as they arise in important eco-systems. We emphasize the importance of keeping a range of research projects in the pipeline – from highly theoretical and generalizable, to working prototypes developed by researchers and resource managers, to deployment experiments.

Acknowledgements

The authors thank those who presented research, development and needs assessment at the workshop, all of which informed this report: Frank Biasi (Nature Conservancy), Alan Borning (Univ. Washington), Geoffrey Bowker (Santa Clara Univ.), C. Marie Denn (USPS), Richard Guldin (US Forest Service), Carol Hert (Univ. Washington-Tacoma), Eduard Hovy (Univ. S. California), Stephen Jensen (European Union), Paul Klarin (State of Oregon), Ron Li (Ohio State Univ.), Molly O'Neil (State EPA Network), Phil Rossignol (Oregon State Univ.) John L. Schnase (NASA), Mark Simonson (Puget Sound Regional Council), Larry Sugarbaker (NaturServe), William Sonntag,   Nancy Tosta (Ross & Assoc Environmental Consulting), Dawn Wright (Oregon State Univ.)., and to János Fülöp, David Roth and Charles Schweik (Univ. Mass-Amherst) and G. Bowker for writing the Decision Making in Eco-Informatics Reports.

We also thank other workshop participants: Bill Backous (Wa. State Dept. of Ecology), Bruce Bargmeyer (University of California, Berkeley), Thomas Beard (USGS), Clifford Duke (Ecological Society of America), Kevin Gergely (USGS), Vivian Hutchison (USGS), Brand Nieman (EPA), Craig Palmer (Univ. Nevada-Las Vegas), Pasky Pascual (USGS), Sherry Pittam (Oregon State Univ.), Louis Sweeny (EPA), Tim Tolle (USFS-Ret.), Andrew Wilson (USDA), Steve Young (EPA), as well as the graduate student panel Steve Abercrombie, Gigi Sanchez, and Hiroshi Sato, and Canopy Database Project staff Anne Fiala, Aaron Crosland, Chris Hardy, and Mike Finch (all of The Evergreen State College) who provided excellent local support. Aaron Ellison of Harvard University and Elaine Hoagland of The National

Council for Science and the Environment for their very helpful comments on an initial short version of this report.

We acknowledge financial and administrative support from the National Science Foundation (NSF IIS 0505790), and emphasize that the opinions stated in this paper are not necessarily those of NSF.

URLs Cited in This Report

dg.o 2005 Panel, Digital Government and the Academy (Delcambre, Giuliano), May 16-18, 2005. http://dgrc.org/dgo2005.

dg.o 2004 Birds of a Feather (Schweis et al), May 24-26, 2004.
http://dgrc.org/dgo2004/disc/bofs/bof_ecoinformatics.pdf

PCAST Panel on Biodiversity and Ecosystems, "Teaming with Life: Investing in Science to Understand and Use America's Living Capital", March 1998.
http://clinton3.nara.gov/WH/EOP/OSTP/Environment/html/teamingcover.html.

BDEI - Biodiversity and Eco-System Informatics Workshops

BDEI-1. NSF, USGS, NASA Workshop (Maier, Landis, Cushing, Frondorf, Silberschatz, Frame, Schnase), NASA (Goddard), June 2000. https://www.evergreen.edu/bdei/2001/

BDEI-2. PI's Meeting Report (Cushing, Beard-Tisdale, Bergen, Clark, Eckman, Henebry, Landis, Maier, Schnase, Stevenson), NSF (Arlington), February 10, 2003. https://www.evergreen.edu/bdei/2003/

BDEI-3. Eco-Informatics for Decision Making (Cushing, Wilson, et al), The Evergreen State College, December 13-15, 2004. http://www.evergreen.edu/bdei/

blm http://www.cse.ogi.edu/forest/papers/blm-briefing.ppt

borning http://www.evergreen.edu/bdei/presentations/borning.pdf

context http://www.evergreen.edu/bdei/presentations/tuesbreakout1_combined.pdf

decision http://www.evergreen.edu/bdei/decisionMaking

gradStudent http://www.evergreen.edu/bdei/ presentations/gradStudentDraft.pdf

hert http://www.evergreen.edu/bdei/presentations/hert-tuesdaylunch1.pdf

hovy http://www.evergreen.edu/bdei/presentations/hovy.pdf

jenson http://www.evergreen.edu/bdei/presentations/jenson.pdf

li http://www.evergreen.edu/bdei/presentations/li.pdf

ontology http://www.evergreen.edu/bdei/presentations/wedbo3ontology.pdf

presentation http://www.evergreen.edu/bdei/presentations/wedbo3summary.pdf

policy http://www.evergreen.edu/bdei/presentations/summaryPolicygroupfinal.pdf

research http://www.evergreen.edu/bdei/presentations/tuesbreakout2_combined.pdf

and http://www.evergreen.edu/bdei/presentations/breakout3_combined.pdf

social http://www.evergreen.edu/bdei/presentations/wedbo3humancenterednessb.doc

sugarbaker http://www.evergreen.edu/bdei/presentations/sugarbaker.pdf

tosta http://www.evergreen.edu/bdei/presentations/GreybeardNT.pdf

wright http://www.evergreen.edu/bdei/presentations/wright.pdf

An Architecture and Application for Integrating Curation Data at the Residue Level for Proteins

Mehmet M. Dalkilic*

Center for Genomics and Bioinformatics,
School of Informatics,
Indiana University, Bloomington,
IN 47405, USA
dalkilic@indiana.edu
http://www.informatics.indiana.edu/dalkilic

Abstract. Understanding protein families requires the bringing together of many different kinds of data. These families are typically derived from multiple sequence alignments. Directed mutagenesis is one of the most common means of inferring which specific amino acid or set of amino acids are important in the function of a protein. Although there are a large number publicly available, protein specific repositories, *e.g.*, PROSITE, UniProt, and Pfam, no tools exist for experimental biologists that provide a means for managing and visualizing the *curation data* of the protein families they study at the individual residue level. We present the development of a novel system designed for experimental biologists, called the Curation Alignment Tool for Protein Analysis (CATPA), that allows for the efficient and effective creation, storage, management, and querying of experimentally curated protein families.

1   Background and Motivation

In the life sciences both the amount of data and its pace of generation is staggering NCBI(1), Pfam(2; 3), UniProt(4), Swiss-Prot(5), TrEMBL(5), and PROSITE(6; 7). Thus, the challenge of integration becomes even greater, since little attention is paid to how all the disparate types of data fit together. Certainly this problem of integration has a good deal to do with the various perspectives of research, *e.g.*, molecular, genomic, proteomic, cellular. Another part of the problem has to do with how all these data are managed. Though integration is a broad problem, some success can be achieved if the focus is sufficiently narrowed. One area that has received virtually no attention is integrating curation information at the individual residue level in protein families. By integration we mean not only unstructured text, but also semi-structured text and images can be associated with sets of residues in a protein of a particular family. Further, this information can be easily managed and visualized.

---

* Partially supported by NSF IIS-0082401.

B. Ludäscher and L. Raschid (Eds.): DILS 2005, LNBI 3615, pp. 335–338, 2005.
© Springer-Verlag Berlin Heidelberg 2005

2   CATPA

We are in the process of creating a portal that provides a library of family curations together with an available application the Curation and Alignment Tool for Protein Analysis (CATPA) that allows for the management and visualization of curation data. In this paper we discuss some elements of CATPA. Figure 1 shows two screen captures. CATPA is designed to

1. used widely recognized formats, for example ClustalX(8);
2. have a GUI (graphical user interface)(9) that allows for most of the typical tasks the biologists performs with respect to visualizing alignments, curated residues, *etc.*;
3. with a rigorous data model and consequently a DBMS to handle the management and security of the stored information; we have implemented CATPA to store in a serialized mode as well.
4. include a querying capability that allows efficient and effective querying of the curations, sequences, and so forth;
5. use the GO vocabulary so that biologists can more easily and correctly search and share information;
6. run as a stand-alone entity which is better suited and can be tailored to the biologists' needs;

A cursory overview of CATPA will be presented here. CATPA recognizes a number of well-known and widely used formats both for importing to and exporting from the system. CATPA utilizes a Java GUI front-end that allows biologists to interact with information in an environment they are accustomed to. Protein families are aligned, and conservation and curation are easily discernable via colors that users can change according to their preferences. Additionally, other kinds of information can be displayed, *e.g.*, entropy, hydrophobicity.

CATPA provides two separate views of the protein family. One is the standard view of an aligned family. The other view is a facility to magnify (increasing or decreasing) over the family called the Dataset View. In addition the Dataset View is used to visualize the results of queries over the protein family. CATPA has extensive query facilities including the ability to query alignments, curations, sequences, and fixed vocabulary. Additionally, CATPA allows for the visualization of query results making perusal easier. CATPA incorporates the Gene ontology (GO) (10) in its curation vocabulary. GO is a collaborative effort to help standardize biological words by providing generalization and component relationships ("is-a" and "is-a-part-of").

CATPA has a small footprint, comprising less than 10MB. The front-end requires Java 1.4.2 and MySQL and can be run on most machines capable of running both. Although not completed in the current release, CATPA is intended as a means of collaboration–biologists wishing to share and query other CATPA databases can do so transparently.

This is ongoing work with Andrew Albrecht, James Costello, Arijit Sengupta, and Peter Cherbas. Sukamol Jakobsson has contributed significantly to the project as well.

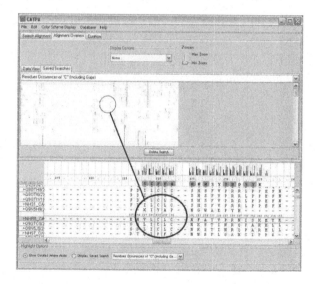

Fig. 1. Two screen captures of CATPA. The top panel shows interaction with curation data: text, URLs, and images. The bottom panel shows one of the querying facilities–the "10K" ft. view of the alignment and the respective residues that it pertains to

References

[1] Benson, D.A., Karsch-Mizrachi, I., Lipman, D.J., et al.: Genbank: update. Nucleic Acids Research **32** (2004) D23–D26

[2] Sonnhammer, E., Eddy, S., Birney, E., Bateman, A., Durbin, R.: Pfam - multiple sequence alignments and hmm-profiles of protein domains. Nucleic Acids Research **26** (1998) 320–322

[3] Bateman, A., Coin, L., Durbin, R., Finn, R., Hollich, V., et al.: The pfam protein families database. Nucleic Acids Research **32** (2004) D138–D141

[4] Apweiler, R., Bairoch, A., Wu, C., Barker, W., Boeckmann, B., Ferro, S., et al.: Uniprot: the universal protein knowledgebase. Nucleic Acids Research **32** (2004) D115–119

[5] Boeckmann, B., Bairoch, A., Apweiler, R., Blatter, M., Estreicher, A., Gasteiger, E., Martin, E., Michoud, K., O'Donovan, C., Phan, I., Pilbout, S., Schneider, M.: The swiss-prot protein knowledgebase and its supplement trembl in 2003. Nucleic Acids Research **31** (2003) 365–370

[6] Sigrist, C., Cerutti, L., Hulo, N., et al.: Prosite: a documented databases using patterns and profiles as motif descriptors. Brief Bioinformatics **3** (2002) 265–274

[7] Falquet, L., Pagni, M., Bucher, P., Hulo, N., Sigrist, C., et al.: The prosite database, its status in 2002. Nucleic Acids Research **30** (2002) 235–238

[8] Thompson, J., Gibson, T., Plewniak, F., Jeanmougin, F., Higgins, D.: The clustal_x windows interface: flexible strategies for multiple sequence alignment aided by quality analysis tools. Nucleic Acids Research **25** (1997) 4876–4882

[9] Norman, D.: The Design of Everyday things. Doubleday Currency (1990)

[10] Gene Ontology Consortium: Creating the gene ontology resource: design and implementation. Genome Research **11** (2001) 1425–1433

The Biozon System for Complex Analysis
of Heterogeneous Interrelated Biological Data
and Discovery of Emergent Structures

Aaron Birkland and Golan Yona*

Department of Computer Science, Cornell University
golan@cs.cornell.edu

1   Introduction

Biological entities are strongly related and mutually dependent on each other.
Therefore, there is a growing need to corroborate and integrate data from different resources and aspects of biological systems in order to analyze them effectively.

To identify entities, existing databases use explicit references by accession
number or a mutual ontology. Some databases relate and cross link elements
from other databases based on these identifiers. However, this information is
very partial and is not readily available in some. Moreover, these links are not
established in coordination with the other linked databases. With the source
databases changing rapidly, this leads to problems of consistency and updatabil-
ity. Furthermore, it is hard to query this wealth of data in ways that can benefit
and exploit the mutual dependency between entities.

Biozon is a unified biological database that integrates heterogeneous data
types and the relationships between them, such as nucleic acid sequences, proteins, structures, protein domains and protein families, protein-protein interactions and cellular pathways, into a single extensive schema. This schema allows
one to see each data instance in its full biological context. More importantly it
allows for complex searches that span multiple data types from a heterogeneous
set of sources and for arbitrary computations on that data. Biozon can also
rank results, the same way Google ranks web documents, and uses similarity
relationships to extend query results to similar biological entities.

2   Data Model

The data in Biozon is composed of two main types: **source data** that are
gleaned from established online databases (such as SWISS-PROT, Genbank,
BIND, KEGG and others), and unique **derived (computed) data** that is
computed in house and includes similarity relationships between objects and

---

* Corresponding author.

B. Ludäscher and L. Raschid (Eds.): DILS 2005, LNBI 3615, pp. 339–342, 2005.

predicted information about their functional role. The derived data introduces another level of complexity to our model, but also allows for even more powerful methods of data querying, management and manipulation to the extent of biological theorem verification and computation.

The information in Biozon is logically represented as graph in which nodes represent some unit of data, and edges indicate a relationship between two nodes. Each graph node or edge is given a classification as part of a hierarchy of data and relation types. Each constituent data set (for example, from a source database such as SwissProt) is mapped onto some subset of this graph.

A special class of nodes is the object class and its descendant subclasses. These are the equivalent of physical objects (such as protein sequences and DNA sequences) and sets thereof (such as interactions and pathways). Instances of fundamental biological objects that are gleaned from source databases are mapped to these object nodes and are required to be non-redundant. As such, physical objects can be viewed as the actual identifiers of the biological entities they represent.

The schemata for different data sets can and do share nodes that represent the same fundamental biological type of object. As a result, our graph ends up becoming highly connected and centered around hubs of such objects. This connectivity allows for efficient formulation and execution of complex queries that span multiple data types. The use of potentially unstable or inconsistent identifiers (such as accession numbers) to indicate relationships or cross-references is eschewed in favor of materializing explicit relations between physical non-redundant Biozon objects.

3   Maintenance

All data in Biozon is integrated and warehoused locally. This provides several benefits over integration methods that rely on independent, distributed sources. The two most critical benefits are speed and data consistency. Speed is obvious, as there are a wealth query plan optimizations and a minimum of network availability and latency issues available over locally warehoused data. Out in the wild, there is no enforced consistency between independent databases that may reference one another. Changes are not coordinated, and one commonly observes dead or misleading references between databases because of this constant, uncoordinated flux. Having local control over the data in the fashion of Biozon affords the ability to detect and mitigate changes that violate consistency.

An additional benefit to locally warehousing the Biozon graph is that the data is readily accessible for large-scale computations, and any resulting derived data can be integrated and maintained in situ. While derived data enriches the understanding of the graph greatly, it comes at the cost of increased maintenance in the face of updates. Indeed, updates on such a tightly integrated graph require that consistency of the data be defined and upheld, especially given the non-redundant model employed by Biozon and domains of knowledge by independent sources overlap.

As a solution, we enforce consistency and freshness of derived data using a framework of rules and actions carried out by small independent subunits that are implemented as graph triggers. We define protocols for the addition, modification, and deletion of data that uphold consistency between Biozon and external data sources as well as consistency of derived data

4   Search and Analysis

Given the non-redundant graph model centered around biological objects, meaning can be inferred by the shapes or topologies observed therein. As a simple example, consider graph nodes representing proteins, interactions, and nucleic acid sequences. Along with the requisite edges, these three objects can form topologies ranging from "nucleic acid sequence encodes for protein which is involved in an interaction" to "protein and nucleic acid sequence are involved in the same interaction".

4.1   Complex Searches

Biozon exploits the graph structure in allowing for complex searches that span multiple data types. In essence, the user initiates by specifying a specific topology to search for, as well as any specific constraints on any documents that should be present in matching topology instances. For example, a valid query could be *"Find all 3D structures of proteins that are involved in phosphorylation interactions and are part of the Prostaglandin and leukotriene metabolism pathway"*. This particular search specifies a topology involving structures, proteins, interactions, and pathways.[1] The online user interface allows these queries to be built in a series of simple steps. Upon execution, the Biozon graph is searched for graph isomorphism in realtime.

4.2   Fuzzy Searches

Fuzzy searches extend complex queries to include similar or homologous objects in the search space. Here Biozon exploits the computed data (such as similarity relationships) that was integrated into the database. Queries may be extended by incorporating materialized similarity data in any appropriate query step. For example, querying for structures of proteins that are in enzyme family 1.1.1.1 and are involved in an interaction returns no results. Incorporating similarity into the search transforms the query into one that searches for structures of proteins that are involved in interactions and *similar* to proteins that are members of the 1.1.1.1 family. This query does return significant results from a very large search space in less than minute.

---

[1] Currently, the specified topology for this search would involve 'enzyme families' as well. Biozon incorporates pathways from KEGG, which defines them in terms of enzyme families. Therefore, a path in the graph between proteins and pathways would have to go through the enzyme family node.

Currently, fuzzy searches use the results of BLAST, expression profile similarity, and structural similarity of proteins. A user may choose which are incorporated into the results, as well as the similarity threshold (such as e-value), where applicable. Because Biozon materializes all similarity data, none of the expensive similarity computations need to be dome when executing such a search.

4.3   Ranking

The graph structure of the Biozon data model and the emergent shape resulting from integration of many sources lends itself well to analysis. Particularly relevant to searches and result sets is the ability to assign ranks to Biozon objects based on the graph structure. Using a ranking system based on a spectral analysis of the data graph (similar to PageRank by Google), search results may be ordered by ranks that reflect the importance of the different entities and is linked to the amount of information associated with them.

5   Current Status

The Biozon database currently stores extensive information about more than 37,000,000 protein and DNA sequences, integrating sequence, structure, protein-protein interactions, pathways and expression data, totaling over 60 million documents from more than 20 different databases. It also stores information about 2.5 billion relations between documents, including explicit relations between objects, and derived or computed relations based on sequence similarities, structural similarities and more.

The Biozon database is accessible now at biozon.org, and serves as a useful proof of concept that the ideas expressed in our approach are practical. Indeed, the following functionality is provided as a direct result of our efforts:

- Browsing and navigating capability that shows the biological context of each object in its own "profile page".
- Interface for building complex and fuzzy queries. By using a step-by-step procedure, users can add objects and relationships to create a search topology, as well as define all search constraints. Results may be subsequently ranked, if desired.
- User accounts system that allows one to attach comments to objects and to materialize the results of complex queries. Queries can be saved, run in the background, and their results may be downloaded as text.
- Online analysis tools for user-supplied data. Currently, Biozon offers BLAST comparisons of submitted sequences with Biozon proteins, EST analysis, expression profile similarity, and domain structure prediction.

All the above features may be used in real time by visiting biozon.org.

Author Index

Lecture Notes in Bioinformatics